LONDON MATHEMATICAL SOCIETY LECTURE NOTE SERIES

Managing Editor: Professor N.J. Hitchin, Mathematical Institute,
University of Oxford, 24–29 St Giles, Oxford OX1 3LB, United Kingdom

The titles below are available from booksellers, or, in case of difficulty, from Cambridge University Press at
www.cambridge.org.

London Mathematical Society Lecture Note Series. 293

Second Order Partial Differential Equations in Hilbert Spaces

Giuseppe Da Prato
Scuola Normale Superiore di Pisa

Jerzy Zabczyk
Polish Academy of Sciences, Warsaw

CAMBRIDGE
UNIVERSITY PRESS

PUBLISHED BY THE PRESS SYNDICATE OF THE UNIVERSITY OF CAMBRIDGE
The Pitt Building, Trumpington Street, Cambridge, United Kingdom

CAMBRIDGE UNIVERSITY PRESS
The Edinburgh Building, Cambridge CB2 2RU, UK
40 West 20th Street, New York, NY 10011-4211, USA
477 Williamstown Road, Port Melbourne, VIC 3207, Australia
Ruiz de Alarcón 13, 28014 Madrid, Spain
Dock House, The Waterfront, Cape Town 8001, South Africa

http://www.cambridge.org

First published 2002

Typeface Computer Modern 10/12pt   *System* $\LaTeX\,2_\varepsilon$ [TB]

A catalogue record for this book is available from the British Library

Library of Congress Cataloguing in Publication data

Da Prato, Giuseppe.
Second order partial differential equations in Hilbert spaces / Giuseppe Da Prato &
Jerzy Zabczyk.
    p.   cm. – (London Mathematical Society lecture note series; 293)
Includes bibliographical references and index.
ISBN 0 521 77729 1 (pbk.)
1. Differential equations, Partial.   2. Hilbert space.   I. Zabczyk, Jerzy.   II. Title.
III. Series.
QA374 .D27   2002
515′.353–dc21   2002022269

ISBN 0 521 77729 1 paperback

Transferred to digital printing 2003

Contents

Preface

The main objects of this book are linear parabolic and elliptic equations of the second order on an infinite dimensional separable Hilbert space H such as

$$\begin{cases} D_t u(t,x) = \frac{1}{2}\text{Tr}[Q(x)D^2 u(t,x)] + \langle F(x), Du(t,x)\rangle, & x \in H,\ t > 0, \\ u(0,x) = \varphi(x), & x \in H, \end{cases}$$

(0.1)

and

$$\lambda\psi(x) - \frac{1}{2}\text{Tr}[Q(x)D^2\psi(x)] - \langle F(x), D\psi(x)\rangle = g(x), \quad x \in H. \qquad (0.2)$$

Here $F : D(F){\subset}H \to H$, $Q : D(Q){\subset}H \to L(H)$ and $\varphi : H \to \mathbb{R}$ are given mappings and λ a given nonnegative number, whereas $u : [0,T]{\times}H \to \mathbb{R}$ and $\psi : H \to \mathbb{R}$ are the unknowns of (0.1) and (0.2) respectively. Moreover D represents derivative and Tr the trace. Some classes of nonlinear equations will be considered as well.

There are several motivations to develop infinite dimensional theory. First of all the theory is a natural part of *functional analysis*. Moreover as in finite dimensions, parabolic equations on Hilbert spaces appear in *mathematical physics* to model systems with infinitely many degrees of freedom. Typical examples are provided by spin configurations in statistical mechanics and by crystals in solid state theory.

Infinite dimensional parabolic equations provide an analytic description of infinite dimensional diffusion processes in such branches of *applied mathematics* as *population biology, fluid dynamics, and mathematical finance*. They are known there under the name of *Kolmogorov equations*.

Nonlinear parabolic problems on Hilbert spaces are present in the control theory of distributed parameter systems. In particular the so called *Bellman-Hamilton-Jacobi* equations for the value functions are intensively studied.

If H is finite dimensional and the coefficients Q and F are continuous and bounded a satisfactory theory is available, see the classical monographs by O. A. Ladyzhenskaja, V. A. Solonnikov and N. N. Ural'ceva [154], and A. Friedman [115]. However, when the coefficients are continuous but unbounded, as in the present book, only a general result on existence, due to S. Itô [143], is available but there is not uniqueness in general, see e.g. [146, page 175].

First attempts to build a theory of partial differential equations on Hilbert spaces were made by R. Gateaux and P. Lévy around 1920. Their approach, based on a specific notion of averaging, was presented by P. Lévy on two books on functional analysis published in 1922 and 1951, see [156].

We adopt here a different approach initiated by L. Gross [138] and Yu. Daleckij [62] about 30 years ago, see also the monograph by Yu. Daleckij and S. V. Fomin [63]. Its main tools are probability measures in Hilbert and Banach spaces, stochastic evolution equations, semigroups of linear operators and interpolation spaces.

In this book we try to present the state of the art of the theory of parabolic or elliptic equations in an infinite dimensional Hilbert space H. Since the theory is rapidly changing and it is far from being complete we shall limit ourselves to basic results referring to more specialized results in notes.

Some results can be extended to general Banach spaces, but these are outside of the scope of the book. Also, for the sake of brevity, we do not treat equations with time dependent coefficients or with an additional potential term $V(x)u(t,x)$, where $V : H \to \mathbb{R}$.

The book is divided into three parts: I. Theory in the space of continuous functions, II. Theory in Sobolev spaces with respect to a Gaussian measure, III. Applications to control theory.

PART I. Here we discuss the case when F and G are continuous and bounded, working on the space $UC_b(H)$ of all uniformly continuous and bounded fuctions from H into $\mathbb{R}$.

A natural starting point is the heat equation:

$$\begin{cases} D_t u(t,x) = \frac{1}{2}\mathrm{Tr}[QD^2 u(t,x)], & t > 0, \ x \in H, \\ u(0,x) = \varphi(x), & x \in H, \end{cases} \tag{0.3}$$

where Q is a given symmetric nonnegative operator of trace class and $\varphi \in UC_b(H)$. The solution of (0.3) is given by the formula

$$u(t,x) = \int_H \varphi(x+y)N_{tQ}(dy), \ x \in H, \ t \geq 0, \tag{0.4}$$

where N_{tQ} is the Gaussian measure with mean 0 and covariance operator tQ.

This problem, initially stated by L. Gross [138], is studied in Chapter 3 where we prove that the requirement that Q is of trace class is necessary to solve problem (0.3) for sufficiently regular initial data φ.

We then study existence, uniqueness and regularity of solutions in $UC_b(H)$. We show that, as noticed by Gross, solutions of (0.3) are smooth only in the directions of the reproducing kernel $Q^{1/2}(H)$.

Finally we study the corresponding strongly continuous semigroup (P_t) and characterize its infinitesimal generator.

In order to make the book self-contained we have devoted Chapter 1 to Gaussian measures and Chapter 2 to properties of continuous functions in an infinite dimensional Hilbert space.

Chapter 4 is the elliptic counterpart of Chapter 3; it is devoted to the Poisson equation:

$$\lambda\psi(x) - \frac{1}{2}\mathrm{Tr}[QD^2\psi(x)] = g(x), \quad x \in H. \tag{0.5}$$

Here, besides existence and uniqueness, *Schauder estimates* are proved.

In Chapter 5, we go to the case of Hölder continuous and bounded coefficients F and G trying to generalize the finite dimensional theory. As in finite dimensions we pass from equations with constant coefficients to equations with variable coefficients by first proving Schauder and interpolatory estimates and then using the classical continuity method.

We notice that the results are not as satisfactory as in the finite dimensional case. In particular they do not characterize the domain of the operator which appears in (0.5). In fact, if g is Hölder continuous, we know that the solution ψ of (0.5) has first and second derivatives Hölder continuous, but we do not have any information about the trace of $QD^2\psi$.

In Chapter 6 we pass to the case when the coefficients F and G are unbounded. The typical important example is the Ornstein-Uhlenbeck operator, that is

$$L\varphi(x) = \frac{1}{2}\mathrm{Tr}[QD^2\varphi(x)] + \langle Ax, D\varphi(x)\rangle, \quad x \in D(A), \; \varphi \in UC_b(H), \tag{0.6}$$

where $A : D(A) \subset H \to H$ generates a C_0 semigroup on H and Q is symmetric and nonnegative.

We show that the problem

$$D_t u = Lu, \quad u(0) = \varphi, \quad \varphi \in UC_b(H),$$

is well posed if and only if

$$\int_0^T \mathrm{Tr}[e^{tA}Qe^{tA^*}]dt < +\infty, \; T > 0. \tag{0.7}$$

In this case, we can construct explicitly the corresponding transition semi-group R_t:

$$R_t\varphi(x) = \int_H \varphi(e^{tA}x + y)N_{Q_t}(dy), \quad x \in H, \;\; \varphi \in UC_b(H), \tag{0.8}$$

where

$$Q_t = \int_0^t e^{sA}Qe^{sA^*}ds. \tag{0.9}$$

It is interesting to notice that now it is not necessary to assume that Q is of trace class as in the case of the heat equation. In fact, there is an important class of Ornstein-Uhlenbeck operators , when

$$e^{tA}(H) \subset Q_t^{1/2}(H), \; t > 0,$$

that behave as elliptic operators in finite dimensions. In this case, R_t is *strong Feller* and the following property, typical of parabolic equations in finite dimensions, holds:

$$\varphi \in C_b(H), \; t > 0 \Rightarrow R_t\varphi \in C_b^\infty(H). \tag{0.10}$$

Notice that (R_t) is not a semigroup of class C_0 in $UC_b(H)$. However, it is possible to define an infinitesimal generator L of (R_t) and study several properties, including Schauder estimates.

Finally, the last two sections are devoted to perturbations of Ornstein-Uhlenbeck operators .

Chapter 7 is concerned with a general Kolmogorov equation under rather strong regularity assumptions on coefficients and on initial functions. We use the method of stochastic characteristics. We recall basic results on stochastic evolution equations and on implicit function theorems which are used to prove regularity of generalized solutions. Existence and uniqueness results are proved in §7.5 and §7.6. Stronger regularity results based on a generalization of the Bismut-Elworthy-Xe formula are presented in §7.7.

In this direction there is still much work to be done to cover more general coefficients; however, for the stochastic reaction-diffusion equations , several results can be found in the monograph by S. Cerrai, [43].

The greater part of the book is devoted to problems on the whole of H. The theory in an open set $\mathcal{O}$ is just starting in the infinite dimensional case, see the papers [92], [207] and [190], [193], [194], [195], [191].

In Chapter 8 we present quite general results on existence and regularity in the interior due to G. Da Prato, B. Goldys and J. Zabczyk [92] and to A. Talarczyk [207].

PART II. To consider equations with very irregular unbounded coefficients arising in different applications such as reaction-diffusion and Ginzburg-Landau systems and stochastic quantization, it is useful to work in spaces $L^2(H, \nu)$ with respect to an invariant measure ν.

Chapter 9 is devoted to basic properties of the space $L^2(H, \mu)$ when μ is a Gaussian measure. In particular the Itô-Wiener decomposition and the compact embedding of $W^{1,2}(H, \mu)$ in $L^2(H, \mu)$ are established.

In Chapter 10 we prove several properties of the Ornstein-Uhlenbeck semigroup R_t on $L^2(H, \nu)$, and of its infinitesimal generator L_2. Here we assume that the operator

$$Q_\infty := \int_0^{+\infty} e^{tA} Q e^{tA^*} dt \tag{0.11}$$

is well defined and of trace class. This implies existence of an invariant Gaussian measure $\mu = N_{Q_\infty}$ of R_t.

Other topics considered are symmetry of R_t and characterization of the domain of L_2. When R_t is strong Feller we show that $\mu = N_{Q_\infty}$ is absolutely continuous with respect to N_{Q_t}, proving that

$$R_t \varphi(x) = \int_H \varphi(e^{tA} x + y) \rho(t, x, y) N_{Q_\infty}(dy). \tag{0.12}$$

Then, using (0.12), we show that R_t is hypercontractive.

Finally, we show Poincaré and log-Sobolev inequalities and some of their consequences such as spectral gap and exponential convergence to equilibrium.

Chapter 11 is devoted to the following perturbation of L:

$$N_0 \varphi(x) = \frac{1}{2} \text{Tr}[QD^2 \varphi(x)] + \langle Ax, D\varphi(x) \rangle + \langle F(x), D\varphi(x) \rangle, \quad x \in D(A), \tag{0.13}$$

where F is bounded or Lipschitz continuous.

More general perturbations of gradient form $F(x) = -DU(x)$ are studied in Chapter 12. In this case, we consider the "Gibbs measure"

$$\nu(dx) = Z^{-1} e^{-2U(x)} \mu(dx), \tag{0.14}$$

where Z is a normalization constant, and try to show that the operator N_0, defined in the space of all exponential functions, is dissipative in $L^2(H, \nu)$ and its closure is m dissipative.

This problem has been extensively studied, using the technique of Dirichlet forms, starting from S. Albeverio and R. Høegh-Krohn [3], see Z. M. Ma and M. Röckner [165].

For the sake of brevity we do not consider perturbations of L_2 that are not Lipschitz continuous and not of gradient form, see comments in Chapter 11 for references to the present literature.

PART III is devoted to applications to control theory. In Chapter 13 we are concerned with a controlled system on a separable Hilbert space H

$$\begin{cases} dX = (AX + G(X) + z(t))dt + Q^{1/2}dW_t, \ t \in [0, T], \\ X(0) = x \in H, \end{cases} \tag{0.15}$$

where $A : D(A) \subset H \to H$ is a linear operator, $G : H \to H$ is a continuous regular mapping, Q is a symme tric nonnegative operator on H, and W is a cylindrical Wiener process. X represents the *state*, z the *control* and $T > 0$ is fixed.

Given $g, \varphi \in UC_b(H)$, and a convex lower semicontinuous function $h : H \to [0, +\infty)$, we want to minimize the cost

$$J(x, z) = \mathbb{E}\left(\int_0^T [g(X(t, x; z)) + h(z(t))] \, dt + \varphi(X(T, x; z)) \right), \tag{0.16}$$

over all $z \in L_W^2(0, T; L^2(\Omega, H))$, the Hilbert space of all square integrable processes adapted to W defined on $[0, T]$ and with values in H.

We solve this problem using the dynamic programming approach, proving existence of a regular solution of the Hamilton-Jacobi equation

$$\begin{cases} D_t u = \frac{1}{2}\mathrm{Tr}[QD^2u] + \langle Ax + G(x), Du \rangle - F(Du) + g \\ u(0) = \varphi, \end{cases} \tag{0.17}$$

where the *Hamiltonian F* is given by the *Legendre transform* of h:

$$F(x) = \sup_{y \in H} \{\langle x, y \rangle - h(y)\}, \ x \in H. \tag{0.18}$$

Finally, Chapter 14 is devoted to Hamilton-Jacobi inequalities which are satisfied by value functions corresponding to optimal stopping problems. In their simplest version they are of the form

$$\begin{cases} u_t(t, x) \in \frac{1}{2}\mathrm{Tr}[QD^2u(t, x)] + \langle Ax + G(x), Du(t, x) \rangle \\ \qquad\qquad + \alpha(x)u(t, x) - \partial I_{K_h}(u(t, x)), \\ u(0, x) = g(x), \ x \in H, \ t \geq 0. \end{cases} \tag{0.19}$$

Here
$$K_h = \left\{ f \in L^2(H,\mu) : f \geq h \right\},$$

where μ is a properly chosen measure and I_{K_h} is the indicator function of K_h :

$$I_{K_h}(f) = \begin{cases} 0 \text{ if } f \in K_h, \\ +\infty \text{ if } f \notin K_h. \end{cases} \tag{0.20}$$

Moreover ∂I_{K_h} is the subgradient of I_{K_h}.

There exist at present four monographs covering some aspects of the infinite dimensional theory, by Z. M. Ma and M. Röckner [165], Yu. Daleckij and S. V. Fomin [63], Y. M. Berezansky and Y. G. Kondratiev [12] and by S. Cerrai, [43]. The overlap between those monographs and our book is however rather small.

The authors acknowledge the financial support of the Italian National Project MURST "Analisi e controllo di equazioni di evoluzione deterministiche e stocastiche", the KBN grant No 2 PO3A 082 08 "Ewolucyjne Równania Stochastyczne" and the Leverhulme Trust, during the preparation of the book.
They also thank F. Gozzi, E. Priola and A. Talarczyk for pointing out some errors and mistakes in earlier versions of the book and S. Cerrai for a careful reading of the whole manuscript.
The authors would like to thank their home institutions Scuola Normale Superiore and the Polish Academy of Sciences for good working conditions.

Part I

THEORY IN SPACES OF CONTINUOUS FUNCTIONS

Chapter 1

Gaussian measures

This chapter is devoted to some basic results on Gaussian measures on separable Hilbert spaces, including the Cameron-Martin and Feldman-Hajek formulae. The greater part of the results are presented with complete proofs.

1.1 Introduction and preliminaries

We are given a real separable Hilbert space H (with norm $|\cdot|$ and inner product $\langle \cdot, \cdot \rangle$). The space of all linear bounded operators from H into H, equipped with the operator norm $\|\cdot\|$, will be denoted by $L(H)$. If $T \in L(H)$, then T^* is the adjoint of T. Moreover, by $L^+(H)$ we shall denote the subset of $L(H)$ consisting of all nonnegative symmetric operators. Finally, we shall denote by $\mathcal{B}(H)$ the σ-algebra of all Borel subsets of H.

Before introducing Gaussian measures we need some results about trace class and Hilbert-Schmidt operators.

A linear bounded operator $R \in L(H)$ is said to be of *trace class* if there exist two sequences (a_k), (b_k) in H such that

$$Ry = \sum_{k=1}^{\infty} \langle y, a_k \rangle b_k, \quad y \in H, \tag{1.1.1}$$

and

$$\sum_{k=1}^{\infty} |a_k| \, |b_k| < +\infty. \tag{1.1.2}$$

Notice that if (1.1.2) holds then the series in (1.1.1) is norm convergent. Moreover, it is not difficult to show that R is compact.

We shall denote by $L_1(H)$ the set of all operators of $L(H)$ of trace class. $L_1(H)$, endowed with the usual linear operations, is a Banach space with the norm

$$\|R\|_{L_1(H)} = \inf \left\{ \sum_{k=1}^{\infty} |a_k| \, |b_k| : \; Ry = \sum_{k=1}^{\infty} \langle y, a_k \rangle b_k, \; y \in H, \; (a_k), (b_k) \subset H \right\}.$$

We set $L_1^+(H) = L^+(H) \cap L_1(H)$. If an operator R is of trace class then its trace, Tr R, is defined by the formula

$$\mathrm{Tr}\, R = \sum_{j=1}^{\infty} \langle Re_j, e_j \rangle,$$

where (e_j) is an orthonormal and complete basis on H. Notice that, if R is given by (1.1.1), we have

$$\mathrm{Tr}\, R = \sum_{j=1}^{\infty} \langle a_j, b_j \rangle.$$

Thus the definition of the trace is independent on the choice of the basis and

$$|\mathrm{Tr}\, R| \le \|R\|_{L_1(H)}.$$

Proposition 1.1.1 *Let $S \in L_1(H)$ and $T \in L(H)$. Then*

(i) $ST, TS \in L_1(H)$ and

$$\|TS\|_{L_1(H)} \le \|S\|_{L_1(H)} \|T\|, \; \|ST\|_{L_1(H)} \le \|S\|_{L_1(H)} \|T\|.$$

(ii) $\mathrm{Tr}(ST) = \mathrm{Tr}(TS)$.

Proof. (i) Assume that $Sy = \sum_{k=1}^{\infty} \langle y, a_k \rangle b_k$, $y \in H$, where $\sum_{k=1}^{\infty} |a_k| |b_k| < +\infty$. Then

$$STy = \sum_{k=1}^{\infty} \langle y, T^* a_k \rangle b_k, \; y \in H,$$

and

$$\sum_{k=1}^{\infty} |T^* a_k| |b_k| \le \|T\| \sum_{k=1}^{\infty} |a_k| |b_k|.$$

It is therefore clear that $ST \in L_1(H)$ and $\|ST\|_{L_1(H)} \leq \|S\|_{L_1(H)}\|T\|$. Similarly we can prove that $\|TS\|_{L_1(H)} \leq \|S\|_{L_1(H)}\|T\|$.

(ii) From part (i) it follows that

$$\mathrm{Tr}(ST) = \sum_{k=1}^{\infty}\langle b_k, T^*a_k\rangle = \sum_{k=1}^{\infty}\langle Tb_k, a_k\rangle.$$

In the same way $\mathrm{Tr}\,(TS) = \sum_{k=1}^{\infty}\langle a_k, Tb_k\rangle$, and the conclusion follows. $\square$

We say that $R \in L(H)$ is of Hilbert-Schmidt class if there exists an orthonormal and complete basis (e_k) in H such that

$$\sum_{k,j=1}^{\infty} |\langle Se_k, e_j\rangle|^2 < +\infty. \tag{1.1.3}$$

If (1.1.3) holds then we have

$$\sum_{k=1}^{\infty} |Se_k|^2 = \sum_{k,j=1}^{\infty} |\langle Se_k, e_j\rangle|^2 = \sum_{k,j=1}^{\infty} |\langle e_k, S^*e_j\rangle|^2 = \sum_{j=1}^{\infty} |S^*e_j|^2. \tag{1.1.4}$$

Now if (f_k) is another complete orthonormal basis in H, we have

$$\sum_{m=1}^{\infty} |Sf_m|^2 = \sum_{m,n=1}^{\infty} |\langle Sf_m, e_n\rangle|^2 = \sum_{m,n=1}^{\infty} |\langle f_m, S^*e_n\rangle|^2 = \sum_{n=1}^{\infty} |S^*e_n|^2.$$

Thus, by (1.1.4) we see that the assertion (1.1.3) is independent of the choice of the complete orthonormal basis (e_k). We shall denote by $L_2(H)$ the space of all Hilbert-Schmidt operators on H. $L_2(H)$, endowed with the norm

$$\|S\|^2_{L_2(H)} = \sum_{k,j=1}^{\infty} |\langle Se_k, e_j\rangle|^2 = \sum_{k=1}^{\infty} |Se_k|^2,$$

is a Banach space.

Proposition 1.1.2 *Let $S, T \in L_2(H)$. Then $ST \in L_1(H)$ and*

$$\|ST\|_{L_1(H)} \leq \|S\|_{L_2(H)}\|T\|_{L_2(H)}. \tag{1.1.5}$$

Proof. Let (e_k) be a complete and orthonormal basis in H, then

$$Ty \;=\; \sum_{k=1}^{\infty}\langle Ty, e_k\rangle e_k = \sum_{k=1}^{\infty}\langle y, T^*e_k\rangle e_k,$$

$$STy \;=\; \sum_{k=1}^{\infty}\langle y, T^*e_k\rangle Se_k.$$

Consequently $ST \in L_1(H)$ and

$$\|ST\|_{L_1(H)} \;\leq\; \sum_{k=1}^{\infty}|T^*e_k|\,|Se_k| \leq \left(\sum_{k=1}^{\infty}|T^*e_k|^2\right)^{1/2}\left(\sum_{k=1}^{\infty}|Se_k|^2\right)^{1/2}$$

$$=\; \|T\|_{L_2(H)}\|S\|_{L_2(H)}.$$

Therefore the conclusion follows. $\square$

    Warning. If S and T are bounded operators, and ST is of trace class then in general TS is not, as the following example, provided by S. Peszat [183], shows.

    Define two linear operators S and T on the product space $H \times H$, by

$$S = \begin{pmatrix} 0 & A \\ B & 0 \end{pmatrix}, \quad T = \begin{pmatrix} I & 0 \\ 0 & 0 \end{pmatrix}.$$

Then

$$ST = \begin{pmatrix} 0 & 0 \\ B & 0 \end{pmatrix}, \quad TS = \begin{pmatrix} 0 & A \\ 0 & 0 \end{pmatrix},$$

and it is enough to take B of trace class and A not of trace class. $\square$

    We have also the following result, see e.g. A. Pietsch [187].

Proposition 1.1.3 *Assume that S is a compact self-adjoint operator, and that (λ_k) are its eigenvalues (repeated according to their multiplicity).*

    (i) $S \in L_1(H)$ if and only if $\sum_{k=1}^{\infty}|\lambda_k| < +\infty$. Moreover $\|S\|_{L_1(H)} = \sum_{k=1}^{\infty}|\lambda_k|$,

and $\operatorname{Tr} S = \sum_{k=1}^{\infty}\lambda_k$.

    (ii) $S \in L_2(H)$ if and only if $\sum_{k=1}^{\infty}|\lambda_k|^2 < +\infty$. Moreover

$$\|S\|_{L_2(H)} = \left(\sum_{k=1}^{\infty}|\lambda_k|^2\right)^{1/2}.$$

More generally let S be a compact operator on H. Denote by (λ_k) the sequence of all positive eigenvalues of the operator $(S^*S)^{1/2}$, repeated according to their multiplicity. Denote by $L_p(H)$, $p > 0$, the set of all operators S such that

$$\|S\|_{L_p(H)} = \left(\sum_{k=1}^{\infty} \lambda_k^p \right)^{1/p} < +\infty. \tag{1.1.6}$$

Operators belonging to $L_1(H)$ and $L_2(H)$ are precisely the trace class and the Hilbert-Schmidt operators.

The following result holds, see N. Dunford and J. T. Schwartz [107].

Proposition 1.1.4 *Let* $S \in L_p(H)$, $T \in L_q(H)$ *with* $p > 0, q > 0$. *Then* $ST \in L_r(H)$ *with* $\frac{1}{r} = \frac{1}{p} + \frac{1}{q}$, *and*

$$\|TS\|_{L_r(H)} \le 2^{1/r} \|S\|_{L_p(H)} \|T\|_{L_q(H)}. \tag{1.1.7}$$

1.2 Definition and first properties of Gaussian measures

1.2.1 Measures in metric spaces

If E is a metric space, then $\mathcal{B}(E)$ will denote the Borel σ-algebra, that is the smallest σ-algebra of subsets of E which contains all closed (open) subsets of E.

Let metric spaces E_1, E_2 be equipped with σ-fields $\mathcal{E}_1, \mathcal{E}_2$ respectively. Measurable mappings $X : E_1 \to E_2$ will often be called *random variables*. If μ is a measure on $(E_1, \mathcal{E}_1)$, then its image by the transformation X will be denoted by $X \circ \mu$:

$$X \circ \mu(A) = \mu(X^{-1}(A)), \quad A \in \mathcal{E}_2.$$

We call $X \circ \mu$ the *law* or the *distribution* of X, and we set $X \circ \mu = \mathcal{L}(X)$.

If ν and μ are two finite measures on $(E, \mathcal{E})$ such that $\Gamma \in \mathcal{E}$, $\mu(\Gamma) = 0$ implies $\nu(\Gamma) = 0$ then one writes $\nu << \mu$ and one says that ν is *absolutely continuous* with respect to μ. If there exist $A, B \in \mathcal{E}$ such that $A \cap B = \emptyset$, $\mu(A) = \nu(B) = 1$, one says that μ and ν are *singular*.

If $\nu << \mu$ then by the Radon-Nikodým theorem there exists $g \in L^1(E, \mathcal{E}, \mu)$ nonnegative such that

$$\nu(\Gamma) = \int_{\Gamma} g(x) \mu(dx), \quad \Gamma \in \mathcal{E}.$$

The function g is denoted by $\frac{d\nu}{d\mu}$.

If $\nu << \mu$ and $\mu << \nu$ then one says that μ and ν are *equivalent* and writes $\mu \sim \nu$.

We have the following change of variable formula. If φ is a nonnegative measurable real function on E_2, then

$$\int_{E_1} \varphi(X(x))\mu(dx) = \int_{E_2} \varphi(y)X \circ \mu(dy). \qquad (1.2.1)$$

Let μ and ν be two measures on a separable Hilbert space H; if $T \circ \mu = T \circ \nu$ for any linear operator $T : H \to \mathbb{R}^n$, $n \in \mathbb{N}$, then $\mu = \nu$.

Random variables $X_1, \dots, X_n$ are said to be *independent* if

$$\mathcal{L}(X_1, \dots, X_n) = \mathcal{L}(X_1) \times \cdots \times \mathcal{L}(X_n).$$

A family of random variables $(X_\alpha)_{\alpha \in A}$ is said to be independent, if any finite subset of the family is independent.

Probability measures on a separable Hilbert space H will always be regarded as defined on $\mathcal{B}(H)$. If μ is a probability measure on H, then its Fourier transform is defined by

$$\hat{\mu}(\lambda) = \int_H e^{i\langle \lambda, x \rangle} \mu(dx), \ \lambda \in H;$$

$\hat{\mu}$ is called the *characteristic function* of μ. One can show that if the characteristic functions of two measures are identical, then the measures are identical as well.

1.2.2 Gaussian measures

We first define Gaussian measures on $\mathbb{R}$. If $a \in \mathbb{R}$ we set

$$N_{a,0}(dx) = \delta_a(dx),$$

where δ_a is the Dirac measure at a. If moreover $\lambda > 0$ we set

$$N_{a,\lambda}(dx) = \frac{1}{\sqrt{2\pi\lambda}} e^{-\frac{(x-a)^2}{2\lambda}} dx.$$

The Fourier transform of $N_{a,\lambda}$ is given by

$$\widehat{N_{a,\lambda}}(h) = \int_\mathbb{R} e^{ihx} N_{a,\lambda}(dx) = e^{iah - \frac{1}{2}\lambda h^2}, \ h \in \mathbb{R}.$$

More generally we show now that in an arbitrary separable Hilbert space and for arbitrary $Q \in L_1^+(H)$ there exists a unique measure $N_{a,Q}$ such that

$$\widehat{N_{a,\lambda}}(h) = \int_H e^{i\langle h,x \rangle} N_{a,Q}(dx) = e^{i\langle h,x \rangle - \frac{1}{2}\langle Qh,h \rangle}, \ h \in H.$$

Let in fact $Q \in L_1^+(H)$. Then there exist a complete orthonormal system (e_k) on H and a sequence of nonnegative numbers (λ_k) such that $Qe_k = \lambda_k e_k$, $k \in \mathbb{N}$. We set $x_h = \langle x, e_h \rangle$, $h \in \mathbb{N}$, and $P_n x = \sum_{k=1}^{n} x_k e_k, x \in H, \ n \in \mathbb{N}$. Let us introduce an isomorphism γ from H into ℓ^2: ([1])

$$x \in H \to \gamma(x) = (x_k) \in \ell^2.$$

In the following we shall always identify H with ℓ^2. In particular we shall write $P_n x = (x_1, ..., x_n)$, $x \in \ell^2$.

A subset I of H of the form $I = \{x \in H : (x_1, \ ... \ , x_n) \in B\}$, where $B \in \mathcal{B}(\mathbb{R}^n)$, is said to be *cylindrical*. It is easy to see that the σ-algebra generated by all cylindrical subsets of H coincides with $\mathcal{B}(H)$.

Theorem 1.2.1 *Let $a \in H$, $Q \in L_1^+(H)$. Then there exists a unique probability measure μ on $(H, \mathcal{B}(H))$ such that*

$$\int_H e^{i\langle h,x \rangle} \mu(dx) = e^{i\langle a,h \rangle} e^{-\frac{1}{2}\langle Qh,h \rangle}, \ h \in H. \tag{1.2.2}$$

Moreover μ is the restriction to H (identified with ℓ^2) of the product measure

$$\bigtimes_{k=1}^{\infty} \mu_k = \bigtimes_{k=1}^{\infty} N_{a_k,\lambda_k},$$

defined on $(\mathbb{R}^\infty, \mathcal{B}(\mathbb{R}^\infty))$. ([2])

We set $\mu = N_{a,Q}$, and call a the *mean* and Q the *covariance operator* of μ. Moreover $N_{0,Q}$ will be denoted by N_Q.

Proof of Theorem 1.2.1. Since a characteristic function uniquely determines the measure, we have only to prove existence.

Let us consider the sequence of Gaussian measures (μ_k) on $\mathbb{R}$ defined as $\mu_k = N_{a_k,\lambda_k}$, $k \in \mathbb{N}$, and the product measure $\mu = \bigtimes_{k=1}^{\infty} \mu_k$ in $\mathbb{R}^\infty$, see e.g

---

[1] For any $p \geq 1$, we denote by ℓ^p the Banach space of all sequences (x_k) of real numbers such that $|x|_p := (\sum_{k=1}^{\infty} |x_k|^p)^{1/p} < +\infty$.

[2] We shall consider $\mathbb{R}^\infty$ as a metric space with the distance $d(x,y) := \sum_{k=1}^{\infty} 2^{-k} \frac{|x_k - y_k|}{1 + |x_k - y_k|}$, $x, y \in \mathbb{R}^\infty$

P. R. Halmos [141, §38.B]. We want to prove that μ is concentrated on ℓ^2, (that it is clearly a Borel subset of $\mathbb{R}^\infty$). For this it is enough to show that

$$\int_{\ell^\infty} |x|^2_{\ell^2}\, \mu(dx) < +\infty. \tag{1.2.3}$$

We have in fact, by the monotone convergence theorem,

$$\int_{\mathbb{R}^\infty} |x|^2_{\ell^2}\mu(dx) \;=\; \sum_{k=1}^\infty \int_{\mathbb{R}^\infty} x_k^2\, \mu(dx) = \sum_{k=1}^\infty \left(\int_{\mathbb{R}} (x_k - a_k)^2 \mu_k(dx) + a_k^2 \right)$$

$$= \sum_{k=1}^\infty (\lambda_k + a_k^2) = \operatorname{Tr} Q + |a|^2 < +\infty.$$

Now we consider the restriction of μ to ℓ^2, which we still denote by μ. We have to prove that (1.2.2) holds. Setting $\nu_n = \prod_{k=1}^n \mu_k$, we have

$$\int_{\ell^2} e^{i\langle x, h \rangle}\mu(dx) = \lim_{n\to\infty} \int_{\ell^2} e^{i\langle P_n h, P_n x \rangle}\mu(dx)$$

$$= \lim_{n\to\infty} \int_{\mathbb{R}^n} e^{i\langle P_n h, P_n x \rangle}\nu_n(dx) = \lim_{n\to\infty} e^{i\langle P_n h, P_n a \rangle - \frac{1}{2}\langle Q P_n h, P_n h \rangle}$$

$$= e^{i\langle h, a \rangle - \frac{1}{2}\langle Q h, h \rangle}. \;\square$$

If the law of a random variable is a Gaussian measure, then the random variable is called *Gaussian*. It easily follows from Theorem 1.2.1 that a random variable X with values in H is Gaussian if and only if for any $h \in H$ the real valued random variable $\langle h, X \rangle$ is Gaussian.

Remark 1.2.2 From the proof of Theorem 1.2.1 it follows that

$$\int_H |x|^2 N_{a,Q}(dx) = \operatorname{Tr} Q + |a|^2. \tag{1.2.4}$$

Proposition 1.2.3 *Let $T \in L(H)$, and $a \in H$, and let $\Gamma x = Tx + a$, $x \in H$. Then $\Gamma \circ N_{m,Q} = N_{Tm+a, TQT^*}$.*

Proof. Notice that, by the change of variables formula (1.2.1), we have

$$\int_H e^{i\langle \lambda, y \rangle} \Gamma \circ N_{m,Q}(dy) = \int_H e^{i\langle \lambda, \Gamma x \rangle} N_{m,Q}(dy)$$

$$= \int_H e^{i\langle \lambda, Tx + a \rangle} N_{m,Q}(dy) = e^{i\langle \lambda, a \rangle} e^{i\langle T^*\lambda, m \rangle - \frac{1}{2}\langle QT^*\lambda, T^*\lambda \rangle}.$$

This shows the result. $\square$

1.2.3 Computation of some Gaussian integrals

We are here given a Gaussian measure $N_{a,Q}$. We set

$$L^2(H, N_{a,Q}) = L^2(H, \mathcal{B}(H), N_{a,Q}).$$

The following identities can be easily proved, using (1.2.2).

Proposition 1.2.4 *We have*

$$\int_H x N_{a,Q}(dx) = a, \qquad (1.2.5)$$

$$\int_H \langle x - a, y\rangle\langle x - a, z\rangle N_{a,Q}(dx) = \langle Qy, z\rangle. \qquad (1.2.6)$$

$$\int_H |x - a|^2 N_{a,Q}(dx) = \operatorname{Tr} Q. \qquad (1.2.7)$$

Proof. We prove as instance (1.2.6). We have

$$\int_H x N_{a,Q}(dx) = \lim_{n\to\infty} \int_H P_n x N_{a,Q}(dx).$$

But

$$\int_H P_n x N_{a,Q}(dx) = (2\pi)^{-n/2} \prod_{k=1}^n \int_{\mathbb{R}} x_k \lambda_k^{-1/2} e^{-\frac{(x_k - a_k)^2}{2\lambda_k}} \, dx_k = a_k,$$

and the conclusion follows. $\square$

Proposition 1.2.5 *For any $h \in H$, the exponential function E_h, defined as*

$$E_h(x) = e^{\langle h, x\rangle}, \quad x \in H,$$

belongs to $L^p(H, N_{a,Q})$, $p \geq 1$, and

$$\int_H e^{\langle h, x\rangle} N_{a,Q}(dx) = e^{\langle u, h\rangle} e^{\frac{1}{2}\langle Qh, h\rangle}. \qquad (1.2.8)$$

Moreover the subspace of $L^2(H, N_{a,Q})$ spanned by all E_h, $h \in H$, is dense on $L^2(H, N_{a,Q})$.

Proof. We have

$$\int_H e^{\langle P_n h, P_n x\rangle} N_{a,Q}(dx) = e^{\langle P_n a, P_n h\rangle} e^{\frac{1}{2}\langle Q P_n h, P_n h\rangle}.$$

Letting n tend to 0 this gives (1.2.8).

Let us prove the last statement. Let $\varphi \in L^2(H, N_{a,Q})$ be such that

$$\int_H e^{\langle h,x \rangle} \varphi(x) N_{a,Q}(dx) = 0, \quad h \in H.$$

Denote by φ^+ and φ^- the positive and negative parts of φ. Then

$$\int_H e^{\langle h,x \rangle} \varphi^+(x) N_{a,Q}(dx) = \int_H e^{\langle h,x \rangle} \varphi^-(x) N_{a,Q}(dx), \quad h \in H.$$

Let us define two measures

$$\mu(dx) = \varphi^+(x) N_{a,Q}(dx), \quad \nu(dx) = \varphi^-(x) N_{a,Q}(dx).$$

Then μ and ν are finite measures such that

$$\int_H e^{\langle h,x \rangle} \mu(dx) = \int_H e^{\langle h,x \rangle} \nu(dx), \quad h \in H.$$

Let T be any linear transformation from H into $\mathbb{R}^n$, $n \in \mathbb{N}$. Then for any $\lambda \in \mathbb{R}^n$

$$\int_{\mathbb{R}^n} e^{\langle \lambda, z \rangle} T \circ \mu(dz) = \int_H e^{\langle \lambda, Tx \rangle} \mu(dx) = \int_H e^{\langle T^*\lambda, \rangle >} \mu(dx)$$

$$= \int_H e^{\langle T^*\lambda, x \rangle} \nu(dx) = \int_{\mathbb{R}^n} e^{\langle \lambda, z \rangle} T \circ \nu(dz).$$

By a well known finite dimensional result $T \circ \mu = T \circ \nu$. Consequently measures μ and ν are identical and so $\varphi = 0$. $\square$

1.2.4 The reproducing kernel

Here we are given an operator $Q \in L_1^+(H)$. We denote as before by (e_k) a complete orthonormal system in H and by (λ_k) a sequence of positive numbers such that $Q e_k = \lambda_k e_k$, $k \in \mathbb{N}$.

The subspace $Q^{1/2}(H)$ is called the *reproducing kernel* of the measure N_Q. If $\text{Ker } Q = \{0\}$, $Q^{1/2}(H)$ is dense on H. In fact, if $x_0 \in H$ is such that $\langle Q^{1/2}h, x_0 \rangle = 0$ for all $h \in H$, we have $Q^{1/2}x_0 = 0$ and so $Qx_0 = 0$, which yields $x_0 = 0$.

Let $\text{Ker } Q = \{0\}$. We are now going to introduce an isomorphism W from H into $L^2(H, N_Q)$ that will play an important rôle in the following. The isomorphism W is defined by

$$f \in Q^{1/2}(H) \rightarrow W_f \in L^2(H, N_Q), \quad W_f(x) = \langle Q^{-1/2}f, x \rangle, \quad x \in H.$$

By (1.2.7) it follows that

$$\int_H W_f(x)W_g(x)N_Q(dx) = \langle f, g \rangle, \; f, g \in H.$$

Thus W is an isometry and it can be uniquely extended to all of H. It will be denoted by the same symbol. For any $f \in H$, W_f is a real Gaussian random variable $N_{|f|^2}$.

More generally, for arbitrary elements $f_1, ..., f_n$, $(W_{f_1}, ..., W_{f_n})$ is a Gaussian vector with mean 0 and covariance matrix $(\langle f_i, f_j \rangle)$. If Ker $Q \neq \{0\}$ then the trasformation $f \to W_f$ can be defined in exactly the same way but only for $f \in H_0 = \overline{Q^{1/2}(H)}$. We will write in some cases $\langle Q^{-1/2}y, f \rangle$ instead of $W_f(y)$.

The proof of the following proposition is left as an exercise to the reader.

Proposition 1.2.6 *For any orthonormal sequence (f_n) in H, the family*

$$1, \; W_{f_n}, \; W_{f_k}W_{f_l}, \; 2^{-1/2}\left(W_{f_m}^2 - 1\right), \; m, n, k, l \in \mathbb{N}, \; k \neq l,$$

is orthonormal in $L^2(H, N_Q)$.

Next we consider the function $f \to e^{W_f}$.

Proposition 1.2.7 *The transformation $f \to e^{W_f}$ acts continuously from H into $L^2(H, N_Q)$, and*

$$\int_H e^{W_f(x)} N_Q(dx) = e^{\frac{1}{2}|f|^2},$$

$$\int_H e^{i\,\lambda W_f(x)} N_Q(dx) = e^{-\frac{1}{2}\lambda^2|f|^2}, \; \lambda \in \mathbb{R}. \qquad (1.2.9)$$

Proof. Since W_f is Gaussian with law $N_{0,|f|^2}$, (1.2.9) follows. Moreover, taking into account (1.2.8) it follows that

$$\int_H \left[e^{W_f} - e^{W_g}\right]^2 dN_Q = \int_H \left[e^{2W_f} - 2e^{W_{f+g}} + e^{2W_g}\right] dN_Q$$

$$= e^{2|f|^2} - 2e^{\frac{1}{2}|f+g|^2} + e^{2|g|^2} = \left[e^{|f|^2} - e^{|g|^2}\right]^2 + 2e^{|f|^2+|g|^2}\left[1 - e^{-\frac{1}{2}|f-g|^2}\right],$$

which shows that W_f is locally uniformly continuous on H. $\square$

Let us define the determinant of $1 + S$ where S is a compact self-adjoint operator in $L_1(H)$:

$$\det(1 + S) = \prod_{k=1}^{\infty}(1 + s_k),$$

where (s_k) is the sequence of eigenvalues of S (repeated according to their multiplicity).

Proposition 1.2.8 *Assume that M is a symmetric operator such that $Q^{1/2}MQ^{1/2} < 1$, (^{3}) and let $b \in H$. Then*

$$\int_H \exp\left\{\frac{1}{2}\langle My, y\rangle + \langle b, y\rangle\right\} N_Q(dy)$$

$$= \left[\det(1 - Q^{1/2}MQ^{1/2})\right]^{-1/2} \exp\left\{\frac{1}{2}|(1 - Q^{1/2}MQ^{1/2})^{-1/2}Q^{1/2}b|^2\right\}.$$

$$(1.2.10)$$

Proof. Let (g_n) be an orthonormal basis for the operator $Q^{1/2}MQ^{1/2}$, and let (γ_n) be the sequence of the corresponding eigenvalues.

Claim 1. We have

$$\langle b, x\rangle = \sum_{k=1}^\infty \langle Q^{1/2}b, g_n\rangle W_{g_n}(x), \quad N_Q\text{-a.e.}$$

Claim 2. We have

$$\langle Mx, x\rangle = \sum_{n=1}^\infty \gamma_n |W_{g_n}(x)|^2, \quad N_Q\text{-a.e},$$

the series being convergent in $L^1(H, N_Q)$.

We shall only prove the more difficult second claim.

Let $P_N = \sum_{k=1}^N e_k \otimes e_k$. (^{4}) Then for any $x \in H$ we have

$$\langle MP_N x, P_N x\rangle = \langle (Q^{1/2}MQ^{1/2})Q^{-1/2}P_N x, Q^{-1/2}P_N x\rangle$$

$$= \sum_{n=1}^\infty \langle (Q^{1/2}MQ^{1/2})Q^{-1/2}P_N x, g_n\rangle\langle Q^{-1/2}P_N x, g_n\rangle$$

$$= \sum_{n=1}^\infty \gamma_n |\langle Q^{-1/2}P_N x, g_n\rangle|^2.$$

Consequently, for each fixed x

$$\langle MP_N x, P_N x\rangle = \sum_{n=1}^\infty \gamma_n |W_{P_N g_n}|^2, \quad N \in \mathbb{N}.$$

---

^{3}This means that $\langle Q^{1/2}MQ^{1/2}x, x\rangle < |x|^2$ for any $x \in H$ different from 0.
^{4}We rember that (e_k) is the sequence of eigenvectors of Q.

Moreover for each $L \in \mathbb{N}$

$$\int_H \left| \langle MP_N x, P_N x \rangle - \sum_{n=1}^{L} \gamma_n |W_{P_N g_n}|^2 \right| N_Q(dx)$$

$$\leq \sum_{n=L+1}^{\infty} |\gamma_n| \int_H |W_{P_N g_n}|^2 N_Q(dx)$$

$$= \sum_{n=L+1}^{\infty} |\gamma_n| \, |P_N g_n|^2 \leq \sum_{n=L+1}^{\infty} |\gamma_n|.$$

As $N \to \infty$ then $P_N x \to x$ and $W_{P_N g_n} \to W_{g_n}$ in $L^2(H, N_Q)$. Passing to subsequences if needed, and using the Fatou lemma, we see that

$$\int_H \left| \langle Mx, x \rangle - \sum_{n=1}^{L} \gamma_n |W_{g_n}|^2 \right| N_Q(dx) \leq \sum_{n=L+1}^{\infty} |\gamma_n|.$$

Therefore the claim is proved.

By the claims it follows that

$$\exp \left\{ \frac{1}{2} \langle Mx, x \rangle + \langle b, x \rangle \right\}$$

$$= \lim_{L \to \infty} \exp \left\{ \sum_{n=1}^{L} \frac{1}{2} \gamma_n |W_{g_n}(x)|^2 + \langle Q^{1/2} b, g_n \rangle W_{g_n}(x) \right\},$$

with a.e. convergence with respect to N_Q for a suitable subsequence. Using the fact that $(W g_n)$ are independent Gaussian random variables, we obtain, by a direct calculation, for $p \geq 1$,

$$\int_H \exp \left\{ p \sum_{n=1}^{L} \frac{1}{2} \gamma_n |W_{g_n}(x)|^2 + p \langle Q^{1/2} b, g_n \rangle W_{g_n}(x) \right\} N_Q(dx)$$

$$= \left[\prod_{n=1}^{L} (1 - p\gamma_n) \right]^{-1/2} \exp \left\{ \frac{1}{2} \sum_{n=1}^{\infty} \frac{|\langle Q^{1/2} b, g_n \rangle|^2}{1 - p\gamma_n} \right\}.$$

Since $\gamma_n < 1$, and $\sum_{n=1}^{\infty} |\gamma_n| < \infty$, there exists $p > 1$ such that $p\gamma_n < 1$, for all $n \in \mathbb{N}$. Therefore

$$\lim_{L \to \infty} \prod_{n=1}^{L} (1 - p\gamma_n)^{-1/2} \exp\left\{\frac{1}{2} \frac{|\langle Q^{1/2}b, g_n\rangle|^2}{1 - p\gamma_n}\right\}$$

$$= \left[\prod_{n=1}^{\infty} (1 - p\gamma_n)\right]^{-1/2} \exp\left\{\frac{1}{2} \sum_{n=1}^{\infty} \frac{|\langle Q^{1/2}b, g_n\rangle|^2}{1 - p\gamma_n}\right\}.$$

So the sequence $\left(\exp\left\{\sum_{n=1}^{L} \left[\frac{1}{2}\gamma_n |W_{g_n}(x)|^2 + \langle Q^{1/2}b, g_n\rangle W_{g_n}(x)\right]\right\}\right)$ is uniformly integrable. Consequently, passing to the limit, we find

$$\int_H \exp\left\{1/2 \langle My, y\rangle + \langle b, y\rangle\right\} N_Q(dy)$$

$$= \lim_{L \to \infty} \int_H \exp\left\{\sum_{n=1}^{L} \left[1/2 \, \gamma_n |W_{g_n}(x)|^2 + \langle Q^{1/2}b, g_n\rangle W_{g_n}(x)\right]\right\} N_Q(dx)$$

$$= \lim_{L \to \infty} \prod_{n=1}^{L} (1 - \gamma_n)^{-1/2} \exp\left\{\frac{1}{2} \frac{|\langle Q^{1/2}b, g_n\rangle|^2}{1 - \gamma_n}\right\}$$

$$= \prod_{n=1}^{\infty} (1 - \gamma_n)^{-1/2} \exp\left\{\frac{1}{2} \frac{|\langle Q^{1/2}b, g_n\rangle|^2}{1 - \gamma_n}\right\}$$

$$= \left(\det(1 - Q^{1/2}MQ^{1/2})\right)^{-1/2} \exp\left\{\frac{1}{2}|(1 - Q^{1/2}MQ^{1/2})^{-1/2}Q^{1/2}b|^2\right\}. \quad \square$$

Remark 1.2.9 It follows from the proof of the proposition that

$$\langle Mx, x\rangle = \sum_{k=1}^{\infty} \gamma_n W_{g_n}^2(x) = \sqrt{2} \sum_{k=1}^{\infty} \gamma_n \left[2^{-1/2}(W_{g_n}^2(x) - 1)\right] + \sum_{k=1}^{\infty} \gamma_n,$$

and so, by Proposition 1.2.6, we have

$$\int_H [\langle Mx, x \rangle]^2 N_Q(dx) = 2 \sum_{k=1}^{\infty} \gamma_n^2 + \left(\sum_{k=1}^{\infty} \gamma_n \right)^2$$

$$= 2 \| Q^{1/2} M Q^{1/2} \|_{L_2(H)}^2 + (\operatorname{Tr} Q^{1/2} M Q^{1/2})^2$$

$$< +\infty.$$

Proposition 1.2.10 *Let $T \in L_1(H)$. Then there exists the limit*

$$\langle T Q^{-1/2} y, Q^{-1/2} y \rangle := \lim_{n \to \infty} \langle T Q^{-1/2} P_n y, Q^{-1/2} P_n y \rangle, \quad N_Q\text{-a.e.},$$

where $P_n = \sum_{k=1}^{n} e_k \otimes e_k$.
Moreover we have the following expansion in $L^2(H, N_Q)$:

$$\langle T Q^{-1/2} y, Q^{-1/2} y \rangle = \sum_{n=1}^{\infty} \langle T g_n, g_n \rangle + \sum_{m \neq n=1}^{\infty} \langle T g_n, g_m \rangle W_{g_n} W_{g_m}$$

$$\times \sqrt{2} \sum_{n=1}^{\infty} \langle T g_n, g_n \rangle \left[2^{-1/2} \left(W_{g_n}^2 - 1 \right) \right]. \quad (1.2.11)$$

The proof of the following result is similar to that of Claim 2 in the proof of Proposition 1.2.8 and it is left to the reader.

Proposition 1.2.11 *Assume that M is a symmetric trace-class operator such that $M < 1$,([5]) and $b \in H$. Then*

$$\int_H \exp \left\{ 1/2 \langle M Q^{-1/2} y, Q^{-1/2} y \rangle + \langle b, Q^{-1/2} y \rangle \right\} N_Q(dy)$$

$$= (\det(1 - M))^{-1/2} \, e^{\frac{1}{2} |(1-M)^{-1/2} b|^2}. \quad (1.2.12)$$

1.3 Absolute continuity of Gaussian measures

We consider here two Gaussian measures μ, ν. We want to prove the Feldman-Hajek theorem , that is they are either singular or equivalent.

---

[5]That is $\langle Mx, x \rangle < |x|^2$ for all $x \neq 0$.

In §1.3.1 we recall some results on equivalence of measures on $\mathbb{R}^\infty$ including the Kakutani theorem. In §1.3.2 we consider the case when $\mu = N_Q$ and $\nu = N_{a,Q}$ with $Q \in L_1^+(H)$ and $a \in H$, proving the Cameron-Martin formula. Finally in §1.3.3 we consider the more difficult case when $\mu = N_Q$ and $\nu = N_R$ with $Q, R \in L_1^+(H)$.

1.3.1   Equivalence of product measures in $\mathbb{R}^\infty$

It is convenient to introduce the notion of *Hellinger* integral.

Let μ, ν be probability measures on a measurable space $(E, \mathcal{E})$. Then $\lambda = \frac{1}{2}(\mu + \nu)$ is also a probability measure on $(E, \mathcal{E})$ and we have obviously

$$\mu << \lambda, \quad \nu << \lambda.$$

We define the *Hellinger integral* by

$$H(\mu, \nu) = \int_E \left[\frac{d\mu}{d\lambda}(x) \frac{d\nu}{d\lambda}(x) \right]^{1/2} \lambda(dx).$$

Instead of $\frac{1}{2}(\mu + \nu)$ one could choose as λ any measure equivalent to $\frac{1}{2}(\mu + \nu)$ without changing the value of $H(\mu, \nu)$.

By using Hölder's inequality we see that

$$|H(\mu, \nu)|^2 \leq \int_E \frac{d\mu}{d\lambda}(x) \lambda(dx) \int_E \frac{d\nu}{d\lambda}(x) \lambda(dx) = 1,$$

so that $0 \leq H(\mu, \nu) \leq 1$.

Exercise 1.3.1 (a) Let $\mu = N_q$ and $\nu = N_{a,q}$, where $a \in \mathbb{R}$ and $q > 0$. Show that we have

$$H(\mu, \nu) = e^{-\frac{a^2}{4q}}. \tag{1.3.1}$$

(b) Let $\mu = N_q$ and $\nu = N_\rho$, where $q, \rho > 0$. Show that we have

$$H(\mu, \nu) = \left[\frac{4q\rho}{(q + \rho)^2} \right]^{1/4}. \tag{1.3.2}$$

Proposition 1.3.2 *Assume that $H(\mu, \nu) = 0$. Then the measures μ and ν are singular.*

Proof. Set $\alpha = \frac{d\mu}{d\lambda}$, $\beta = \frac{d\nu}{d\lambda}$. Since $H(\mu, \nu) = \int_\Omega \sqrt{\alpha\beta} \, d\lambda = 0$, we have $\alpha\beta = 0$, λ-a.e. Consequently, setting

$$A = \{\omega \in \Omega : \alpha(\omega) = 0\}, \quad B = \{\omega \in \Omega : \beta(\omega) = 0\},$$

we have $\lambda(A \cup B) = 1$. This means that $\lambda(C) = 0$ where $C = \Omega \setminus (A \cup B)$, and hence $\mu(C) = \nu(C) = 0$. Then, as

$$\mu(A) = \int_A \alpha \, d\lambda = 0, \quad \nu(B) = \int_B \beta \, d\lambda = 0,$$

we have that μ and ν are singular since

$$\mu(A \cup C) = \nu(B) = 0, \quad (A \cup C) \cap B = \emptyset. \quad \square$$

Proposition 1.3.3 *Let* $\mathcal{G} \subset \mathcal{E}$ *be a* σ-*algebra, and let* $\mu_\mathcal{G}$ *and* $\nu_\mathcal{G}$ *be the restrictions of* μ *and* ν *to* $(E, \mathcal{G})$. *Then we have* $H(\mu, \nu) \leq H(\mu_\mathcal{G}, \nu_\mathcal{G})$.

Proof. Let $\lambda_\mathcal{G}$ be the restriction of λ to $(E, \mathcal{G})$. It is easy to check that

$$\frac{d\mu_\mathcal{G}}{d\lambda_\mathcal{G}} = E_\lambda \left(\frac{d\mu}{d\lambda} \Big| \mathcal{G} \right) \quad \frac{d\nu_\mathcal{G}}{d\lambda_\mathcal{G}} = E_\lambda \left(\frac{d\nu}{d\lambda} \Big| \mathcal{G} \right), \quad \lambda\text{-a.e.}(^6)$$

Consequently we have $(^7)$

$$H(\mu_\mathcal{G}, \nu_\mathcal{G}) = \int_E \left[E_\lambda \left(\frac{d\mu}{d\lambda} \Big| \mathcal{G} \right) E_\lambda \left(\frac{d\nu}{d\lambda} \Big| \mathcal{G} \right) \right]^{1/2} d\lambda.$$

Since λ-a.e.

$$\frac{\left[\frac{d\mu}{d\lambda} \frac{d\nu}{d\lambda} \right]^{1/2}}{\left[E_\lambda \left(\frac{d\mu}{d\lambda} \big| \mathcal{G} \right) E_\lambda \left(\frac{d\nu}{d\lambda} \big| \mathcal{G} \right) \right]^{1/2}} \leq \frac{1}{2} \left(\frac{\frac{d\mu}{d\lambda}}{E_\lambda \left(\frac{d\mu}{d\lambda} \big| \mathcal{G} \right)} + \frac{\frac{d\nu}{d\lambda}}{E_\lambda \left(\frac{d\nu}{d\lambda} \big| \mathcal{G} \right)} \right),$$

taking conditional expectations of both sides one finds, λ-a.e.,

$$\left[E_\lambda \left(\frac{d\mu}{d\lambda} \Big| \mathcal{G} \right) E_\lambda \left(\frac{d\nu}{d\lambda} \Big| \mathcal{G} \right) \right]^{1/2} \geq E_\lambda \left(\left(\frac{d\mu}{d\lambda} \right)^{1/2} \left(\frac{d\nu}{d\lambda} \right)^{1/2} \Big| \mathcal{G} \right). \quad (1.3.3)$$

$^6 E_\lambda(\eta | \mathcal{G})$ is the conditional expectation of the random variable η with respect to $\mathcal{G}$ and measure λ.

^{7}For positive numbers a, b, c, d, $\sqrt{\frac{ab}{cd}} \leq \frac{1}{2} \left(\frac{a}{c} + \frac{b}{d} \right)$.

Integrating with respect to λ both sides of (1.3.3), the required result follows.
$\square$

Now let us consider two sequences of measures (μ_k) and (ν_k) on $(\mathbb{R}, \mathcal{B}(\mathbb{R}))$ such that $\nu_k \sim \mu_k$ for all $k \in \mathbb{N}$. We set $\lambda_k = \frac{1}{2}(\mu_k + \nu_k)$, and we consider the Hellinger integral

$$H(\mu_k, \nu_k) = \int_{\mathbb{R}} \left[\frac{d\mu_k}{d\lambda_k}(x) \frac{d\nu_k}{d\lambda_k}(x) \right]^{1/2} \lambda_k(dx), \ k \in \mathbb{N}.$$

Remark 1.3.4 Since (μ_k) and (ν_k) are equivalent, we have

$$\frac{d\mu_k}{d\lambda_k} \frac{d\nu_k}{d\lambda_k} = \frac{d\mu_k}{d\lambda_k} \frac{d\nu_k}{d\mu_k} \frac{d\mu_k}{d\lambda_k} = \frac{d\nu_k}{d\mu_k} \left(\frac{d\mu_k}{d\lambda_k} \right)^2.$$

Thus

$$H(\mu_k, \nu_k) = \int_{\mathbb{R}} \left[\frac{d\nu_k}{d\mu_k}(x) \right]^{1/2} \mu_k(dx). \tag{1.3.4}$$

We also consider the product measures on $\mathbb{R}^\infty$

$$\mu = \prod_{k=1}^\infty \mu_k, \ \nu = \prod_{k=1}^\infty \nu_k,$$

and the corresponding Hellinger integral $H(\mu, \nu)$. As is easily checked we have

$$H(\mu, \nu) = \prod_{k=1}^\infty H(\mu_k, \nu_k).$$

Proposition 1.3.5 (Kakutani) *If $H(\mu, \nu) > 0$ then μ and ν are equivalent. Moreover*

$$f(x) := \frac{d\nu}{d\mu}(x) = \prod_{k=1}^\infty \frac{d\nu_k}{d\mu_k}(x_k), \ x \in \mathbb{R}^\infty, \ \mu\text{-a.e.} \tag{1.3.5}$$

Proof. We set

$$f_n(x) = \prod_{k=1}^n \frac{d\nu_k}{d\mu_k}(x_k), \ x \in \mathbb{R}^\infty, \ n \in \mathbb{N}.$$

We are going to prove that the sequence (f_n) is convergent on $L^1(\mathbb{R}^\infty, \mathcal{B}(\mathbb{R}^\infty), \mu)$. Let $m, n \in \mathbb{N}$, then we have

$$\int_{\mathbb{R}^\infty} \left| f_{n+m}^{1/2}(x) - f_n^{1/2}(x) \right|^2 \mu(dx)$$

$$= \int_{\mathbb{R}^\infty} \prod_{k=1}^n \frac{d\nu_k}{d\mu_k}(x_k) \left| \prod_{k=n+1}^{n+m} \left(\frac{d\nu_k}{d\mu_k}(x_k) \right)^{1/2} - 1 \right|^2 \mu(dx)$$

$$= \prod_{k=1}^n \int_{\mathbb{R}^\infty} \frac{d\nu_k}{d\mu_k}(x_k) \mu(dx) \int_{\mathbb{R}^\infty} \left| \prod_{k=n+1}^{n+m} \left(\frac{d\nu_k}{d\mu_k}(x_k) \right)^{1/2} - 1 \right|^2 \mu(dx).$$

Consequently

$$\int_{\mathbb{R}^\infty} |f_{n+p}^{1/2}(x) - f_n^{1/2}(x)|^2 \mu(dx)$$

$$= \int_{\mathbb{R}^\infty} \left[\prod_{k=n+1}^{n+p} \frac{d\nu_k}{d\mu_k}(x_k) - 2 \prod_{k=n+1}^{n+p} \left(\frac{d\nu_k}{d\mu_k}(x_k) \right)^{1/2} + 1 \right] \mu(dx)$$

$$= 2 \left(1 - \prod_{k=n+1}^{n+p} \int_{\mathbb{R}} \left(\frac{d\nu_k}{d\mu_k}(x_k) \right)^{1/2} \mu_k(dx_k) \right)$$

$$= 2 \left(1 - \prod_{k=n+1}^{n+p} H(\mu_k, \nu_k) \right). \tag{1.3.6}$$

On the other hand we know by assumption that

$$H(\mu, \nu) = \prod_{k=1}^\infty H(\mu_k, \nu_k) > 0,$$

or, equivalently, that

$$-\log H(\mu, \nu) = -\sum_{k=1}^\infty \log[H(\mu_k, \nu_k)] < +\infty.$$

Consequently, for any $\varepsilon > 0$ there exists $n_\varepsilon \in \mathbb{N}$ such that if $n > n_\varepsilon$ and $p \in \mathbb{N}$, we have

$$- \sum_{k=n+1}^{n+p} \log[H(\mu_k, \nu_k)] < \varepsilon.$$

By (1.3.6) if $n > n_\varepsilon$ we have

$$\int_{\mathbb{R}^\infty} |\sqrt{f_{n+p}} - \sqrt{f_n}|^2 d\mu \le 2(1 - e^{-\varepsilon}).$$

Thus the sequence $(f_n^{1/2})$ is convergent on $L^2(\mathbb{R}^\infty, \mathcal{B}(\mathbb{R}^\infty), \mu)$ to some function $f^{1/2}$. Therefore $f_n \to f$ in $L^1(\mathbb{R}^\infty, \mathcal{B}(\mathbb{R}^\infty), \mu)$.

Finally, we prove that $\nu << \mu$ and $f = \frac{d\nu}{d\mu}$. Let φ be a continuous bounded Borel function on $\mathbb{R}^\infty$, and set $\varphi_n(x) = \varphi(P_n(x))$, $x \in \mathbb{R}^\infty$, where $P_n x = \{x_1, \ldots, x_n, 0, 0, \ldots\}$. Then we have

$$\int_{\mathbb{R}^\infty} \varphi(P_n x)\nu(dx) = \int_{\mathbb{R}^n} \varphi(P_n x)\, \nu_1(dx_1) \ldots \nu_n(dx_n)$$

$$= \int_{\mathbb{R}^n} \varphi(P_n x) \frac{d\nu_1}{d\mu_1}(x_1) \ldots \frac{d\nu_n}{d\mu_n}(x_n)\, \mu_1(dx_1) \ldots \mu_n(dx_n)$$

$$= \int_{\mathbb{R}^\infty} \varphi(P_n x) f_n(x)\mu(dx).$$

Letting n tend to infinity, we find

$$\int_{\mathbb{R}^\infty} \varphi(x)\nu(dx) = \int_{\mathbb{R}^\infty} \varphi(x)f(x)\mu(dx),$$

so that $\nu << \mu$. Finally, by exchanging the rôles of μ and ν, we find $\mu << \nu$. $\square$

1.3.2 The Cameron-Martin formula

We consider here the measures $\mu = N_{a,Q}$ and $\nu = N_Q$, and for any $a \in Q^{1/2}(H)$ we set

$$\rho_a(x) = \exp\left\{-\frac{1}{2}|Q^{-1/2}a|^2 + \langle Q^{-1/2}a, Q^{-1/2}x\rangle\right\}, \quad x \in H. \qquad (1.3.7)$$

Let us recall, see §1.2.4, that $W_f(x) = \langle f, Q^{-1/2}x\rangle$ was defined for all $f \in \overline{Q^{1/2}(H)}$. Since $Q^{-1/2}a \in Q^{1/2}(H)$ the definition (1.3.7) is meaningful.

Theorem 1.3.6 *(i) If $a \in Q^{1/2}(H)$ then the measures μ and ν are equivalent and*

$$\frac{d\mu}{d\nu}(x) = \rho_a(x), \; x \in H. \tag{1.3.8}$$

(ii) If $a \notin Q^{1/2}(H)$ then the measures μ and ν are singular.

Proof. To prove (i) it is enough to show that for any real function φ in H bounded and Borel we have

$$\int_H \varphi(x) N_{a,Q}(dx) = \int_H \varphi(x)\rho_a(x) N_Q(dx). \tag{1.3.9}$$

It is enough to show (1.3.9) for $\varphi = E_h$, $h \in H$. In this case we have in fact

$$\int_H E_h(x) N_{a,Q}(dx) = e^{\langle a,h \rangle + \frac{1}{2}\langle Qh,h \rangle},$$

and

$$\int_H E_a(x)\rho_a(x) N_Q(dx)$$

$$= \exp\left\{ -\frac{1}{2}|Q^{-1/2}a|^2 \right\} \int_H \exp\left\{ \langle h,x \rangle + \langle Q^{-1/2}a, Q^{-1/2}x \rangle \right\} N_Q(dx)$$

$$= \exp\left\{ \langle a,h \rangle + \frac{1}{2}\langle Qh,h \rangle \right\}.$$

Therefore (i) is proved. Let us prove (ii). By identifying as before H with ℓ^2 we can write

$$\mu = \prod_{k=1}^{\infty} N_{\lambda_k}, \; \nu = \prod_{k=1}^{\infty} N_{a_k,\lambda_k}.$$

Then by (1.3.1) we can compute the Hellinger integral $H(\mu,\nu)$:

$$H(\mu,\nu) = \prod_{k=1}^{\infty} e^{-\frac{a_k^2}{4\lambda_k}}.$$

Now if $a \notin Q^{1/2}(H)$ we have $H(\mu,\nu) = 0$. $\square$

1.3.3 The Feldman-Hajek theorem

We consider here two linear operators $Q, R \in L_1^+(H)$ such that

$$\text{Ker } Q = \text{Ker } R = \{0\},$$

and set $\mu = N_Q$ and $\nu = N_R$. We denote by (e_k) a complete orthonormal system in H such that $Qe_k = \lambda_k e_k$, $k \in \mathbb{N}$, where (λ_k) is the sequence of eigenvalues of Q. We set $x_k = \langle x, e_k \rangle$, $x \in H$, $k \in \mathbb{N}$.

The following result is a consequence of the general theorem 1.3.9. We prefer however to prove it directly.

Theorem 1.3.7 *Assume that Q and R commute and that for a sequence of positive numbers (ρ_k) we have $Re_k = \rho_k e_k$, $k \in \mathbb{N}$.*

(i) If $\displaystyle\sum_{k=1}^{\infty} \frac{(\lambda_k - \rho_k)^2}{(\lambda_k + \rho_k)^2} < +\infty$, *$\mu$ and ν are equivalent and we have*

$$\frac{d\nu}{d\mu}(x) = \prod_{k=1}^{\infty} \exp\left\{-\frac{(\lambda_k - \rho_k)x_k^2}{2\lambda_k\rho_k}\right\}. \qquad (1.3.10)$$

(ii) If $\displaystyle\sum_{k=1}^{\infty} \frac{(\lambda_k - \rho_k)^2}{(\lambda_k + \rho_k)^2} = +\infty$, *$\mu$ and ν are singular.*

Proof. Let us compute the Hellinger integral $H(\mu, \nu) = \displaystyle\prod_{k=1}^{\infty} H(\mu_k, \nu_k)$, where $\mu_k = N_{\lambda_k}$, $\nu_k = N_{\rho_k}$. By (1.3.2) we have $H(\mu, \nu) = \prod_{k=1}^{\infty} \left[\frac{4\lambda_k\rho_k}{(\lambda_k+\rho_k)^2}\right]^{1/4}$, therefore the conclusion follows from Propositions 1.3.2 and 1.3.5. $\square$

Example 1.3.8 (i) Let $R = \alpha Q$, for some $\alpha > 0$ different from 1. Then μ and ν are singular.

(ii) Let $\lambda_k = \frac{1}{k^2}$, $\rho_k = \frac{1}{k^2} + \frac{1}{k^3}$, $k \in \mathbb{N}$. Then μ and ν are equivalent.

Theorem 1.3.9 *Assume that μ and ν are not singular. Then there exists a symmetric Hilbert-Schmidt operator S such that $R = Q^{1/2}(1 - S)Q^{1/2}$.*

Proof. Since μ and ν are not singular we have $H(\mu, \nu) =: \delta > 0$ by Proposition 1.3.2. Let us consider the mapping:

$$\gamma_n : H \to \mathbb{R}^n, \ x \to \left(\frac{x_1}{\sqrt{\lambda_1}}, \dots, \frac{x_n}{\sqrt{\lambda_n}}\right).$$

Set $\tilde{\mu}_n = \gamma_n \circ \mu$, and $\tilde{\nu}_n = \gamma_n \circ \nu$. Then $\tilde{\mu}_n$ and $\tilde{\nu}_n$ are Gaussian measures with mean 0 and covariance operators $\gamma_n Q \gamma_n^*$ and $\gamma_n R \gamma_n^*$ respectively. Moreover

$$\gamma_n^* : \mathbb{R}^n \to H, \ (\xi_1, \, \dots, \xi_n) \to \sum_{k=1}^{n} \frac{\xi_k e_k}{\sqrt{\lambda_k}}.$$

It follows $\gamma_n Q \gamma_n^* = I_n$, where I_n is the identity in $\mathbb{R}^n$, and for $T_n = \gamma_n R \gamma_n^*$,

$$(T_n)_{h,k} = \frac{\langle Re_h, e_k \rangle}{\sqrt{\lambda_h \lambda_k}}, \ h, k = 1, \dots, n.$$

Let $\mathcal{G}_n$ be the mimimal σ-field such that $x_1, \dots, x_n$ are measurable. Then by Proposition 1.3.3 we have $0 < \delta \le H(\mu, \nu) \le H(\tilde{\mu}_n, \tilde{\nu}_n)$. Now, by an elementary computation, we find

$$\delta^2 \le H^2(\tilde{\mu}_n, \tilde{\nu}_n) = \det \left[\frac{2T_n^{1/2}}{1 + T_n} \right] = \prod_{j=1}^{n} 2 \frac{\theta_{n,j}^{1/2}}{1 + \theta_{n,j}},$$

where $\theta_{n,j}, \ j = 1, \dots, n$, are the eigenvalues of T_n. Consequently

$$\sum_{j=1}^{n} \left[2 \log \left(\frac{1 + \theta_{n,i}}{2} \right) - \log \theta_{n,j} \right] = -4 \log H(\tilde{\mu}_n \tilde{\nu}_n) \le -4 \log \delta.$$

Moreover, since $\|T_n\| \le \|1 + R\|$, there exists $c > 0$ such that

$$2 \log \frac{1 + x}{2} - \log x \ge c(1 - x)^2, \ x > 0, \text{ and } 0 < x < \|1 + R\|,$$

we have $\sum_{j=1}^{n} (1 - \theta_{n,j})^2 \le -4 \log \delta$. By the arbitrariness of n we infer

$$\sum_{j=1}^{\infty} (1 - \theta_{n,j})^2 \le -4 \log \delta. \tag{1.3.11}$$

Now we can conclude the proof. Set in fact

$$T = \sum_{i,j=1}^{\infty} \frac{\langle Re_i, e_j \rangle}{\sqrt{\lambda_i \lambda_j}} e_i \otimes e_j,$$

and $S = 1 - T$. Then S is of Hilbert-Schmidt class. Moreover

$$Q^{1/2} T Q^{1/2} e_i = Re_i, \ i \in \mathbb{N},$$

so that $R = Q^{1/2} T Q^{1/2}$ as required. $\square$

We now prove a converse of Theorem 1.3.9.

Theorem 1.3.10 *Assume that there exists $S \in L(H)$ symmetric and of Hilbert-Schmidt class such that $R = Q^{1/2}(1 - S)Q^{1/2}$. Then μ and ν are equivalent.*

Proof. Let (g_k) be a complete orthonormal system that diagonalizes $1 + S$, that is $(1 + S)g_k = \tau_k g_k$, $k \in \mathbb{N}$, where (τ_k) are the eigenvalues of $1 + S$.

Let us consider two measures $\hat{\mu}$ and $\hat{\nu}$ on $\mathbb{R}^\infty$

$$\hat{\mu} = \prod_{k=1}^{\infty} N_1, \quad \hat{\nu} = \prod_{k=1}^{\infty} N_{\tau_k},$$

and the mapping ψ

$$\psi : \mathbb{R}^\infty \to H, \quad c = (c_k) \to \psi(c) = \sum_{k=1}^{\infty} c_k Q^{1/2} g_k,$$

with value 0 if the series is not convergent in H. Then we have $\mu = \psi \circ \hat{\mu}$, $\nu = \psi \circ \hat{\nu}$. Moreover, by Proposition 1.3.5 $\hat{\mu}$ and $\hat{\nu}$ are equivalent since

$$H(\hat{\mu}, \hat{\nu}) = \prod_{k=1}^{\infty} \left[\frac{4\tau_k}{(1 + \tau_k)^2} \right] > 0,$$

and so μ and ν are equivalent as well. $\square$

Assume that $\mu = N_Q$ and $\nu = N_R$ are equivalent. Then we know by Theorem 1.3.9 that there exists $S \in L_2(H)$ such that $R = Q^{1/2}(1 - S)Q^{1/2}$. If $S \in L_1^+(H)$ we can give a simple formula for the density $d\nu/d\mu$.

Proposition 1.3.11 *Let $Q, R \in L_1^+(H)$ and $R = Q^{1/2}(1 - S)Q^{1/2}$ with $S \in L_1(H)$ and $S < 1$. Then $\mu = N_Q$ and $\nu = N_R$ are equivalent and*

$$\frac{d\nu}{d\mu}(x) = [\det(1 - S)]^{-1/2} \exp \left\{ -\frac{1}{2} \langle S(1 - S)^{-1} Q^{-1/2} x, Q^{-1/2} x \rangle \right\}, \quad x \in H.$$
$$(1.3.12)$$

Proof. We set

$$\rho(x) = [\det(1 - S)]^{-1/2} \exp \left\{ -\frac{1}{2} \langle S(1 - S)^{-1} Q^{-1/2} x, Q^{-1/2} x \rangle \right\}, \quad x \in H,$$

and prove that the characteristic function of the measure $\zeta : \zeta(dx) = \rho(x)\mu(dx)$ coincide with that of ν. In fact, by applying a slight modification of Proposition 1.2.8 with $M = -Q^{-1/2} S(1 - S)^{-1} Q^{-1/2}$, $b = ih$, we find

$$\int_H e^{i \langle h, x \rangle} \rho(x)\mu(dx) = \exp \left\{ -\frac{1}{2} |(1 - S)^{1/2} Q^{1/2} h|^2 \right\}.$$

Since
$$|(1 - S)^{1/2}Q^{1/2}h|^2 = \langle Q^{1/2}(1 - S)Q^{1/2}h \rangle = \langle Rh, h \rangle,$$

we have
$$\int_H e^{i\langle h, x \rangle} \rho(x)\mu(dx) = \exp\left\{-\frac{1}{2}\langle Rh, h \rangle\right\},$$

as required. □

1.4 Brownian motion

Let $(\Omega, \mathcal{F}, \mathbb{P})$ be a probability space. A family of real random variables $X = (X(t))_{t \geq 0}$ is called a real *stochastic process* in $[0, +\infty)$. The stochastic process X is said to be *continuous* if the function $X(\cdot)(\omega)$ is continuous $\mathbb{P}$-a.s.

A real *Brownian motion* B is a continuous real stochastic process in $[0, +\infty)$ such that

  (i) $B(0) = 0$ and if $0 \leq s < t$, $B(t) - B(s)$ is a real Gaussian random variable with law N_{t-s},

  (ii) if $n \in \mathbb{N}$ and $0 < t_1 < \cdots < t_n$, the random variables

$$B(t_1), \ B(t_2) - B(t_1), \dots, B(t_n) - B(t_{n-1})$$

  are independent.

We are now going to construct a Brownian motion. To this purpose, let us consider the probability space $(H, \mathcal{B}(H), \mu)$, where $H = L^2(0, +\infty)$, and $\mu = N_Q$, where Q is any operator in $L_1^+(H)$ such that $\text{Ker } Q = \{0\}$.

Theorem 1.4.1 *Let* $B(t) = W_{\chi_{[0,t]}}$, $t \geq 0$, *where* $\chi_{[0,t]}$ *is the characteristic function of the interval* $[0, t]$. *Then* B *is a real Brownian motion on* $(H, \mathcal{B}(H), \mu)$. ([8])

Proof. Clearly $B(0) = 0$. Since for $t > s$,

$$B(t) - B(s) = W_{\chi_{[0,t]}} - W_{\chi_{[0,s]}} = W_{\chi_{(s,t]}},$$

we have that $B(t) - B(s)$ is a real Gaussian random variable N_{t-s} and (i) is proved. Let us prove (ii). Since the system of elements of H,

$$(\chi_{[0,t_1]}, \chi_{(t_1,t_2]}, \cdots, \chi_{(t_{n-1},t_n]}),$$

---

[8] More precisely B has a modification which is a continuous path Brownian motion.

is orthogonal, we see that the random variables

$$B(t_1),\ B(t_2) - B(t_1), \ldots, B(t_n) - B(t_{n-1})$$

are independent. Thus (ii) is proved.

It remains to show the continuity of B. For this we shall use the so called *factorization method*, see [93], based on the following elementary identity:

$$\int_s^t (t-\sigma)^{\alpha-1}(\sigma-s)^{-\alpha}d\sigma = \frac{\pi}{\sin \pi\alpha}, \quad 0 \le s \le \sigma \le t \le 1, \qquad (1.4.1)$$

where $\alpha \in (0,1)$.

From now on we take $\alpha < 1/2$. Then identity (1.4.1) can be written as

$$\chi_{[0,t]}(s) = \frac{\sin \pi\alpha}{\pi} \int_0^t (t-\sigma)^{\alpha-1}\chi_{[0,\sigma]}(s)(\sigma-s)^{-\alpha}d\sigma, \quad t,s \in [0,+\infty).$$

We can also write

$$\chi_{[0,t]} = \frac{\sin \pi\alpha}{\pi} \int_0^t (t-\sigma)^{\alpha-1}g_\sigma d\sigma, \qquad (1.4.2)$$

where $g_\sigma(s) = \chi_{[0,\sigma]}(s)(\sigma-s)^{-\alpha}$. Since $\alpha < 1/2$, $g_\sigma \in H$ and $|g_\sigma|^2 = \frac{\sigma^{1-2\alpha}}{1-2\alpha}$. We note that the integral in (1.4.2) is a Riemann integral of the H-valued function $(t-\sigma)^{\alpha-1}g_\sigma$. Using the fact that the mapping $H \to L^2(H,\mu)$, $f \to W_f$, is continuous, we obtain the following representation formula for B:

$$B(t) = \frac{\sin \pi\alpha}{\pi} \int_0^t (t-\sigma)^{\alpha-1}W_{g_\sigma}d\sigma. \qquad (1.4.3)$$

Now it is enough to prove that if $m > 1/(2\alpha)$, then $W_{g_{(\cdot)}}(x) \in L^{2m}(0,T)$ for μ-almost all $x \in H$ and for any $T > 0$; in fact this implies that B is continuous by the elementary Lemma 1.4.2 below.

To show summability of $W_{g_{(\cdot)}}(x)$, we notice that, since W_{g_σ} is a real Gaussian random variable with law $N_{\frac{\sigma^{1-2\alpha}}{1-2\alpha}}$, we have

$$\int_H |W_{g_\sigma}(x)|^{2m}\mu(dx) = \frac{(2m)!}{2^m m!}(1-2\alpha)^{-m}\sigma^{m(1-2\alpha)}. \qquad (1.4.4)$$

Since $\alpha < 1/2$ we have, by the Fubini theorem,

$$\int_H \left[\int_0^T |W_{g_\sigma}(x)|^{2m}d\sigma\right]\mu(dx) = \int_0^T \left[\int_H |W_{g_\sigma}(x)|^{2m}\mu(dx)\right]d\sigma < +\infty.$$
$$\qquad (1.4.5)$$

It follows that $W_{g_\cdot}(x) \in L^{2m}(0,T)$ for μ-almost all $x \in H$ and the conclusion follows. $\square$

Lemma 1.4.2 *Let $m > 1, \alpha \in (1/(2m), 1)$ and $f \in L^{2m}(0, T)$. Set*

$$F(t) = \int_0^t (t - \sigma)^{\alpha - 1} f(\sigma) d\sigma, \ t \in [0, 1].$$

Then $F \in C([0, T])$.

Proof. Let $t \in [0, T]$, then by Hölder's inequality we have (notice that $2m\alpha - 1 > 0$)

$$|F(t)| \leq \left(\int_0^t (t - \sigma)^{(\alpha - 1) \frac{2m}{2m - 1}} d\sigma \right)^{\frac{2m - 1}{2m}} |f|_{L^{2m}(0,1)}. \tag{1.4.6}$$

Therefore $F \in L^\infty(0, 1)$. It remains to show continuity of F. Continuity at 0 follows from (1.4.6). Let $t_0 \in (0, T]$. We are going to prove that F is continuous on $[\frac{t_0}{2}, T]$. Let us set for $\varepsilon < \frac{t_0}{2}$,

$$F_\varepsilon(t) = \int_0^{t - \varepsilon} (t - \sigma)^{\alpha - 1} f(\sigma) d\sigma, \ t \in [0, 1].$$

F_ε is obviously continuous on $[\frac{t_0}{2}, T]$. Moreover, using once again Hölder's inequality, we find

$$|F(t) - F_\varepsilon(t)| \leq M \left(\frac{2m - 1}{2m\alpha - 1} \right)^{\frac{2m - 1}{2m}} \varepsilon^{\alpha - \frac{1}{2m}} |f|_{L^{2m}(0,1)}.$$

Thus $\lim_{\varepsilon \to 0} F_\varepsilon(t) = F(t)$, uniformly on $[\frac{t_0}{2}, T]$, and F is continuous as required. $\square$

Exercise 1.4.3 Prove that $B(\cdot)x$ is Hölder continuous with any exponent $\beta < 1/2$ for almost all $x \in H$.

Chapter 2

Spaces of continuous functions

In this chapter we deal with spaces of continuous functions on a separable Hilbert space H (with norm $|\cdot|$ and inner product $\langle \cdot, \cdot \rangle$). After some definitions in §2.1, we prove approximation results for continuous functions in terms of smoother ones. When H is finite dimensional this kind of results is usually proved with the help of convolutions. In the infinite dimensional case, since a reference measure playing the rôle of the Lebesgue measure is not available, we will use the so called *inf-sup convolutions* introduced by J. M. Lasry and P. L. Lions [155], see §2.2.

Notice that in infinite dimensions some classical approximation results, valid when the dimension of H is finite, fail. For instance the space of real bounded functions of class C^2 is not dense in the space of continuous and bounded functions.

Finally, §2.3 is devoted to interpolation between space of continuous functions .

2.1 Preliminary results

The basic space we will consider in this section is the space $C_b(H)$ (resp. $UC_b(H)$), consisting of all mappings $\varphi : H \to \mathbb{R}$ that are continuous (resp. uniformly continuous) and bounded. As is easily checked $C_b(H)$ (resp. $UC_b(H)$), endowed with the norm

$$\|\varphi\|_0 = \sup_{x \in H} |\varphi(x)|, \ \varphi \in C_b(H),$$

is a Banach space.

Remark 2.1.1 $UC_b(H)$ is not a separable space even if $H = \mathbb{R}$. Here is an example of an uncountable family of functions such that any pair of functions of the family have distance greater than 1.

Fix a nonnegative function $\rho \in C_0^\infty(\mathbb{R})$ with support in $[0, 1]$ and maximum equal to $1/2$. For arbitrary sequences $\varepsilon = (\varepsilon_k)$, with $\varepsilon_k = 1$ or -1, we set

$$\varphi_\varepsilon(x) = \sum_{k=1}^\infty \rho(x + k)\varepsilon_k.$$

Then if $\varepsilon \neq \varepsilon'$ we have $\|\varphi_\varepsilon - \varphi_{\varepsilon'}\|_0 \geq 1$.

We shall also consider the space $C_b(H; E)$ (resp. $UC_b(H; E)$), where E is a Banach space, consisting of all mappings $F : H \to E$ that are continuous (resp. uniformly continuous) and bounded. $C_b(H; E)$ (resp. $UC_b(H; E)$), endowed with the norm

$$\|F\|_0 = \sup_{x \in H} |F(x)|_E, \ F \in UC_b(H; E),$$

is a Banach space.

Let us define some important subspaces of $C_b(H)$ (resp. $UC_b(H)$).

(i) $C_b^1(H)$ (resp. $UC_b^1(H)$) is the space of all continuous (resp. uniformly continuous) and bounded functions $\varphi : H \to \mathbb{R}$ which are Fréchet differentiable on H with a continuous (resp. uniformly continuous) and bounded derivative $D\varphi$. We set

$$[\varphi]_1 := \sup_{x \in H} |D\varphi(x)|, \ \ \|\varphi\|_1 = \|\varphi\|_0 + [\varphi]_1, \ \varphi \in C_b^1(H).$$

If $\varphi \in C_b^1(H)$ and $x \in H$, we shall identify $D\varphi(x)$ with the unique element h of H such that

$$D\varphi(x)y = \langle h, y \rangle, \ y \in H.$$

(ii) $C_b^2(H)$ (resp. $UC_b^2(H)$) is the subspace of $UC_b^1(H)$ of all functions $\varphi : H \to \mathbb{R}$ which are twice Fréchet differentiable on H with a continuous (resp. uniformly continuous) and bounded second derivative $D^2\varphi$. We set

$$[\varphi]_2 := \sup_{x \in H} \|D^2\varphi(x)\|, \ \ \|\varphi\|_2 := \|\varphi\|_1 + [\varphi]_2, \ \varphi \in C_b^2(H).$$

If $\varphi \in C_b^2(H)$ and $x \in H$, we shall identify $D^2\varphi(x)$ with the unique linear operator $T \in L(H)$ such that

$$D\varphi(x)(y, z) = \langle Ty, z \rangle, \ y, z \in H.$$

(iii) For any $k \in \mathbb{N}$, $C_b^k(H)$ (resp. $UC_b^k(H)$) is the subspace of $UC_b(H)$ of all functions $\varphi : H \to \mathbb{R}$ which are k times Fréchet differentiable on H with continuous (resp. uniformly continuous) and bounded derivatives $D^h\varphi$ with h less than or equal to k. We set

$$[\varphi]_k := \sup_{x \in H} \|D^k \varphi(x)\|, \quad \|\varphi\|_k = \|\varphi\|_1 + [\varphi]_2, \ \varphi \in C_b^k(H).$$

We set moreover

$$C_b^\infty(H) = \bigcap_{k=1}^\infty C_b^k(H).$$

(iv) $C_b^{0,1}(H)$ is the subspace of $UC_b(H)$ of all Lipschitz continuous functions. If $\varphi \in C_b^{0,1}(H)$ we set

$$[\varphi]_1 := \sup_{\substack{x,y \in H \\ x \neq y}} \frac{|\varphi(x) - \varphi(y)|}{|x - y|}, \quad \varphi \in C_b^{0,1}(H).$$

$C_b^{0,1}(H)$ is a Banach space with the norm

$$\|\varphi\|_1 := \|\varphi\|_0 + [\varphi]_1, \ \varphi \in C_b^{0,1}(H).$$

(v) $C_b^{1,1}(H)$ is the space of all functions $\varphi \in C_b^1(H)$ such that $D\varphi$ is Lipschitz continuous. We set

$$[\varphi]_{1,1} := \sup_{x \neq y} \frac{|D\varphi(x) - D\varphi(y)|}{|x - y|}, \ \varphi \in C_b^{1,1}(H).$$

$C_b^{1,1}(H)$ is a Banach space with the norm

$$\|\varphi\|_{1,1} = \|f\|_1 + [\varphi]_{1,1}, \ \varphi \in C_b^{1,1}(H).$$

(vi) $C_b^\alpha(H)$, $\alpha \in (0,1)$, is the subspace of $C_b(H)$ of all functions $\varphi : H \to \mathbb{R}$ such that

$$[\varphi]_1 := \sup_{\substack{x,y \in H \\ x \neq y}} \frac{|\varphi(x) - \varphi(y)|}{|x - y|^\alpha}, \quad \varphi \in C_b^{0,1}(H).$$

$C_b^\alpha(H)$ is a Banach space with the norm

$$\|\varphi\|_\alpha := \|\varphi\|_0 + [\varphi]_\alpha, \ \varphi \in C_b^\alpha(H).$$

(vii) $C_b^{1,\alpha}(H)$, $\alpha \in (0,1)$, is the space of all functions $\varphi \in C_b^1(H)$ such that Df is α-Hölder continuous. We set

$$[\varphi]_{1,\alpha} := \sup_{x \neq y} \frac{|Df(x) - Df(y)|}{|x-y|^\alpha}, \ \varphi \in C_b^{1,\alpha}(H).$$

$C_b^{1,\alpha}(H)$ is a Banach space with the norm

$$\|\varphi\|_{1,\alpha} := \|f\|_1 + [\varphi]_{1,\alpha}, \ \varphi \in C_b^{1,\alpha}(H).$$

(viii) $C_b^{2,\alpha}(H)$ $\alpha \in (0,1)$, is the space of all functions $\varphi \in C_b^2(H)$ such that $D^2 f$ is α-Hölder continuous. We set

$$[\varphi]_{2,\alpha} := \sup_{x \neq y} \frac{\|D^2 f(x) - D^2 f(y)\|}{|x-y|^\alpha}, \ \varphi \in C_b^{2,\alpha}(H).$$

$C_b^{2,\alpha}(H)$ is a Banach space with the norm

$$\|\varphi\|_{2,\alpha} := \|f\|_1 + [\varphi]_{1,\alpha}, \ \varphi \in C_b^{2,\alpha}(H).$$

2.2 Approximation of continuous functions

In this section we want to show that a given function $\varphi \in UC_b(H)$ can be uniformly approximated by functions on $UC_b^1(H)$. This fact can be easily proved, using convolutions, when dim $H < \infty$. Let in fact $\rho \in C_b^\infty(H)$ be nonnegative, with compact support and such that $\int_H \rho(x)dx = 1$. Then it is easy to see that setting

$$\varphi_t(x) = t^{-n} \int_H \rho\left(\frac{x-y}{t}\right) \varphi(y)dy, \ t > 0,$$

we have $\varphi_t \in C_b^\infty(H)$ and $\lim_{t \to 0} \|\varphi - \varphi_t\|_0 = 0$.

If dim $H = \infty$, it was proved in 1954 by J. Kurtzweil [152] that there exists a sequence $(\varphi_\varepsilon) \subset UC_b(H) \cap C^\infty(H)$ such that $\lim_{\varepsilon \to 0} \sup_{x \in H} |\varphi(x) - \varphi_\varepsilon(x)| = 0$. However in 1973 A. S. Nemirowski and S. M. Semenov [176] proved that $UC_b^2(H)$ is not dense in $UC_b(H)$, whereas $UC_b^{1,1}(H)$ is.

We will give now a proof of this last result following J. M. Lasry and P. L. Lions [155]. For any $t > 0$ and any $\varphi \in UC_b(H)$, we set $\varphi_t = V_{t/2}U_t\varphi$, where U, V are defined by

$$U_t\varphi(x) = \inf_{y \in H} \left\{ \varphi(y) + \frac{|x-y|^2}{2t} \right\} = \inf_{y \in H} \left\{ \varphi(x-y) + \frac{|y|^2}{2t} \right\} \qquad (2.2.1)$$

and

$$V_t\varphi(x) = \sup_{y\in H}\left\{\varphi(y) - \frac{|x-y|^2}{2t}\right\} = \sup_{y\in H}\left\{\varphi(x-y) - \frac{|y|^2}{2t}\right\}. \qquad (2.2.2)$$

We set moreover

$$U_0\varphi = V_0\varphi = \varphi, \quad \varphi \in UC_b(H). \qquad (2.2.3)$$

Several properties of U_t and V_t are studied in Appendix C.

Note that

$$\varphi_t(x) = \sup_{z\in H}\left\{\inf_{y\in H}\left[\varphi(y) + \frac{|z-y|^2}{2t}\right] - \frac{|z-x|^2}{t}\right\}. \qquad (2.2.4)$$

For any $\varphi \in UC_b(H)$ we define the *uniform continuity modulus* ω_φ of φ by setting

$$\omega_\varphi(t) = \sup\{|\varphi(x) - \varphi(y)| : x, y \in H, |x-y| \le t\}, \ t \ge 0.$$

Theorem 2.2.1 *Let $\varphi \in UC_b(H)$ and let φ_t, $t \in [0,1]$, be defined by (2.2.4). Then the following statements hold.*

(i) *If $\varphi, \psi \in UC_b(H)$ and $\varphi(x) \le \psi(x)$ for all $x \in H$, then $\varphi_t(x) \le \psi_t(x)$ for all $x \in H$.*

(ii) *$\varphi_t \in UC_b^{1,1}(H)$ and the following estimates hold.*

$$\|\varphi_t\|_0 \ \le \ \|\varphi\|_0, \ [\varphi_t]_1 \le \frac{2}{\sqrt{t}}\sqrt{\|\varphi\|_0}, \quad t \in [0,1].$$

$$[\varphi_t]_{1,1} \ \le \ \frac{1}{t}, \ \|\varphi - \varphi_t\|_0 \le \omega_\varphi\left(2\sqrt{t\|\varphi\|_0}\right), \quad t \in [0,1].$$

(iii) *Let $\alpha \in (0,1)$. If $\varphi \in UC_b^\alpha(H)$ then there exists $C_\alpha > 0$ such that*

$$\|\varphi - \varphi_t\|_0 \ \le \ C_\alpha[\varphi]_\alpha^{\frac{2}{2-\alpha}} t^{\frac{\alpha}{2-\alpha}}, \qquad (2.2.5)$$

$$[\varphi_t]_1 \ \le \ C_\alpha[\varphi]_\alpha^{\frac{2}{2-\alpha}} t^{\frac{\alpha-1}{2-\alpha}}. \qquad (2.2.6)$$

From the theorem it follows that $UC_b^{1,1}(H)$ is dense in $UC_b(H)$.

Proof of Theorem 2.2.1. Part (i) follows from the definition (2.2.4). Let us prove (ii). Let $\varphi \in UC_b(H), t > 0$, and $\varphi_t = V_{t/2}U_t\varphi$. Then φ_t is Lipschitz continuous. From Proposition C.3.4 the mapping $x \to \varphi_t(x) + \frac{|x|^2}{t}$ is convex.

Moreover by Proposition C.3.6, since the mapping $x \to U_t\varphi(x) - \frac{|x|^2}{2t}$ is concave, we have that the mapping $x \to \varphi_t(x) - \frac{|x|^2}{t}$ is concave too. So the conclusion follows from Proposition C.2.1.

Finally, we prove (iii). We start with (2.2.5). We have

$$|\varphi(x) - \varphi_t(x)| = |\varphi(x) - V_{t/2}U_t\varphi(x)|$$

$$\leq |\varphi(x) - V_{t/2}\varphi(x)| + |V_{t/2}\varphi(x) - V_{t/2}U_t\varphi(x)|.$$

Using (C.3.6) and the fact that V_t is a contraction, we find

$$|\varphi(x) - \varphi_t(x)| \leq C_\alpha[\varphi]_\alpha^{\frac{1}{2-\alpha}} (t/2)^{\frac{\alpha}{2-\alpha}} + \|\varphi - U_{t/2}\varphi\|_0.$$

Using (C.3.6) again, the estimate (2.2.5) follows. Finally we prove (2.2.6). We first remark that for any $\varepsilon > 0$ and $x \in H$, there exists $z_{\varepsilon,x} \in H$ such that

$$\varphi_t(x) < \inf_{y\in H}\left[\varphi(y) + \frac{|z_{\varepsilon,x} - y|^2}{2t}\right] - \frac{|z_{\varepsilon,x} - x|^2}{t} + \varepsilon. \qquad (2.2.7)$$

It follows that

$$\varphi_t(x) \leq \varphi(x) + \frac{|z_{\varepsilon,x} - x|^2}{2t} - \frac{|z_{\varepsilon,x} - x|^2}{t} + \varepsilon = \varphi(x) - \frac{|z_{\varepsilon,x} - x|^2}{2t} + \varepsilon,$$

which implies, by (2.2.5),

$$\frac{|z_{\varepsilon,x} - x|^2}{2t} \leq \varphi(x) - \varphi_t(x) + \varepsilon \leq C_\alpha[\varphi]_\alpha^{\frac{2}{2-\alpha}} t^{\frac{\alpha}{2-\alpha}} + \varepsilon. \qquad (2.2.8)$$

Now, if $x, \overline{x} \in H$ then by (2.2.7) it follows that

$$\varphi_t(x) - \varphi_t(\overline{x}) < \inf_{y\in H}\left[\varphi(y) + \frac{|z_{\varepsilon,x} - y|^2}{2t}\right] - \frac{|z_{\varepsilon,x} - x|^2}{t} - \varphi_t(\overline{x}) + \varepsilon. \qquad (2.2.9)$$

On the other hand by (2.2.4) it follows that

$$\varphi_t(\overline{x}) \geq \inf_{y\in H}\left[\varphi(y) + \frac{|z_{\varepsilon,x} - y|^2}{2t}\right] - \frac{|z_{\varepsilon,x} - \overline{x}|^2}{t},$$

and so, taking into account (2.2.8), we get

$$\varphi_t(x) - \varphi_t(\overline{x}) \leq \frac{|z_{\varepsilon,x} - \overline{x}|^2}{t} - \frac{|z_{\varepsilon,x} - x|^2}{t} + \varepsilon$$

$$= \varepsilon + \frac{1}{t}\left\{|x - \overline{x}|^2 + 2\langle x - \overline{x}, z_{\varepsilon,x} - x\rangle\right\}$$

$$\leq \varepsilon + \frac{1}{t}\left\{|x - \overline{x}|^2 + 2|x - \overline{x}|\sqrt{C_\alpha}[\varphi]_\alpha^{\frac{1}{2-\alpha}} t^{\frac{1}{2-\alpha}}\right\},$$

and the conclusion follows. □

2.3   Interpolation spaces

This section is devoted to a characterization of some interpolation spaces between $UC_b(H)$ and $UC_b^1(H)$, and to some interpolatory estimates which will be needed in the following.

Let us recall the definition of interpolation spaces. We shall use the so called K-method, see e.g. H. Triebel [212].

Let X and Y be Banach spaces such that $Y \subset X$ with continuous embedding. We denote by M a number greater than 1 such that $\|x\|_X \leq M\|x\|_Y$, $x \in Y$. Then we define

$$K(t,x) = \inf\{\|a\|_X + t\|b\|_Y \; : \; x = a + b, \; a \in X, \; b \in Y\},$$

and for arbitrary $\theta \in [0,1]$ we set

$$[x]_{(X,Y)_{\theta,\infty}} = \sup_{t \in (0,1]} t^{-\theta} K(t,x),$$

$$(X,Y)_{\theta,\infty} = \{x \in X : [x]_{(X,Y)_{\theta,\infty}} < +\infty\}.$$

As is easily seen $(X,Y)_{\theta,\infty}$, endowed with the norm

$$\|x\|_{(X,Y)_{\theta,\infty}} = \|x\|_X + [x]_{(X,Y)_{\theta,\infty}}, \; x \in (X,Y)_{\theta,\infty},$$

is a Banach space.

Remark 2.3.1 It is not difficult to check that the following statements are equivalent.

(i) $x \in (X,Y)_{\theta,\infty}$ and $[x]_{(X,Y)_{\theta,\infty}} \leq L$.

(ii) For all $t \in (0,1]$ there exist $a_t \in X$ and $b_t \in Y$ such that $x = a_t + b_t$ and
$$\|a_t\|_X + t\|b_t\|_Y \leq Lt^\theta.$$

2.3.1   Interpolation between $UC_b(H)$ and $UC_b^1(H)$

Let us start with the one dimensional case.

Proposition 2.3.2 *We have*

$$\left(UC_b(\mathbb{R}), UC_b^1(\mathbb{R})\right)_{\theta,\infty} = C_b^\theta(\mathbb{R}), \; \theta \in (0,1). \tag{2.3.1}$$

Proof.

Step 1. $\left(UC_b(\mathbb{R}), UC_b^1(\mathbb{R})\right)_{\theta,\infty}$ is continuously embedded in $C_b^\theta(\mathbb{R})$.

Let $x \in \left(UC_b(\mathbb{R}), UC_b^1(\mathbb{R})\right)_{\theta,\infty}$, and set $L = [x]_{\left(UC_b(\mathbb{R}), UC_b^1(\mathbb{R})\right)_{\theta,\infty}}$. Then by definition, for any $t \in (0,1]$ there exist $a_t \in UC_b(\mathbb{R})$, $b_t \in C_b^1(\mathbb{R})$, such that $\varphi = a_t + b_t$ and

$$\|a_t\|_0 + t\|b_t\|_1 \leq Lt^\theta, \ t \in (0,1].$$

Then, if $x, y \in \mathbb{R}$, we have

$$\varphi(x) - \varphi(y) = a_t(x) - a_t(y) + b_t(x) - b_t(y)$$

$$= a_t(x) - a_t(y) + \int_0^1 \langle Db_t(\xi x + (1-\xi)y), x - y \rangle \, d\xi.$$

It follows that

$$|\varphi(x) - \varphi(y)| \leq 2Lt^\theta + Lt^{\theta-1}|x - y|, \ t \in (0,1], \ x, y \in \mathbb{R}.$$

Now, if $|x - y| \leq 1$, setting $t = |x - y|$, we have

$$|\varphi(x) - \varphi(y)| \leq 3L|x - y|^\theta, \ x, y \in \mathbb{R},$$

whereas if $|x - y| \geq 1$, we have

$$|\varphi(x) - \varphi(y)| \leq 2\|\varphi\|_0 |x - y|^\theta, \ x, y \in \mathbb{R}.$$

Consequently

$$|\varphi(x) - \varphi(y)| \leq (3L + 2\|\varphi\|_0)|x - y|^\theta, \ x, y \in \mathbb{R},$$

and the statement is proved.

Step 2. $C_b^\theta(\mathbb{R})$ is continuously embedded in $\left(UC_b(\mathbb{R}), UC_b^1(\mathbb{R})\right)_{\theta,\infty}$.

Let $\rho \in UC_b^\infty(\mathbb{R})$ be nonnegative, with compact support and such that

$$\int_{-\infty}^\infty \rho(x)dx = 1,$$

and let $\rho_t(x) = \frac{1}{t}\rho\left(\frac{x}{t}\right)$ for $t > 0$ and $x \in \mathbb{R}$. Let $\varphi \in C_b^\theta(\mathbb{R})$ be fixed, and denote by φ_t the convolution $\varphi_t = \varphi * \rho_t$. Set moreover $a_t = \varphi - \varphi_t$, $b_t = \varphi_t$, $t \in (0,1]$. We are going to show that, for some constant $C > 0$, we have

$$\|a_t\|_0 \leq C\|\varphi\|_\theta \, t^\theta, \ \|Db_t\|_0 \leq C\|\varphi\|_\theta \, t^{\theta-1}, \ t \in (0,1]. \tag{2.3.2}$$

We have in fact

$$a_t(x) = \varphi(x) - \varphi_t(x) = \varphi(x) - \frac{1}{t} \int_{-\infty}^{+\infty} \varphi(x-y)\rho\left(\frac{y}{t}\right) dy$$

$$= \varphi(x) - \int_{-\infty}^{+\infty} \varphi(x-tz)\rho(z)dz = \int_{\infty}^{+\infty} [\varphi(x) - \varphi(x-tz)]\rho(z)dz.$$

It follows that

$$|a_t(x)| \le [\varphi]_\theta t^\theta \int_{-\infty}^{+\infty} |z|^\theta \rho(z)\, dz. \tag{2.3.3}$$

Moreover

$$Db_t(x) = \frac{1}{t^2} \int_{-\infty}^{+\infty} \varphi(y) D\rho\left(\frac{x-y}{t}\right) dy$$

$$= \frac{1}{t} \int_{-\infty}^{+\infty} \varphi(x-tz) D\rho(z)\, dz$$

$$= \frac{1}{t} \int_{-\infty}^{+\infty} [\varphi(x) - \varphi(x-tz)] D\rho(z)\, dz.$$

Consequently

$$|Db_t(x)| \le \frac{1}{t}[\varphi]_\theta t^\theta \int_{-\infty}^{+\infty} |z|^\theta D\rho(z)\, dz. \tag{2.3.4}$$

Finally, (2.3.2) follows from (2.3.3) and (2.3.4). □
Now we go to the infinite dimensional case proving a result due to P. Cannarsa and G. Da Prato [31].

Theorem 2.3.3 *Let H be a separable Hilbert space. Then we have*

$$\left(UC_b(H), UC_b^1(H)\right)_{\theta,\infty} = C_b^\theta(H),\ \theta \in (0,1). \tag{2.3.5}$$

Proof.
Step 1. $\left(UC_b(H), UC_b^1(H)\right)_{\theta,\infty}$ is continuously embedded in $UC_b^\theta(H)$.
The proof is completely similar to that of Proposition 2.3.2 and it is left as an exercise to the reader.
Step 2. $C_b^\theta(H)$ is continuously embedded in $\left(UC_b(H), UC_b^1(H)\right)_{\theta,\infty}$.

Let $\varphi \in UC_b^\theta(H)$ and let φ_t be defined by (2.2.4). Set

$$a_t = \varphi - \varphi_{t^{2-\theta}}, \quad b_t = \varphi_{t^{2-\theta}}.$$

Then by (2.2.5)-(2.2.6) we see that, for some constant $C > 0$ depending on $\|\varphi\|_\theta$, we have

$$\|a_t\|_0 \le Ct^\theta, \quad \|b_t\|_1 \le Ct^{\theta-1}, \ t \in (0,1]$$

so the conclusion follows. $\square$

2.3.2 Interpolatory estimates

Let X, Y, E be Banach spaces with $Y \subset E \subset X$, all embeddings being continuous.

Let $\theta \in [0,1]$. We say that E belongs to the class $J_\theta(X,Y)$ if there exists $C > 0$ such that

$$\|x\|_E \le C\|x\|_X^{1-\theta} \|x\|_Y^\theta, \quad x \in Y.$$

Example 2.3.4 The interpolation space $(X,Y)_{\theta,\infty}$ belongs to the class $J_\theta(X,Y)$. In fact if $x \in Y$ we have

$$K(t,x) \le \|x\|_X, \quad K(t,x) \le t\|x\|_Y.$$

It follows that $K(t,x) \le t^\theta \|x\|_X^{1-\theta} \|x\|_Y^\theta$, and so $[x]_{\theta,\infty} \le \|x\|_X^{1-\theta} \|x\|_Y^\theta$.

Now we want to extend some classical interpolatory estimates between different spaces of continuous functions to the case of an infinite dimensional Hilbert space.

Theorem 2.3.5 *Let* $0 \le a < b < c$. *Then we have* ([1])

$$UC_b^b(H) \in J_{\frac{b-a}{c-a}}\left(UC_b^a(H), UC_b^c(H)\right), \tag{2.3.6}$$

and there exists a constant $C_{a,b,c} > 0$ *such that*

$$\|u\|_b \le C_{a,b,c} \|u\|_a^{\frac{c-b}{c-a}} \|u\|_c^{\frac{b-a}{c-a}}, \quad u \in UC_b^c(H). \tag{2.3.7}$$

Proof. The proof is very similar to the finite dimensional one: it consists in several steps following the different possible choices of a, b, c. We will consider only those cases that we will use in the following.

---

[1] If $a \ge 0$ is integer we set $UC_b^a(H) = C_b^a(H)$.

Case 1. $a = 0$, $b = 1$, $c = 2$.

Formula (2.3.6) reads $UC_b^1(H) \in J_{1/2}\left(UC_b(H), UC_b^2(H)\right)$, and formula (2.3.7) becomes

$$\|u\|_1 \leq C_{0,1,2} \|u\|_0^{1/2} \|u\|_2^{1/2}, \quad u \in UC_b^2(H). \tag{2.3.8}$$

Let $u \in UC_b^2(H)$, $h > 0$, $x, z \in H$ with $|z| = 1$. From the equality

$$
\begin{aligned}
u(x + hz) \;=\; & u(x) + h\langle Du(x), z\rangle \\
& + h^2 \int_0^1 (1 - \sigma)\langle D^2 u(x + \sigma hz) \cdot z, z\rangle d\sigma, \tag{2.3.9}
\end{aligned}
$$

it follows that

$$|\langle Du(x), z\rangle| \leq \frac{2}{h} \|u\|_0 + \frac{h}{2} \|D^2 u\|_0, \quad h > 0.$$

Since $\min_{h>0}\left(\frac{a}{h} + bh\right) = 2\sqrt{ab}$, we have

$$|\langle Du(x), z\rangle| \leq 2\|u\|_0^{1/2} \|D^2 u\|_0^{1/2}.$$

Then, from the arbitrariness of z we have $\|Du\|_0 \leq 2\|u\|_0^{1/2} \|u\|_2^{1/2}$. On the other hand $\|u\|_0 = \|u\|_0^{1/2} \|u\|_0^{1/2} \leq \|u\|_0^{1/2} \|u\|_2^{1/2}$, which yields $\|u\|_1 \leq 3\|u\|_0^{1/2} \|u\|_2^{1/2}$.

Case 2. $a = 1$, $b = 2$, $c = 2 + \alpha$, $\alpha \in (0, 1)$.

Formula (2.3.6) reads

$$UC_b^2(H) \in J_{\frac{1}{1+\alpha}}\left(UC_b^1(H), UC_b^{2+\alpha}(H)\right),$$

and (2.3.7)

$$\|u\|_2 \leq C_{1,2,2+\alpha} \|u\|_1^{\frac{\alpha}{1+\alpha}} \|u\|_{2+\alpha}^{\frac{1}{1+\alpha}}, \quad u \in C_b^{2+\alpha}(H). \tag{2.3.10}$$

Let $u \in UC_b^{2+\alpha}(H)$ and let $h > 0$, $x, z \in H$ with $|z| = 1$. Then from the equality

$$
\begin{aligned}
u(x + hz) \;=\; & u(x) + h\langle Du(x), z\rangle + \frac{1}{2}h^2\langle D^2 u(x) \cdot z, z\rangle \\
& + h^2 \int_0^1 (1 - \sigma)\langle (D^2 u(x + \sigma hz) - D^2 u(x)) \cdot z, z\rangle d\sigma,
\end{aligned}
$$

there exists $C_\alpha > 0$ such that

$$\frac{1}{2}h^2|\langle D^2u(x) \cdot z, z\rangle| \leq |u(x+hz) - u(x) - hDu(x) \cdot z|$$

$$+h^2 [D^2u]_\alpha \int_0^1 (1 - \sigma)\sigma^\alpha h^\alpha \, d\sigma$$

$$\leq 2h\|Du\|_0 + C_\alpha h^{2+\alpha}[D^2u]_\alpha.$$

It follows that

$$\|D^2u(x)\| \leq \frac{1}{h}\|u\|_0 + 2C_\alpha h^\alpha[D^2u]_\alpha.$$

Taking the minimum for $h > 0$ it follows, for a suitable constant $C_{1,\alpha}$, that

$$\|D^2u\|_0 \leq C \, \|Du\|_0^{\frac{\alpha}{1+\alpha}} \, [D^2u]_\alpha^{\frac{1}{1+\alpha}},$$

and the conclusion follows.

Case 3. $a = \alpha$, $b = 1$, $c = 2$, $\alpha \in (0,1)$.

Formula (2.3.6) becomes $UC_b^1(H) \in J_{\frac{1-\alpha}{2-\alpha}}\left(UC_b^\alpha(H), UC_b^2(H)\right)$, and (2.3.7)

$$\|u\|_1 \leq C_{\alpha,1,2} \, \|u\|_\alpha^{\frac{1}{2-\alpha}} \, \|u\|_2^{\frac{1-\alpha}{2-\alpha}}, \quad u \in C_b^2(H). \tag{2.3.11}$$

Let $u \in UC_b^2(H)$ and let $h > 0$, $x, z \in H$ with $|z| = 1$. From (2.3.9) it follows that

$$\|Du\|_0 \leq h^{\alpha-1}[u]_\alpha + h^2\|D^2u\|_0.$$

The conclusion follows again taking minimum for $h > 0$.

Case 4 $a = \alpha$, $b = 2$, $c = 2 + \alpha$, $\alpha \in (0,1)$.

Formula (2.3.6) becomes $UC_b^2(H) \in J_{1-\alpha/2}\left(UC_b^\alpha(H), UC_b^{2+\alpha}(H)\right)$, and (2.3.7)

$$\|u\|_2 \leq C_{\alpha,2,2+\alpha} \, \|u\|_{2+\alpha}^{1-\alpha/2} \, \|u\|_\alpha^{\alpha/2}, \quad u \in C_b^{2+\alpha}(H). \tag{2.3.12}$$

From (2.3.10) we have

$$\|u\|_2 \leq C\|u\|_{2+\alpha}^{\frac{1}{1+\alpha}} \, \|u\|_1^{\frac{\alpha}{1+\alpha}},$$

from which, using (2.3.11), we have, for suitable constants C_1 and C_2,

$$\|u\|_2 \leq C_1\|u\|_{2+\alpha}^{\frac{1}{1+\alpha}} \, \|u\|_\alpha^{\frac{\alpha}{(2-\alpha)(1+\alpha)}} \, \|u\|_2^{\frac{\alpha(1-\alpha)}{(2-\alpha)(1+\alpha)}},$$

which yields

$$\|u\|_2^{\frac{2}{(1+\alpha)(2-\alpha)}} \leq C_2 \|u\|_{2+\alpha}^{\frac{1}{1+\alpha}} \|u\|_\alpha^{\frac{\alpha}{(2-\alpha)(1+\alpha)}},$$

and the conclusion follows.

Case 5. $a = 0$, $b = 2$, $c = 2 + \alpha$, $\alpha \in (0,1)$.

Now (2.3.6) becomes $UC_b^2(H) \in J_{\frac{2}{2+\alpha}}\left(UC_b(H), UC_b^{2+\alpha}(H)\right)$, and (2.3.7)

$$\|u\|_2 \leq C_{0,2,2+\alpha} \|u\|_0^{\frac{\alpha}{2+\alpha}} \|u\|_{2+\alpha}^{\frac{2}{2+\alpha}}, \quad u \in C_b^{2+\alpha}(H). \tag{2.3.13}$$

The conclusion follows from (2.3.9) and (2.3.8). $\square$

2.3.3   Additional interpolation results

Let us first recall the classical *reiteration theorem*, Theorems 2.3.6 and 2.3.7 below. For a proof see e.g A. Lunardi [162].

Let X, Y, E, F be Banach spaces with $Y \subset F \subset E \subset X$, all embeddings being continuous.

Theorem 2.3.6 *Assume that*
(i) there exists $\alpha \in (0,1)$ such that $E \in J_\alpha(X,Y)$,
(ii) there exists $\beta \in (0,1)$ with $\alpha \neq \beta$, such that $F \in J_\beta(X,Y)$.
Then, given $\theta \in (0,1)$, and setting $\omega = (1-\theta)\alpha + \theta\beta$, we have

$$(X,Y)_{\omega,\infty} \subset (E,F)_{\theta,\infty}, \tag{2.3.14}$$

the inclusion being continuous.

Theorem 2.3.7 *Assume that*
(i) there exists $\alpha \in (0,1)$ such that $E \subset (X,Y)_{\alpha,\infty}$,
(ii) there exists $\beta \in (0,1)$ with $\alpha \neq \beta$, such that $F \subset (X,Y)_{\beta,\infty}$.
Then, given $\theta \in (0,1)$ and setting $\omega = (1-\theta)\alpha + \theta\beta$, we have

$$(X,Y)_{\omega,\infty} \supset (E,F)_{\theta,\infty}, \tag{2.3.15}$$

the inclusion being continuous.

We now give two applications of the reiteration theorem, useful later.

Proposition 2.3.8 *Let $\beta, \theta \in (0,1)$. Then we have*

$$\left(UC_b(H), C_b^\beta(H)\right)_{\theta,\infty} = C_b^{\theta\beta}(H), \tag{2.3.16}$$

$$\left(C_b^\beta(H), C_b^1(H)\right)_{\theta,\infty} = C_b^{\beta+\theta(1-\beta)}(H). \tag{2.3.17}$$

Proof. To prove (2.3.16) it suffices to apply Theorems 2.3.6 and 2.3.7 with

$$Y = UC_b^1(H), \ F = C_b^\beta(H), \ E = X = UC_b(H),$$

and with $\alpha = 0$. The proof of (2.3.17) is similar. $\square$

Proposition 2.3.9 *Let $\delta, \gamma \in (0,1)$. Then we have*

$$\left(C_b^\delta(H), C_b^{2+\delta}(H)\right)_{1-\frac{\delta}{2}(1-\gamma),\infty} \subset C_b^{2+\delta\gamma}(H). \tag{2.3.18}$$

Proof.

Step 1. We have

$$\left(C_b^\delta(H), C_b^{2+\delta}(H)\right)_{1-\frac{\delta}{2}(1-\gamma),\infty} \subset \left(UC_b^2(H), C_b^{2+\delta}(H)\right)_{\gamma,\infty}.$$

Set $Y = F = C_b^{2+\delta}(H), \ E = UC_b^2(H), \ X = C_b^\delta(H)$. Then from Theorem 2.3.5 it follows that $E \in J_{1-\frac{\delta}{2}}(X,Y)$, and obviously $F \in J_1(X,Y)$. So we can apply Theorem 2.3.6 with $\alpha = 1 - \frac{\delta}{2}$, $\beta = 1$, $\theta = \gamma$. In this case, setting $\omega = (1-\theta)\alpha + \theta\beta = 1 - \frac{\delta}{2}(1-\gamma)$, we have

$$(X,Y)_{1-\frac{\delta}{2}(1-\gamma),\infty} \subset (E,F)_{\gamma,\infty}.$$

Step 2. $\left(UC_b^2(H), C_b^{2+\delta}(H)\right)_{\gamma,\infty} \subset C_b^{2+\delta\gamma}(H).$

Let $\varphi \in \left(UC_b^2(H), C_b^{2+\delta}(H)\right)_{\gamma,\infty}$. From the very definition of an interpolation space, there exists $C > 0$, $a_t \in UC_b^2(H)$, $b_t \in C_b^{2+\delta}(H)$, such that

$$a_t + b_t = \varphi, \ \|D^2 a_t\|_0 \le Ct^\gamma, \ [D^2 b_t]_\delta \le Ct^\gamma, \ t \in (0,1].$$

It follows that

$$\|D^2\varphi(x) - D^2\varphi(y)\| \le 2Ct^\gamma + Ct^{\gamma-1}|x-y|^\delta, \ t \in]0,1]. \tag{2.3.19}$$

Setting $t = |x-y|^\delta$ in (2.3.19) we have the conclusion. $\square$

Chapter 3

The heat equation

The chapter is devoted to the existence, uniqueness and regularity of solutions of the following heat equation on a separable Hilbert space H (norm $|\cdot|$, inner product $\langle \cdot, \cdot \rangle$):

$$
\begin{cases}
D_t u(t, x) &= \frac{1}{2} \operatorname{Tr}[Q D^2 u(t, x)], \quad t > 0, \ x \in H, \\
u(0, x) &= \varphi(x), \quad x \in H,
\end{cases}
$$

where $Q \in L^+(H)$. We first show in §3.1 that, in order to have well-posedness of this problem for all $\varphi \in UC_b(H)$, one has to assume that the operator Q is of trace class. In §3.2 and §3.3 we study existence, uniqueness and regularity of solutions in $UC_b(H)$. In §3.5 we give several properties of the corresponding strongly continuous semigroup (P_t) and characterize its infinitesimal generator.

The heat equation was first studied in a pioneering paper of L. Gross [138], see also [62]. His setting was different, as can be seen in §3.4.

We shall use all notations introduced in Chapters 1 and 2.

3.1 Preliminaries

Let us start with the heat equations in $H = \mathbb{R}^d$, $d \in \mathbb{N}$:

$$
\begin{cases}
D_t u(t, x) &= \frac{1}{2} \sum_{i,j=1}^d q_{ij} D_i D_j u(t, x), \quad t > 0, \ x = (x_1, \dots, x_d) \in \mathbb{R}^d, \\
u(0, x) &= \varphi(x), \quad x \in \mathbb{R}^d,
\end{cases}
$$

$$(3.1.1)$$

44

where $Q = (q_{ij})$ is a symmetric nonnegative definite $d \times d$ real matrix. Note that (3.1.1) can be written in more compact form as

$$\begin{cases} D_t u(t, x) & = \frac{1}{2} \operatorname{Tr}[Q D^2 u(t, x)], \quad t > 0, \ x \in \mathbb{R}^d, \\ u(0, x) & = \varphi(x), \quad x \in \mathbb{R}^d. \end{cases}$$

It is well known, see e.g. A. Friedman [115], or N. V. Krylov [147], that if $\det Q > 0$, and $\varphi \in C_b(\mathbb{R}^d)$, then there exists a unique continuous and bounded function $u \colon [0, +\infty) \times \mathbb{R}^d \to \mathbb{R}$ such that the partial derivatives

$$D_t u(x, t), \quad D_i D_j u(t, x), \quad i, j = 1, \dots, d,$$

are well defined and continuous on $(0, +\infty) \times \mathbb{R}^d$ and (3.1.1) holds. Moreover the solution u is given by the following formula:

$$\begin{aligned} u(t, x) & = \frac{1}{\sqrt{(2\pi t)^d \det Q}} \int_{\mathbb{R}^d} e^{-\frac{1}{2t} \langle Q^{-1}(y-x), y-x \rangle} \varphi(y) \, dy \\ & = \int_{\mathbb{R}^d} \varphi(x + y) n_t(y) dy, \quad t > 0, \ x \in \mathbb{R}^d, \end{aligned} \tag{3.1.2}$$

where

$$n_t(y) = \frac{1}{\sqrt{(2\pi t)^d \det Q}} e^{-\frac{1}{2t} \langle Q^{-1} y, y \rangle}, \ y \in \mathbb{R}^d.$$

It follows from (3.1.2) that the solution u is a C^∞ function for each $t > 0$, and it depends linearly on the initial function φ. In fact for each $t > 0$, the formula

$$u(t, x) = P_t \varphi(x) = \int_{\mathbb{R}^d} \varphi(x + y) n_t(y) dy, \quad x \in \mathbb{R}^d, \tag{3.1.3}$$

defines a linear operator $P_t \in L(C_b(\mathbb{R}^d)) \cap L(UC_b(\mathbb{R}^d))$. From the fact that the formula (3.1.3) defines a unique solution to (3.1.1), it follows that the family P_t, $t > 0$, has the semigroup property:

$$P_{t+s} = P_t P_s, \quad t, s > 0. \tag{3.1.4}$$

If one sets $P_0 = I$, then (3.1.4) is valid for all $t, s \geq 0$.

Basic properties of the solution u can be expressed in terms of the semigroup (P_t) as follows.

Theorem 3.1.1 *Assume that the matrix Q is positive definite. Then the formula (3.1.3) defines a semigroup of bounded operators on $C_b(\mathbb{R}^d)$ and on*

$UC_b(\mathbb{R}^d)$. *The semigroup (P_t) is strongly continuous on $UC_b(\mathbb{R}^d)$ but not on $C_b(\mathbb{R}^d)$.* (^{1}) *Moreover, for arbitrary $\varphi \in C_b(\mathbb{R}^d)$ the function $P_t\varphi$ is of class C^∞ with respect to variables $t > 0$ and $x \in \mathbb{R}^d$.*

We leave the elementary proof to the reader.

Let Q be a nonnegative symmetric matrix with $\det Q = 0$, and denote by $e_1, \ldots, e_d$ its eigenvectors and by $\gamma_1, \ldots, \gamma_d$ its eigenvalues. Assume that $m < d$ and $\gamma_1, \ldots, \gamma_m$ are positive whereas the remaining ones are equal to 0. Therefore, in the coordinates determined by the eigenvectors of Q, we have

$$\mathrm{Tr}[QD^2u(x)] = \sum_{i=1}^{m} \gamma_i D_i^2 u(x), \quad x \in H.$$

The solution u of

$$\begin{cases} D_t u(t,x) & = \quad \frac{1}{2}\sum_{i=1}^{m} \gamma_i D_i^2 u(t,x), \quad t > 0, \ x \in \mathbb{R}^d, \\ u(0,x) & = \quad \varphi(x) \end{cases}$$

is given, for $x = (x_1, \ldots, x_m, x_{m+1}, \ldots, x_d)$ and $t > 0$, by the formula

$$u(t,x) = \frac{1}{\sqrt{(2t\pi)^m}\, \gamma_1 \cdots \gamma_m}$$

$$\times \int_{\mathbb{R}^m} e^{-\frac{1}{2t}\sum_{i=1}^{m} \frac{1}{\gamma_i}(x_i - y_i)^2} \varphi(y_1, \ldots, y_m, x_{m+1}, \ldots, x_d)\, dy_1 \ldots dy_m.$$

$$(3.1.5)$$

If $\varphi \in C_b(\mathbb{R}^d)$ then u is not, in general, a C^∞ function. This is clear if, for instance, φ depends only on the variables $x_{m+1}, \ldots, x_d$, say $\varphi(x_1, \ldots, x_d) = \varphi_0(x_{m+1}, \ldots, x_d)$, as then

$$u(t,x) = \varphi_0(x_{d_1+1}, \ldots, x_d), \quad x \in \mathbb{R}^d,$$

and therefore the solution u has the same regularity as the initial function φ_0. We will see that *the lack of the regularizing power* for a degenerate heat equation in $\mathbb{R}^d$ will be shared by all heat equations on infinite dimensional Hilbert spaces.

Let now H be a separable infinite dimensional Hilbert space and Q a bounded self-adjoint nonnegative operator on H. Consider the following problem:

$$\begin{cases} D_t u(t,x) & = \quad \frac{1}{2}\mathrm{Tr}[QD^2u(t,x)], \quad t > 0, \ x \in H, \\ u(0,x) & = \quad \varphi(x). \end{cases} \qquad (3.1.6)$$

---

^{1}since the closure of the domain of the infinitesimal generator of (P_t) is $UC_b(H)$ which is a proper closed subspace of $C_b(H)$.

Note that the formula (3.1.3) for the solutions loses its meaning if dim $H = \infty$. It depends in fact on the dimension d of the space $\mathbb{R}^d$ and is written in terms of the Lebesgue measure for which an exact counterpart in infinite dimensions does not exist. One way to construct a solution to (3.1.6) in the infinite dimensional situation would be to consider a sequence of equations

$$\begin{cases} D_t u_n(t,x) &= \frac{1}{2}\mathrm{Tr}[Q_n D^2 u_n(t,x)], \quad t > 0, \; x \in H, \\ u_n(0,x) &= \varphi(x), \end{cases} \tag{3.1.7}$$

with finite rank nonnegative operators Q_n strongly converging to Q. For each $n \in \mathbb{N}$ this equation has a unique solution u_n given by formula (3.1.5). If the sequence (u_n) of solutions were convergent, then its limit could be taken as a candidate for the solution to (3.1.6).

However, to define the solution in this way, some *restrictions* on the operator Q have to be imposed, see P. Cannarsa and G. Da Prato [31]. We have in fact the following result due to J. Zabczyk [220].

Proposition 3.1.2 *Assume that* $\varphi \in C_b(H)$ *and* $\lim_{|y| \to \infty} \varphi(y) = 0$. *If* $\mathrm{Tr}\, Q = \infty$ *and* (Q_n) *is a sequence of finite rank nonnegative operators converging strongly to* Q, *then* $\lim_{n \to \infty} u_n(t,x) = 0$, *for all* $t > 0$ *and* $x \in H$.

Proof. It follows easily that $\lim_{n \to \infty} \mathrm{Tr}\, Q_n = \infty$. Without any loss of generality one can assume that $x = 0$. Let us consider first the special function

$$\hat{\varphi}(y) := e^{-|y|^2}, \quad y \in H.$$

If $\gamma_1^n, \ldots, \gamma_{m_n}^n$ are all the positive eigenvalues of Q_n, by (3.1.5) we find for the corresponding solutions $\hat{u}_n$,

$$\hat{u}_n(t,0) = \left(\prod_{k=1}^{m_n} (1 + 2t\gamma_k^n) \right)^{-\frac{1}{2}}.$$

One can assume that $2t\gamma_k^n \leq \alpha$, $k = 1, 2, \ldots, m_n$, $n \in \mathbb{N}$, for some $\alpha > 0$. Then, for a constant $\beta > 0$, $\hat{u}_n(t,0) \leq e^{-\beta \sum_{k=1}^{m_n} \gamma_k^n}$. Since $\mathrm{Tr}\, Q_n = \sum_{k=1}^{m_n} \gamma_k^n \to \infty$, the result is true for the special function $\hat{\varphi}$. If now φ is any function satisfying the conditions of the proposition, for any $\epsilon > 0$ one can find $\delta > 0$ and a decomposition $\varphi = \varphi_0 + \varphi_1$, with $\varphi_0, \varphi_1 \in C_b(H)$ such that

$$|\varphi_0(x)| \leq \delta\, e^{-|x|^2}, \quad |\varphi_1(x)| \leq \epsilon, \quad x \in H.$$

Consequently, for the solutions u_n,

$$|u_n(t,0)| \leq \delta\, \hat{u}_n(t,0) + \epsilon,$$

and $\limsup_{n\to\infty} |u_n(t,0)| \leq \epsilon$. Since ϵ is an arbitrary positive number the proof of the proposition is complete. $\square$

Proposition 3.1.2 indicates that if $\mathrm{Tr}\, Q = \infty$ then, for a majority of initial functions φ, the equation (3.1.6) *does not have a continuous solution* on $[0, +\infty) \times H$. This is why we will assume that the symmetric nonnegative operator Q is of trace class.

It is easy to see that if $\varphi \in C_b(H)$ then the approximating solutions u_n from Proposition 3.1.2 are given by the formula

$$u_n(t,x) = \int_H \varphi(x+y) N_{tQ_n}(dy), \qquad x \in H, \ t \geq 0. \qquad (3.1.8)$$

Now if $\mathrm{Tr}\, Q < +\infty$, $u_n \to u$ as $n \to \infty$, where

$$u(t,x) = \int_H \varphi(x+y) N_{tQ}(dy), \qquad x \in H, \ t \geq 0. \qquad (3.1.9)$$

Moreover the operators P_t,

$$P_t\varphi(x) = \int_H \varphi(x+y) N_{tQ}(dy), \qquad x \in H, \ \varphi \in C_b(H), \qquad (3.1.10)$$

form a semigroup of bounded operators (since it is a limit of semigroups) in $C_b(H)$ and in $UC_b(H)$. The function

$$u(t,x) = P_t\varphi(x), \qquad t \geq 0, \ x \in H,$$

where (P_t) is given by (3.1.10), will be called the *generalized solution to* (3.1.6). The semigroup (P_t) is strongly continuous on $UC_b(H)$ but not on $C_b(H)$. We will call (P_t) the *heat semigroup*.

Remark 3.1.3 It is instructive to realize that the heat equation (3.1.6) is the Kolmogorov equation corresponding to the simplest Itô equation

$$dX(t) = dW(t), \quad X(0) = x, \ t \geq 0, \qquad (3.1.11)$$

on a Hilbert space H, where W is a Wiener process on some probability space $(\Omega, \mathcal{F}, \mathbb{P})$ taking values in H, with covariance operator Q.

3.2   Strict solutions

A function $u : [0, +\infty) \times H \to \mathbb{R}$ is said to be a *strict solution* to (3.1.6) if the derivatives $D_t u(t,x)$ and $D^2 u(t,x)$ ([2]) exist for all $t \geq 0$ and $x \in H$, are continuous and bounded on $[0, +\infty) \times H$ and satisfy (3.1.6).

---

[2]We shall denote by D (resp. D^2) the first (resp. second) Fréchet derivative with respect to x.

In this section we show that if the function φ is sufficiently regular then (3.1.10) defines a strict solution to (3.1.6), and that strict solutions are unique.

Remark 3.2.1 With the same proof one can show existence of solutions under local boundedness and polynomial growth at infinity of the initial condition φ. The proof is left to the reader.

We have the following lemma.

Lemma 3.2.2 *If $\varphi \in UC_b^2(H)$ (resp. $C_b^2(H)$), then for arbitrary $t \geq 0$, $u(t, \cdot) \in UC_b^2(H)$ (resp. $C_b^2(H)$) and*

$$Du(t,x) = \int_H D\varphi(x + \sqrt{t}\, y) N_Q(dy), \qquad t \geq 0, \ x \in H, \quad (3.2.1)$$

$$D^2 u(t,x) = \int_H D^2\varphi(x + \sqrt{t}\, y) N_Q(dy), \qquad t \geq 0, \ x \in H. \quad (3.2.2)$$

Proof. The integral in (3.2.1) is of Bochner type and in (3.2.2) it is strong Bochner, see e.g. [102]. Let $g, h \in H$; then

$$\langle Du(t,x), g \rangle = \int_H \langle D\varphi(x + \sqrt{t}\, y), g \rangle N_Q(dy),$$

$$\langle D^2 u(t,x)h, g \rangle = \int_H \langle D^2\varphi(x + \sqrt{t}\, y)g, h \rangle N_Q(dy).$$

Therefore (3.2.1), (3.2.2) follow. The uniform continuity of the functions $u(t, \cdot)$, $Du(t, \cdot)$, $D^2 u(t, \cdot)$ is an easy consequence of the formulae (3.2.1), (3.2.2). $\square$

The following theorem is taken from [102].

Theorem 3.2.3 *If $\varphi \in C_b^2(H)$, then the function u given by (3.1.9) is a strict solution to (3.1.6).*

Proof. By the mean value theorem

$$u(t,x) = \varphi(x) + \sqrt{t} \int_H \langle D\varphi(x), y \rangle N_Q(dy)$$

$$+ \frac{1}{2} t \int_H \langle D^2\varphi(x + \sigma(t,y)\sqrt{t}\, y)y, y \rangle N_Q(dy),$$

where σ is a Borel function from $[0, +\infty) \times H$ into $[0, 1]$. Note that by Proposition 1.2.4 we have

$$\int_H \langle D\varphi(x), y \rangle N_Q(dy) = 0,$$

$$\int_H \langle D^2\varphi(x)y, y \rangle N_Q(dy) = \mathrm{Tr}[QD^2\varphi(x)], \quad x \in H,$$

and consequently

$$\frac{1}{t}\left(u(t, x) - u(0, x)\right) - \frac{1}{2}\,\mathrm{Tr}[QD^2 u(0, x)]$$

$$= \frac{1}{2}\int_H \langle [D^2\varphi(x + \sigma(t, y)\sqrt{t}\,y) - D^2\varphi(x)]y, y\rangle N_Q(dy).$$

Therefore

$$\left| \frac{u(t, x) - u(0, x)}{t} - \frac{1}{2}\,\mathrm{Tr}[QD^2 u(0, x)] \right|$$

$$\leq \frac{1}{2}\left[\int_H |D^2\varphi(x + \sigma(t, y)\sqrt{t}y) - D^2\varphi(x)|^2 N_Q(dy)\right]^{1/2} \left[\int_H |y|^4 N_Q(dy)\right]^{1/2}.$$

By Proposition 1.2.8

$$\int_H e^{-\frac{\xi^2}{2}|y|^2} N_Q(dy) = [\det(1 + \xi Q)]^{-1/2}, \quad \xi \in \mathbb{R}.$$

Differentiating this identity twice with respect to ξ and setting $\xi = 0$ yields

$$\int_H |y|^4 N_Q(dy) = [\mathrm{Tr}\,Q]^2 + 2\,\mathrm{Tr}(Q^2).$$

Thus

$$\left| \frac{u(t, x) - u(0, x)}{t} - \frac{1}{2}\,\mathrm{Tr}[QD^2 u(0, x)] \right| \leq \frac{1}{2}\left([\mathrm{Tr}\,Q]^2 + 2\,\mathrm{Tr}(Q^2)\right)$$

$$\times \left[\int_H |D^2\varphi(x + \sigma(t, y)\sqrt{t}y) - D^2\varphi(x)|^2 N_Q(dy)\right]^{1/2}.$$

Consequently

$$\frac{1}{t}\left(u(t, x) - u(0, x)\right) \to \frac{1}{2}\,\mathrm{Tr}[QD^2 u(0, x)], \quad \text{as } t \to 0,$$

uniformly in $x \in H$.

In this way we have shown that the right derivative of $u(t, x)$ at $t = 0$ is given by

$$D_t^+ u(0, x) = \frac{1}{2} \operatorname{Tr}[QD^2 u(0, x)], \quad x \in H.$$

Now fix $s > 0$. Since $P_{s+t}\varphi = P_t(P_s\varphi)$, applying the previous argument with φ replaced by $P_s\varphi$, we obtain for all $x \in H$,

$$D_s^+ u(s, x) = \frac{1}{2} \operatorname{Tr}[QD^2 u(s, x)], \quad s \geq 0, \; x \in H. \tag{3.2.3}$$

The right hand side of (3.2.3) is continuous on $[0, +\infty) \times H$. In particular, by Lemma 3.2.2, for every x, the right derivative in time of u is a bounded and continuous function in $s \in [0, +\infty)$. From elementary calculus, see Lemma 3.2.4 below, $u(\cdot, x)$ is continuously differentiable in s and the result follows. $\square$

Lemma 3.2.4 *If a continuous real function f has a continuous right derivative $D^+ f$ on an interval $[0, a)$ then it is of class C^1 on $[0, a)$.*

Proof. Define

$$g(t) = f(0) + \int_0^t D^+ f(s) ds, \quad \varphi(t) = g(t) - f(t), \; t \in [0, a),$$

and assume that for some $t_0 \in (0, a)$, $\varphi(t_0) > 0$. Let us choose $\varepsilon > 0$ such that $\psi_\varepsilon(t_0) < \varphi(t_0)$ where $\psi_\varepsilon(t) = \varepsilon + \varepsilon t$, $t \in [0, a)$, and let $t_1 = \sup\{t \in [0, t_0] : \psi_\varepsilon(t) = \varphi(t)\}$. Then $t_1 < t_0$ and $\psi_\varepsilon(t_1) = \varphi(t_1)$. However

$$\varepsilon = D\psi_\varepsilon(t_1) = D^+\psi_\varepsilon(t_1) \leq D^+\varphi(t_1) = 0,$$

a contradiction. So $g(t) \leq f(t)$ for all $t \in [0, a)$. In a similar way, considering $\varphi(t) = f(t) - g(t)$, $t \in [0, a)$, one shows that $g(t) \geq f(t)$ for all $t \in [0, a)$. Consequently $f(t) = g(t)$, $t \in [0, a)$ and the result follows. $\square$

We are going to prove, using the maximum principle, that there exists at most one strict solution of (3.1.6). For this we will need the following Asplund's theorem, for a proof see for instance J. P. Aubin [7].

Theorem 3.2.5 *Let X be a Hilbert space, K a closed bounded subset of X, and ζ a bounded function on K. Then there exists a dense subset Σ of X such that the mapping $K \to \mathbb{R}$, $x \to \zeta(x) + \langle x, y \rangle$, attains a maximum at K for all $y \in \Sigma$.*

Lemma 3.2.6 *Assume that K is a closed subset of a Hilbert space X, and that u is a bounded and continuous function on K. For arbitrary $\varepsilon > 0$ and $n \in \mathbb{N}$, there exists $p \in C_0^\infty(X)$ such that*

(i) $u + p$ attains its maximum on K,

(ii) $\displaystyle\sup_{x \in X} \sum_{k=0}^{n} \|D^k p(x)\| \leq C\varepsilon.$

Proof. It is enough to prove the lemma for a nonnegative u. For $\varepsilon > 0$ let $x_\varepsilon \in K$ be such that $u(x_\varepsilon) \geq \sup_{x \in K} u(x) - \varepsilon$. Consider a new function $u_\varepsilon(x) = u(x) + 2\varepsilon\eta(|x - x_\varepsilon|^2)$, $x \in K$, where $\eta \in C^\infty([0, +\infty); \mathbb{R})$ such that

$$0 \leq \eta \leq 1, \ \eta(0) = 1, \ \eta(r) = 0, \ \forall r \geq 1.$$

Then $u_\varepsilon(x_\varepsilon) = u(x_\varepsilon) + 2\varepsilon \geq \sup_{x \in K} u(x) + \varepsilon$, and

$$u_\varepsilon(x) = u(x) \leq \sup_{x \in K} u(x), \ |x - x_\varepsilon| \geq 1, \ x \in K.$$

Consequently, setting $K_\varepsilon = K \cap B(x_\varepsilon, 1)$, we have

$$\sup_{x \in K} u_\varepsilon(x) = \sup_{x \in K_\varepsilon} u_\varepsilon(x).$$

From Asplund's theorem, for a dense set of $q \in X$, the function $u_\varepsilon(x) + \langle q, x \rangle$, $x \in K_\varepsilon$, attains its maximum at some point of K_ε. We finally set

$$p(x) = 2\varepsilon\eta(|x - x_\varepsilon|^2 + \rho(|x - x_\varepsilon|^2))\langle q, x \rangle,$$

where $\rho \in C^\infty([0, +\infty); \mathbb{R})$ is such that

$$0 \leq \rho \leq 1, \ \rho(r) = 1, \ \forall r \in [0, 1], \ \rho(r) = 0, \ \forall r \geq 2,$$

and $|q|$ is sufficiently small. One can easily check that p has the required properties. $\square$

Now we are in position to prove uniqueness.

Theorem 3.2.7 *Let $\varphi \in UC_b^2(H)$. Then there exists at most one strict solution to (3.1.6).*

Proof. Let $T > 0$ be fixed, and let u be a strict solution to equation (3.1.6). It is enough to show that

$$\sup_{t \in [0,T]} e^{-t} \|u(t, \cdot)\|_0 \leq \|\varphi\|_0. \tag{3.2.4}$$

Assume that u is not identically equal to 0 and set

$$\lambda := \sup_{(t,x)\in[0,T]\times H} u(t,x),$$

and $v(t,x) = e^{-t}u(t,x)$, $t \in [0,T]$, $x \in H$. Then $\lambda > 0$ and we have

$$D_t v = \frac{1}{2}\mathrm{Tr}[QD^2 v] - v.$$

We apply now Lemma 3.2.6 to the Hilbert space $X = \mathbb{R} \times H$ and to the set $K = [0,T] \times H$. Consequently for any $\varepsilon > 0$ there exists p_ε such that the function $v_\varepsilon := v + p_\varepsilon$ attains maximum at a point $(t_\varepsilon, x_\varepsilon)$ in $[0,T] \times H$ and

$$\sup_{(t,x)\in[0,T]\times H} \left(|p_\varepsilon(t,x)| + |D_t p_\varepsilon(t,x)| + \|D^2 p_\varepsilon(t,x)\|\right) \leq \varepsilon.$$

We have

$$D_t v_\varepsilon - \frac{1}{2}\mathrm{Tr}[QD^2 v_\varepsilon] = -v_\varepsilon + g_\varepsilon,$$

where

$$g_\varepsilon = D_t p_\varepsilon - \frac{1}{2}\mathrm{Tr}[QD^2 p_\varepsilon] + p_\varepsilon,$$

so that

$$\sup_{(t,x)\in[0,T]\times H} |g_\varepsilon(t,x)| \leq \varepsilon\left(1 + \frac{1}{2}\mathrm{Tr}\,Q\right).$$

We show now that for sufficiently small $\varepsilon > 0$ we have $t_\varepsilon = 0$. Denote by (λ_k) the eigenvalues of Q and assume in contradiction that $t_\varepsilon \in (0,T]$. Since $D_t v_\varepsilon(t_\varepsilon, x_\varepsilon) \geq 0$ and $D^2 v_\varepsilon(t_\varepsilon, x_\varepsilon) \leq 0$, we have

$$D_t v_\varepsilon(t_\varepsilon, x_\varepsilon) - \frac{1}{2}\sum_{k=1}^{\infty} \lambda_k D_k^2 v_\varepsilon(t_\varepsilon, x_\varepsilon) \geq 0,$$

and then $v_\varepsilon(t_\varepsilon, x_\varepsilon) \leq g_\varepsilon(t_\varepsilon, x_\varepsilon)$. This means that

$$u(t_\varepsilon, x_\varepsilon) = e^{t_\varepsilon} v(t_\varepsilon, x_\varepsilon) \leq \varepsilon e^{t_\varepsilon}\left(2 + \frac{1}{2}\mathrm{Tr}\,Q\right).$$

Therefore if ε is small enough, we get a contradiction with the positivity of λ. In conclusion the maximum of v_ε is attained at a point $(0, x_\varepsilon)$ and (3.2.4) easily follows. $\square$

3.3   Regularity of generalized solutions

Let $\varphi \in C_b(H)$ and let $u(t,x) = P_t\varphi(x)$. We know that if H is finite dimensional and Q nondegenerate then $u(t,x)$ is of class C^∞ in t and x when
$t > 0$. This result is not true in infinite dimensions. We cannot say even
that $u(t,\cdot) \in C_b^1(H)$, as follows from Proposition 3.3.9 below. However,
as discovered by L. Gross [138] $u(t,x)$ is infinitely differentiable in x in all
directions of the Cameron-Martin space $Q^{1/2}(H)$. To show this we will introduce Q-derivatives in the following subsection. The original setting of L.
Gross [138] will be recalled in the last section of this chapter.

3.3.1 Q-derivatives

Let φ be a mapping from H into a Banach space E, then φ is called Q-
differentiable at x if the function $F(y) = \varphi(x + Q^{1/2}y)$, $y \in H$, is differentiable at 0. We set $D_Q\varphi(x) = DF(0)$, and call $D_Q\varphi(x)$ the Q-*derivative* of
φ at x.

   If F is differentiable in a neighborhood of 0 and DF is differentiable
at 0, then we say that φ is twice Q-*differentiable* at x and its second Q-
derivative is given by definition by $D_Q^2\varphi(x) = D^2F(0)$. If φ is a real valued
function then $D_Q\varphi(x)$ (resp. $D_Q^2\varphi(x)$) will be identified with an appropriate
element of H (resp. $L(H)$). Derivatives $D_Q^k\varphi(x)$ of higher orders are defined
similarly.

   We notice that if $\varphi \in C_b^1(H)$ we have $D_Q\varphi(x) = Q^{1/2}D\varphi(x)$, and if
$\varphi \in C_b^2(H)$, $D_Q^2\varphi(x) = Q^{1/2}D^2\varphi(x)Q^{1/2}$.

   Let us define some functional subspaces of $C_b(H)$ and $UC_b(H)$ related
to Q-derivatives.

   For any $k \in \mathbb{N}$ we denote by $C_Q^k(H)$ (resp. $UC_Q^k(H)$) the set of all $\varphi \in$
$C_b(H)$ (resp. $UC_b(H)$) that possess continuous (resp. uniformly continuous)
Q-derivatives of order smaller than or equal to k.

   If $\varphi \in C_Q^k(H)$ we set

$$[\varphi]_{k,Q} = \sup_{x \in H} \|D_Q^k\varphi\|_0, \ k \in \mathbb{N}.$$

It is easy to check that $C_Q^k(H)$ (resp. $UC_Q^k(H)$), endowed with the norm

$$\|\varphi\|_{C_Q^k(H)} = \|\varphi\|_0 + \sum_{j=1}^{k}[\varphi]_{j,Q},$$

is a Banach space. Notice that $C_b^k(H) \subset UC_Q^k(H) \subset C_Q^k(H)$, $k \in \mathbb{N}$.

For any $k \in \mathbb{N} \cup \{0\}$ and $\theta \in (0,1)$ we set (^{3})

$$C_Q^{k+\theta}(H) = \left\{ \varphi \in C_Q^k(H) : D_Q^k \varphi(Q^{1/2} \cdot) \in C_b^\theta(H, H^{\otimes k}) \right\}.$$

$C_Q^{k+\theta}(H)$, endowed with the norm

$$\|\varphi\|_{k+\theta, Q} = \|\varphi\|_{k, Q} + [\varphi]_{k+\theta, Q},$$

where

$$[\varphi]_{k+\theta, Q} = \sup_{\substack{x, y \in H \\ x \neq y}} \frac{\|D_Q^k \varphi(Q^{1/2} x) - D_Q^k \varphi(Q^{1/2} y)\|}{|x - y|^\theta},$$

is a Banach space.

The following lemma is an easy consequence of the mean value theorem.

Lemma 3.3.1 *Define for $x, y, g, h \in H$ the function*

$$\psi_y(s, t) = \varphi(x + Q^{1/2} y + s Q^{1/2} g + t Q^{1/2} h), \quad s, t \in \mathbb{R}.$$

(i) Assume that for all y from a neighborhood of 0 we have $D_s \psi_y(0,0) = \langle a(y), g \rangle$, where a is continuous at 0. Then φ is Q-differentiable at x and $D_Q \varphi(x) = a(0)$.

(ii) Assume in addition that a is continuous in a neighborhood of 0 and for each y in such neighborhood and for each g and h, $D_s D_t \psi_y(0,0) = \langle A(y) h, g \rangle$, where A is continuous at 0. Then φ is twice Q-differentiable at x and $D_Q^2 \varphi(x) = A(0)$.

Similar characterizations hold for higher derivatives.

We end this subsection by giving some interpolatory estimates , needed in what follows. We use notations introduced in §2.3.2. The proofs are similar to that of Theorem 2.3.5.

Theorem 3.3.2 *Let $0 \leq a < b < c$. Then we have*

$$UC_Q^b(H) \in J_{\frac{b-a}{c-a}} \left(UC_Q^a(H), UC_Q^c(H) \right), \tag{3.3.1}$$

and there is a constant $C_{a,b,c} > 0$ such that

$$\|u\|_{b, Q} \leq C_{a,b,c} \|u\|_{a, Q}^{\frac{c-b}{c-a}} \|u\|_{c, Q}^{\frac{b-a}{c-a}}, \quad u \in UC_Q^c(H). \tag{3.3.2}$$

$^3 H^{\otimes k}$ denotes the linear space of all continuous k-linear real valued mappings on H.

Proof. We prove the result in some cases we shall use later.

Case 1. $a = 0$, $b = 1$, $c = 2$.

Take $y \in H$, $|y| = 1$. By Taylor's formula with integral remainder, we have

$$u(x+tQ^{1/2}y) = u(x)+t\langle D_Q u(x), y\rangle+t^2\int_0^1 (1-\sigma)\langle D_Q^2 u(x+\sigma t Q^{1/2}y)y, y\rangle d\sigma.$$

Then

$$|D_Q u(x)| \le |\langle D_Q u(x), y\rangle| \le \frac{2}{t}\|u\|_0 + \frac{t}{2}\|D_Q^2 u\|_0,$$

so that the conclusion follows by taking the supremum with respect to t. ([4])

Case 2. $a = 0$, $b = \theta$, $c = 1$.

Let $\psi(t) = u(Q^{1/2}x + tQ^{1/2}(y-x))$. Then

$$\psi'(t) = \langle D_Q u(Q^{1/2}x + tQ^{1/2}(y-x)), y-x\rangle.$$

So

$$|u(Q^{1/2}x) - u(Q^{1/2}y)| = |\psi(1) - \psi(0)| \le \|D_Q u\|_0\, |x-y|.$$

Fix $M > 0$. If $|x-y| \le M$ then $\frac{|x-y|}{M} \le \left(\frac{|x-y|}{M}\right)^\theta$. So

$$|u(Q^{1/2}x) - u(Q^{1/2}y)| \le \|D_Q u\|_0 |x-y|^\theta M^{1-\theta}.$$

If $|x-y| \ge M$ then $(\frac{|x-y|}{M})^\theta \ge 1$, and consequently

$$|u(Q^{1/2}x) - u(Q^{1/2}y)| \le 2\|u\|_0 M^{-\theta}|x-y|^\theta.$$

Summing up the two inequalities yields

$$|u(Q^{1/2}x) - u(Q^{1/2}y)| \le \left[\|D_Q u\|_0 M^{1-\theta} + 2\|u\|_0 M^{-\theta}\right]|x-y|^\theta.$$

Taking the minimum with respect to M the conclusion follows.

Case 3. $a = 1$, $b = 1+\theta$, $c = 2$.

Let $\psi(t) = \langle D_Q u(Q^{1/2}x + tQ^{1/2}(y-x)), h\rangle$. Then

$$\psi'(t) = \langle D_Q^2 u(Q^{1/2}x + t(y-x))(y-x), h\rangle.$$

So

$$|D_Q u(Q^{1/2}x) - D_Q u(Q^{1/2}y)| \le \|D_Q^2 u\|_0\, |x-y|.$$

Now the proof is exactly the same as that of Case 2. $\square$

---

[4] For any $a, b, \gamma > 0$ we have $\min_{\sigma>0}(a\sigma + b\sigma^{-\gamma}) = c(\gamma)a^{\frac{\gamma}{1+\gamma}}b^{\frac{1}{1+\gamma}}$, where $c(\gamma) = \gamma^{\frac{1}{1+\gamma}} + \gamma^{-\frac{\gamma}{1+\gamma}}$.

3.3.2 Q-derivatives of generalized solutions

We are going to give several regularity properties of the generalized solution of (3.1.6). For this we need some notations.

For $\varphi \in L^2(H, N_Q)$ we define ([5])

$$\int_H Q^{-1/2}y \, \varphi(y) N_Q(dy) := \sum_{k=1}^{\infty} g_n \langle W_{g_n}, \varphi \rangle_{L^2(H, N_Q)} \in H,$$

$$\int_H [Q^{-1/2}y\varphi(y) \otimes Q^{-1/2}y\varphi(y) - 1] N_Q(dy)$$

$$:= \sum_{n,n=1}^{\infty} g_n \otimes g_m \langle W_{g_n} W_{g_m} - \delta_{n.m}, \varphi \rangle_{L^2(H, N_Q)} \in L_2(H),$$

where (g_n) is any orthonormal complete basis on H. The reader can easily check that these definitions are meaningful and independent of the choice of the basis (g_n).

Note that

$$\left\langle \int_H Q^{-1/2}y \, \varphi(y) N_Q(dy), g \right\rangle = \int_H \langle Q^{-1/2}y \, \varphi(y), g \rangle N_Q(dy), \ g \in H,$$

and

$$\left\langle \int_H (Q^{-1/2}y \otimes Q^{-1/2}y - 1)\varphi(y) N_Q(dy)g, h \right\rangle$$

$$= \int_H [\langle Q^{-1/2}y, g \rangle \langle Q^{-1/2}y, h \rangle - \langle g, h \rangle]\varphi(y) N_Q(dy), \ g, h \in H.$$

Theorem 3.3.3 *Let $\varphi \in C_b(H)$ and $u(t, \cdot) = P_t\varphi$. Then for all $t > 0$, and $x \in H$ we have $u(t, \cdot) \in C_Q^2(H)$ and*

$$D_Q u(t, x) = \frac{1}{\sqrt{t}} \int_H (tQ)^{-1/2}y \, \varphi(x + y) N_{tQ}(dy), \qquad (3.3.3)$$

and

$$D_Q^2 u(t, x) = \frac{1}{t} \int_H ((tQ)^{-1/2}y \otimes (tQ)^{-1/2}y - 1)\varphi(x + y) N_{tQ}(dy). \quad (3.3.4)$$

---

[5]We recall that $W_x(y) = \langle Q^{-1/2}x, y \rangle$, see Chapter 1.

Moreover

$$|D_Q u(t,x)| \le \frac{1}{\sqrt{t}} \|\varphi\|_0, \quad t > 0, \ x \in H, \qquad (3.3.5)$$

$$\|D_Q^2 u(t,x)\| \le \frac{\sqrt{2}}{t} \|\varphi\|_0, \quad t > 0, \ x \in H. \qquad (3.3.6)$$

If $\varphi \in UC_b(H)$ then $u(t,\cdot) \in UC_Q^2(H)$ for all $t > 0$

Proof. We apply Lemma 3.3.1. Let us prove (3.3.3). We have for $g \in H$

$$u(t, x + \alpha Q^{1/2} g) = \int_H \varphi(x+y) N_{\alpha Q^{1/2} g, tQ}(dy).$$

By the Cameron-Martin formula (Theorem 1.3.6) it follows that

$$\frac{dN_{\alpha Q^{1/2} g, tQ}}{dN_{tQ}}(y) = e^{-\frac{\alpha^2}{2t}|g|^2 + \frac{\alpha}{\sqrt{t}}\langle g, (tQ)^{-1/2} y\rangle}.$$

Therefore

$$u(t, x + \alpha Q^{1/2} g) = \int_H \varphi(x+y) e^{-\frac{\alpha^2}{2t}|g|^2 + \frac{\alpha}{\sqrt{t}}\langle g, (tQ)^{-1/2} y\rangle} N_Q(dy).$$

By the very definition we obtain (3.3.3) taking derivatives with respect to α at $\alpha = 0$. Equation (3.3.4) can be proved in the same way.

To prove (3.3.5) it is enough to note that by the Hölder inequality we have

$$|\langle D_Q u(t,x), g\rangle|^2 \le \frac{1}{t} \|\varphi\|_0^2 \int_H |\langle g, (tQ)^{-1/2} y\rangle|^2 N_{tQ}(dy) = \frac{1}{t} \|\varphi\|_0^2 |g|^2.$$

Inequality (3.3.6) can be proved similarly. $\square$

We now prove that the solution of the heat equation has Q-derivatives of all orders.

Theorem 3.3.4 *Let φ be Borel and bounded and $u(t,x) = P_t\varphi(x)$, $t \ge 0$, $x \in H$. Then $u(t,\cdot)$ has Q-derivatives of all orders for any $t > 0$.*

Proof. We have for $g_1, \ldots, g_n \in H$

$$u(t, x + Q^{1/2}(\alpha_1 g_1 + \cdots + \alpha_n g_n)) = \int_H \varphi(x+z) N_{Q^{1/2}(\alpha_1 g_1 + \cdots + \alpha_n g_n), tQ}(dz)$$

$$= e^{-\frac{1}{2t}|\alpha_1 g_1 + \cdots + \alpha_n g_n|^2} \int_H \varphi(x+z) e^{\frac{1}{\sqrt{t}}\langle \alpha_1 g_1 + \cdots + \alpha_n g_n, (tQ)^{-1/2} z\rangle} N_{tQ}(dz)$$

$$= e^{-\frac{1}{2}|\alpha_1 g_1 + \cdots + \alpha_n g_n|^2} \int_H \varphi(x+z) e^{\alpha_1 W_{a_1}(z) + \cdots + \alpha_n W_{a_n}(z)} N_{tQ}(dz),$$

where $a_j = t^{-1/2} g_j$, $j = 1, \ldots, n$. Moreover $(W_{a_1}, \cdots, W_{a_n})$ is a Gaussian vector on (H, N_{tQ}) with mean vector 0 and covariance matrix $(\langle a_i, a_j \rangle,\ i, j = 1, \ldots, n)$. Thus

$$u(t, x + Q^{1/2}(\alpha_1 g_1 + \cdots + \alpha_n g_n)) = \Psi_0(\alpha_1 + \cdots + \alpha_n)\Psi_1(\alpha_1 + \cdots + \alpha_n; x),$$

where

$$\Psi_0(\alpha_1 + \cdots + \alpha_n) = e^{-\frac{1}{2}|\alpha_1 g_1 + \cdots + \alpha_n g_n|^2},$$

and

$$\Psi_1(\alpha_1 + \cdots + \alpha_n; x) = \int_H \varphi(x + z) e^{\alpha_1 W_{a_1}(z) + \cdots + \alpha_n W_{a_n}(z)} N_{tQ}(dz).$$

Our aim is to calculate

$$\frac{\partial^{k_1 + \cdots + k_n}}{\partial \alpha_1^{k_1} \cdots \partial \alpha_n^{k_n}} \left(\Psi_0(\alpha_1 + \cdots + \alpha_n)\Psi_1(\alpha_1 + \cdots + \alpha_n; x)\right)\Bigg|_{\alpha_1 = \cdots = \alpha_n = 0}$$

$$= \sum_{j_1=0}^{k_1} \cdots \sum_{j_n=0}^{k_n} \binom{k_1}{j_1} \cdots \binom{k_n}{j_n} \left[\frac{\partial^{j_1 + \cdots + j_n}}{\partial \alpha_1^{j_1} \cdots \partial \alpha_n^{j_n}} \Psi_0(0, \ldots, 0)\right]$$

$$\times \left[\frac{\partial^{(k_1 + \cdots + k_n) - (j_1 + \cdots + j_n)}}{\partial \alpha_1^{k_1 - j_1} \cdots \partial \alpha_n^{k_n - j_n}} \Psi_1(0, \ldots, 0, x)\right] = S_1 + S_2 + S_3,$$

where

$$S_1 = \frac{\partial^{k_1 + \cdots + k_n}}{\partial \alpha_1^{k_1} \cdots \partial \alpha_n^{k_n}} \Psi_1(0, \ldots, 0, x),$$

$$S_2 = \sum_{\substack{m=1 \\ k_m \geq 2}}^{n} \binom{k_m}{2}$$

$$\times \frac{\partial^2}{\partial \alpha_m^2} \Psi_0(0, \ldots, 0) \frac{\partial^{k_1}}{\partial \alpha_1^{k_1}} \cdots \frac{\partial^{k_{m-1}}}{\partial \alpha_{m-1}^{k_{m-1}}} \frac{\partial^{k_m - 2}}{\partial \alpha_m^{k_m - 2}} \cdots \frac{\partial^{k_n}}{\partial \alpha_n^{k_n}} \Psi_1(0, \ldots, 0; x),$$

and

$$S_3 = \sum_{1 \leq l < m \leq n} \binom{k_l}{1}\binom{k_m}{1} \frac{\partial^2}{\partial \alpha_l \partial \alpha_m} \Psi_0(0, \ldots, 0)$$

$$\times \frac{\partial^{k_1}}{\partial \alpha_1^{k_1}} \cdots \frac{\partial^{k_{l-1}}}{\partial \alpha_{l-1}^{k_{l-1}}} \frac{\partial^{k_l - 1}}{\partial \alpha_l^{k_l - 1}} \cdots \frac{\partial^{k_{m-1}}}{\partial \alpha_{m-1}^{k_{m-1}}} \frac{\partial^{k_m - 1}}{\partial \alpha_m^{k_m - 1}} \cdots \frac{\partial^{k_n}}{\partial \alpha_n^{k_n}} \Psi_1(0, \ldots, 0, x).$$

Other partial derivatives at $(0,\ldots,0)$ vanish.

We have

$$\frac{\partial^{k_1+\cdots+k_n}}{\partial\alpha_1^{k_1}\cdots\partial\alpha_n^{k_n}}\Psi_1(\alpha_1,\ldots,\alpha_n;x)$$

$$=\int_H \varphi(x+z)\frac{\partial^{k_1+\cdots+k_n}}{\partial\alpha_1^{k_1}\cdots\partial\alpha_n^{k_n}}e^{\alpha_1\beta_1+\cdots+\alpha_n\beta_n}N_{tQ}(dz),$$

where $\beta_j=W_{a_j}(z)$, $j=1,\ldots,n$. Consequently

$$\frac{\partial^{k_1+\cdots+k_n}}{\partial\alpha_1^{k_1}\cdots\partial\alpha_n^{k_n}}\Psi_1(0,\ldots,0;x)=\int_H \varphi(x+z)W_{a_1}^{k_1}(z)\cdots W_{a_n}^{k_n}(z)N_{tQ}(dz).$$

Therefore

$$\frac{\partial^{k_1+\cdots+k_n}}{\partial\alpha_1^{k_1}\cdots\partial\alpha_n^{k_n}}u(t,x+Q^{1/2}(\alpha_1g_1+\cdots+\alpha_ng_n))\bigg|_{\alpha_1=\cdots=\alpha_n=0}$$

$$=\Sigma_1+\Sigma_2+\Sigma_3,$$

where

$$\Sigma_1=\int_H \varphi(x+z)W_{a_1}^{k_1}(z)\ldots W_{a_n}^{k_n}(z)N_{tQ}(dz),$$

$$\Sigma_2=-\sum_{m=1}^{n}\sum_{k_m\geq 2}\binom{k_m}{2}|a_m|^2$$

$$\times\int_H \varphi(x+z)W_{a_1}^{k_1}(z)\ldots W_{a_{m-1}}^{k_{m-1}}(z)W_{a_m}^{k_m-2}(z)\ldots W_{a_n}^{k_n}(z)N_{tQ}(dz),$$

$$\Sigma_3=-\sum_{1\leq l<m\leq n}\binom{k_l}{1}\binom{k_m}{1}\langle a_l,a_m\rangle$$

$$\times\int_H \varphi(x+z)W_{a_1}^{k_1}(z)\ldots W_{a_{l-1}}^{k_{l-1}}(z)W_{a_l}^{k_l-1}(z)W_{a_{l+1}}^{k_{l+1}}(z)$$

$$\times\ldots W_{a_{m-1}}^{k_{m-1}}(z)\ldots W_{a_m}^{k_m-1}(z)W_{a_{m+1}}^{k_{m+1}}(z)\ldots W_{a_n}^{k_n}(z)N_{tQ}(dz).$$

Consequently

$$
\frac{\partial^{k_1+\cdots+k_n}}{\partial \alpha_1^{k_1} \cdots \partial \alpha_n^{k_n}} u(t, x + \alpha_1 Q^{1/2} g_1 + \cdots + \alpha_n Q^{1/2} g_n) \Big|_{\alpha_1=\cdots=\alpha_n=0}
$$

$$
= \int_H \varphi(x+z) \Bigg[W_{a_1}^{k_1}(z) \ldots W_{a_n}^{k_n}(z)
$$

$$
- \sum_{m=1}^n |a_m|^2 \binom{k_m}{2} \sum_{k_m \geq 2} W_{a_1}^{k_1}(z) \ldots W_{a_{m-1}}^{k_{m-1}}(z) W_{a_m}^{k_m-2}(z) \ldots W_{a_n}^{k_n}(z)
$$

$$
- \sum_{1 \leq l < m \leq n} \binom{k_l}{1} \binom{k_m}{1} \langle a_l, a_m \rangle W_{a_1}^{k_1}(z) \ldots W_{a_{l-1}}^{k_{l-1}}(z) W_{a_l}^{k_l-1}(z) W_{a_{l+1}}^{k_{l+1}}(z)
$$

$$
\times \ldots W_{a_{m-1}}^{k_{m-1}}(z) \cdots W_{a_m}^{k_m-1}(z) W_{a_{m+1}}^{k_{m+1}}(z) W_{a_n}^{k_n}(z) \Bigg] N_{tQ}(dz)
$$

$$
= \frac{\partial^{k_1+\cdots+k_n}}{\partial \alpha_1^{k_1} \cdots \partial \alpha_n^{k_n}} P_t \varphi(x + \alpha_1 Q^{1/2} g_1 + \cdots + \alpha_n Q^{1/2} g_n) \Big|_{\alpha_1=\cdots=\alpha_n=0} .
$$

It follows from these formulae that Q-derivatives exist of all orders and that they are continuous in x. $\square$

Given $\varphi \in C_b(H)$, we should expect that $D_Q^2 P_t \varphi$ is of trace class for any $t > 0$. Unfortunately this is not the case in general, see Proposition 3.3.9 below. However, the weaker result that $D_Q^2 P_t \varphi$ is Hilbert-Schmidt holds, as was shown by L. Gross [138] (see also [197]). The next proof is taken from J. Zabczyk [220].

Theorem 3.3.5 *For arbitrary $\varphi \in C_b(H)$, $t > 0$ and $x \in H$, the operator $D_Q^2 u(t, x)$ is of Hilbert-Schmidt type and*

$$
\|D_Q^2 u(t, x)\|_{L^2(H)} \leq \frac{\sqrt{2}}{t} \|\varphi\|_0, \ t > 0, \ x \in H. \tag{3.3.7}
$$

Proof. Let (e_n) be an orthonormal and complete basis in H. For $n, m \in \mathbb{N}$ define

$$
f_{n,m}(y) = \langle e_n, (tQ)^{-1/2} y \rangle \langle e_m, (tQ)^{-1/2} y \rangle - \langle e_n, e_m \rangle, \quad y \in H.
$$

Then

$$\|D_Q^2 u(t,x)\|_{L^2(H)}^2 = \frac{1}{t^2} \sum_{n,m \in \mathbb{N}} \left| \int_H f_{n,m}(y) \varphi(x+y) N_{tQ}(dy) \right|^2$$

$$= \frac{1}{t^2} \sum_{n \in \mathbb{N}} \left| \int_H f_{n,n}(y) \varphi(x+y) N_{tQ}(dy) \right|^2$$

$$+ \frac{1}{t^2} \sum_{n \neq m} \left| \int_H f_{n,m}(y) \varphi(x+y) N_{tQ}(dy) \right|^2.$$

By Proposition 1.2.6 the system of functions $\tilde{f}_{n,m}$, $n, m \in \mathbb{N}$,

$$\tilde{f}_{n,n} = \frac{f_{n,n}}{\sqrt{2}}, \ n \in \mathbb{N}, \quad \tilde{f}_{n,m} = f_{n,m}, n, m \in \mathbb{N}, \ n \neq m,$$

is orthonormal in $L^2(H, N_{tQ})$. Since

$$\|D_Q^2 u(t,x)\|_{L^2(H)}^2 \leq \frac{2}{t^2} \sum_{n,m} |\langle \tilde{f}_{n,m}, \varphi(x+\cdot)\rangle_{L^2(H,N_{tQ})}|^2,$$

therefore

$$\|D_Q^2 u(t,x)\|_{L^2(H)}^2 \leq \frac{2}{t^2} \int_H \varphi^2(x+y) N_{tQ}(dy) \leq \frac{2}{t^2} \|\varphi\|_0^2. \quad \square$$

We now consider the case when $\varphi \in UC_b^1(H)$.

Proposition 3.3.6 *Assume that* $\varphi \in C_b^1(H)$. *Then*

$$\|D_Q^2 P_t \varphi(x)\| \leq t^{-1/2} \|\varphi\|_{1,Q}, \ x \in H. \qquad (3.3.8)$$

Proof. By (3.3.3) for $k \in H$,

$$\langle D_Q P_t \varphi(x), k \rangle = t^{-1/2} \int_H \langle k, (tQ)^{-1/2} y \rangle \varphi(x+y) N_{tQ}(dy), \qquad (3.3.9)$$

and therefore, differentiating (3.3.9) in the direction $Q^{1/2}h$ gives for any $h, k \in H$,

$$\langle D_Q^2 P_t \varphi(x) h, k \rangle = t^{-1/2} \int_H \langle k, (tQ)^{-1/2} y \rangle \langle D\varphi(x+y), Q^{1/2}h \rangle N_{tQ}(dy).$$

$$(3.3.10)$$

By using the Hölder inequality we find

$$|\langle D_Q^2 P_t\varphi(x)h, k\rangle|^2 \;\leq\; t^{-1}\,\|\varphi\|_{1,Q}^2 \int_H |\langle k, (tQ)^{-1/2}y\rangle|^2 N_{tQ}(dy)$$

$$= \; t^{-1}\,\|\varphi\|_{1,Q}^2|h|^2|k|^2,$$

which yields (3.3.8). $\square$

We now prove that $D_Q^2 u(t,x)$ is of trace class, see L. Gross [138].

Theorem 3.3.7 *Assume that $\varphi \in C_b^1(H)$. Then the operator $D_Q^2 u(t,x)$ is of trace class for any $t > 0$, $x \in H$, and*

$$\mathrm{Tr}[D_Q^2 u(t,x)] = \frac{1}{t}\int_H \langle y, D\varphi(x+y)\rangle N_{tQ}(dy). \tag{3.3.11}$$

Moreover

$$|\,\mathrm{Tr}[D_Q^2 u(t,x)]| \leq \frac{1}{\sqrt{t}}\,(\mathrm{Tr}\,Q)^{1/2}\|D\varphi\|_0 \tag{3.3.12}$$

and the heat equation holds:

$$D_t u(t,x) = \frac{1}{2}\,\mathrm{Tr}[D_Q^2 u(t,x)], \qquad t > 0, \; x \in H. \tag{3.3.13}$$

Proof. It follows from (3.3.10) that for $k \in H$

$$D_Q^2 u(t,x)k = t^{-1/2}Q^{1/2}\int_H D\varphi(x+y)\langle k, (tQ)^{-1/2}y\rangle N_{tQ}(dy) := Q^{1/2}G\varphi(x).$$

Since $Q^{1/2}$ is a Hilbert-Schmidt operator, in order to show that $D_Q^2 u(t,x)$ is of trace class it is enough to prove that G is also a Hilbert-Schmidt operator. Note that

$$\|G\|_{L^2(H)}^2 = \frac{1}{t}\sum_{i,j=1}^{\infty}\left|\int_H \langle e_j, D\varphi(x+y)\rangle\langle e_i, (tQ)^{-1/2}y\rangle N_{tQ}(dy)\right|^2,$$

where (e_k) is an orthonormal and complete basis in H. The sequence $(\langle e_k, (tQ)^{-1/2}(\cdot)\rangle)$ is orthonormal in $L^2(H, N_{tQ})$ and therefore by the Parseval identity,

$$\|G\|_{L_2(H)}^2 \leq \frac{1}{t}\sum_{j=1}^{\infty}\int_H \langle e_j, D\varphi(x+y)\rangle^2 N_{tQ}(dy).$$

Consequently

$$\|G\|_{L^2(H)}^2 \le \frac{1}{t} \int_H |D\varphi(x+y)|^2 N_{tQ}(dy),$$

and

$$\|D_Q^2 u(t,x)\|_{L^1(H)} \le \|Q^{1/2}\|_{L^2(H)} \, \|G\|_{L^2(H)} \le \frac{1}{\sqrt{t}} \, (\mathrm{Tr}\ Q)^{1/2} \|D\varphi\|_0.$$

In this way inequality (3.3.12) has been established and in particular $D_Q^2 u(t, x)$ is a trace class operator. To show that (3.3.11) holds we use (3.3.9). If (e_k, λ_k) is an orthonormal sequence determined by Q, then

$$\sum_{k=1}^\infty \langle D_Q^2 u(t,x)e_k, e_k \rangle$$

$$= \frac{1}{\sqrt{t}} \int_H \sum_{k=1}^\infty \langle (tQ)^{-1/2} e_k, y \rangle \langle D\varphi(x+y), Q^{1/2} e_k \rangle] N_{tQ}(dy)$$

$$= \frac{1}{\sqrt{t}} \int_H \sum_{k=1}^\infty \frac{1}{\sqrt{t}} \frac{1}{\sqrt{\lambda_k}} \langle e_k, y \rangle \langle D\varphi(x+y), \sqrt{\lambda_k} e_k \rangle N_{tQ}(dy)$$

$$= \frac{1}{t} \int_H \langle y, D\varphi(x+y) \rangle N_{tQ}(dy)$$

and (3.3.11) holds. Since $u(t,x) = \int_H \varphi(x+\sqrt{t}y)N_Q(dy)$, we have

$$D_t u(t,x) = \int_H \left\langle D\varphi(x+\sqrt{t}y), \frac{1}{2\sqrt{t}} y \right\rangle N_Q(dy)$$

$$= \frac{1}{2t} \int_H \left\langle D\varphi(x+\sqrt{t}y), \sqrt{t}y \right\rangle N_Q(dy) = \frac{1}{2t} \int_H \langle D\varphi(x+y), y \rangle N_{tQ}(dy),$$

and then the equation (3.3.13) holds by (3.3.11). $\square$

Remark 3.3.8 It is known, see R. R. Phelps [185], that the set of all points where a Lipschitz continuous function on a Hilbert space is not Gateaux differentiable is of measure zero with respect to any nondegenerate Gaussian measure. Using this result one can extend Theorem 3.3.7 to Lipschitz continuous φ, see L. Gross [138] for a different proof.

We conclude this section by giving examples showing that the generalized solution is not regular if the dimension of H is infinite. The following result, due to J. Zabczyk [220], which extends a previous result of E. Priola [190], shows that if $\dim H = \infty$ and $t > 0$, then $P_t\varphi \notin UC_b^1(H)$ for some $\varphi \in UC_b(H)$. If $\psi \in UC_b(H)$ then $D_v\psi(x)$ will denote the directional derivative of ψ at x in the direction v.

Proposition 3.3.9 *Assume that* $\dim H = \infty$, *and that* $Q \in L_1^+(H)$. *Then for arbitrary sequences* (t_k) *in* $(0, +\infty)$ *and* (x_m) *in* H, *there exists* $\varphi \in UC_b(H)$ *such that functions* $P_{t_k}\varphi$ *are not Lipschitz continuous in any neighborhood of any point* x_m, $m \in \mathbb{N}$.

Proof. Let $\varphi \in UC_b(H)$, let $x \in H$, and $v \in Q^{1/2}h$. Then we have

$$D_v P_t\varphi(x) = \frac{1}{\sqrt{t}} \int_H \varphi(x+y)\langle Q^{-1/2}v, (tQ)^{-1/2}y\rangle N_{tQ}(dy).$$

Let (v_n) be a sequence in $Q^{1/2}(H)$ such that $|v_n| = 1$ and $\lim_{n\to\infty} |Q^{-1/2}v_n| = \infty$. For each $t > 0$ and $x \in H$ define the following linear functionals $F_n^{t,x}$ from $UC_b(H)$ into $\mathbb{R}$:

$$F_n^{t,x}(\varphi) = \frac{1}{\sqrt{t}} \int_H \varphi(x+y)\langle Q^{-1/2}v_n, (tQ)^{-1/2}y\rangle N_{tQ}(dy).$$

Then we have

$$\|F_n^{t,x}\| = \frac{1}{\sqrt{t}} \int_H |\langle Q^{-1/2}v_n, (tQ)^{-1/2}y\rangle| N_{tQ}(dy).$$

Since, for each $h \in H$, $y \to \langle h, Q^{-1/2}y\rangle$ is a Gaussian random variable on $L^2(H, N_Q)$ with mean 0 and covariance $|h|^2$, one easily gets that

$$\|F_n^{t,x}\| = \frac{\pi t}{2} |Q^{-1/2}v_n|.$$

Therefore, for arbitrary t and x, it results that $\lim_{n\to\infty} \|F_n^{t,x}\| = \infty$. Consequently by the Banach-Steinhaus principle of the condensation of singularities , there exists $\varphi \in UC_b(H)$ such that for any t_k and x_m,

$$\sup_{n\in\mathbb{N}} |D_{v_n} P_{t_k}\varphi(x_m)| = \sup_{n\in\mathbb{N}} |F_n^{t_k,x_m}(\varphi)| = \infty. \tag{3.3.14}$$

Assume that for some t_k, x_m the function $P_{t_k}\varphi$ is Lipschitz continuous in a neighborhood of x_m. Then there exists $\delta > 0$ such that for $\psi = P_{t_k}\varphi$

$$|\psi(x_m + y) - \psi(x_m)| \le C|y| \quad \text{if} \quad |y| \le \delta.$$

Let $\sigma_n > 0$ be such that $\sigma_n < \delta$ and

$$\frac{|\psi(x_m + \sigma_n v_n) - \psi(x_m)|}{\sigma_n} \geq \frac{1}{2}|D_{v_n}\psi(x_m)|.$$

Then

$$\frac{1}{2}|D_{v_n}\psi(x_m))| \leq \frac{C|\sigma_n v_n|}{\sigma_n} \leq C, \quad n \in \mathbb{N},$$

a contradiction with (3.3.14). □

We will show that for arbitrary $t > 0$ and $x \in H$ there exist initial functions $\varphi \in UC_b(H)$ such that the generalized solution u of the heat equation in the form (3.3.13) is not satisfied at (t, x).

The following result is taken from J. Zabczyk [220].

Proposition 3.3.10 *Assume that* $\dim(H) = \infty$, *and* $Q \in L_1^+(H)$. *Then for arbitrary sequences* $(t_k) \in (0, +\infty)$ *and* $(x_m) \subset H$, *there exists* $\varphi \in UC_b(H)$ *such that operators* $D_Q^2 P_{t_k}\varphi(x_m)$ *are not of trace class.*

Proof. Let (g_n) be a complete orthonormal basis in H. Then

$$\langle D_Q^2 P_t\varphi(x)g_k, g_k \rangle = \frac{1}{t}\int_H \left[|\langle g_k, (tQ)^{-1/2}y\rangle|^2 - 1\right]\varphi(x+y)N_{tQ}(dy).$$

For $\varphi \in UC_b(H), t > 0, x \in H$, and $n \in \mathbb{N}$, we define

$$F_n^{t,x}(\varphi) := \sum_{k=1}^{n}\langle D_Q^2 P_t\varphi(x)g_k, g_k \rangle$$

$$= \frac{1}{t}\int_H \sum_{k=1}^{n}\left[|\langle g_k, (tQ)^{-1/2}y\rangle|^2 - 1\right]\varphi(x+y)N_{tQ}(dy).$$

Then for fixed $t > 0, x \in H$, and $n \in \mathbb{N}$, the norms of the functionals $F_n^{t,x}$ on $UC_b(H)$ can be easily computed as

$$\|F_n^{t,x}\| = \frac{1}{t}\int_H \left|\sum_{k=1}^{n}\left[|\langle g_k, (tQ)^{-1/2}y\rangle|^2 - 1\right]\right|N_{tQ}(dy).$$

Let $\eta_j(y) = \frac{\langle y, g_j\rangle^2}{\lambda_j} - 1, j = 1, 2, \ldots, y \in H$. Then the sequence (η_j) consists of independent random variables on $L^2(H, N_Q)$, identically distributed, with mean value 0 and finite second moment. Therefore it follows easily, for instance by the central limit theorem, that

$$\lim_{n\to\infty} \|F_n^{t,x}\| = \lim_{n\to\infty} \int_H |\eta_1 + \cdots + \eta_n| N_{tQ}(dy) = \infty.$$

Consequently for arbitrary sequences (t_k) and (x_m), there exists a function $\varphi \in UC_b(H)$ such that $\sup_n |F_n^{t_k,x_m}| = +\infty$. This means that for all such (t_k) and (x_m), the operators $D_Q^2 P_{t_k}\varphi(x_m)$ cannot be of trace class. $\square$

3.4 Comments on the Gross Laplacian

We describe here the relationship between the Q-derivatives and the Gross Laplacian introduced in [138]. The results of this section are not used in what follows.

Let $(B, \|\cdot\|_B)$ be a separable Banach space and G a linear dense subspace of B equipped with a scalar product $\langle\cdot,\cdot\rangle_G$ and the Hilbertian norm: $\|g\|_G = \sqrt{\langle g,g\rangle_G}$, $g \in G$. It is assumed that $(G, \|\cdot\|_G)$ is a separable Hilbert space and for some $c > 0$

$$\|g\|_B \le c\|g\|_G, \qquad g \in G. \tag{3.4.1}$$

Identifying G with its dual G^* and taking into account that the embedding $G \subset B$ is continuous one can identify B^* with a subset of G. Thus

$$B^* \subset G^* = G \subset B. \tag{3.4.2}$$

Let E be another Banach space and u a transformation from B into E. If there exists $T \in L(G, E)$ such that

$$\lim_{\|g\|_G \to 0} \frac{\|u(x+g) - u(x) - Tg\|_E}{\|g\|_G} = 0$$

then T is called the *G-derivative* of u at x and it is denoted by $D_G u(x)$. Replacing the space E with $L(G, E)$, one can define in the same way $D_G^2 u(x)$ as an element of $L(G, L(G, E))$. Identifying $L(G, L(G, E))$ in the usual way, with the Banach space $L^2(G, E)$ of all bilinear transformations from G into E, one gets that $D_G^2 u(x) \in L^2(G, E)$. In a similar way, $D_G^n u(x)$ can be defined and if $D_G^n u(x)$ exists then $D_G^n u(x) \in L^n(G, E)$.

If $E = \mathbb{R}$, then $D_G u(x)$ and $D_G^2 u(x)$ will be identified with elements of G and $L(G)$ respectively.

If $D^2_G u(x)$ is an operator of trace class on G, then the Gross Laplacian Δ_G is defined by the formula

$$\Delta_G u(x) = \text{ Tr } D^2_G u(x) = \sum_{m=1}^{\infty} \langle D^2_G u(x) g_m, g_m \rangle_G,$$

where (g_n) is an orthonormal basis on G.

Let now B be a separable Hilbert space H and $Q : H \to H$ a self-adjoint nonnegative operator of trace class. Define $G = Q^{1/2}(H)$ and

$$\langle g_1, g_2 \rangle_G = \langle Q^{-1/2} g_1, Q^{-1/2} g_2 \rangle_H, \qquad g_1, g_2 \in G.$$

Then $H^* \subset G^* = G \subset H$ and $H^* = Q(H) = Q^{1/2}(G)$ with the induced norms. Assume that $u : H \to R$ is twice Fréchet differentiable at $x \in H$ with $Du(x)$, $D^2 u(x)$ its first and second Fréchet derivatives. Then for arbitrary $g, g_1, g_2 \in G$

$$\langle Du(x), g \rangle_H = \langle D_G u(x), g \rangle_G,$$

and

$$\langle D^2 u(x) g_1, g_2 \rangle_H = \langle D^2_G u(z) g_1, g_2 \rangle_G,$$

where the bilinear forms $D^2 u(x)$, $D^2_G u(x)$ were identified with linear operators on H and G respectively. Since

$$\langle D_G u(x), g \rangle_G = \langle Q^{-1/2} D_G u(x), Q^{-1/2} g \rangle_H,$$

and

$$\langle D^2_G u(x) g_1, g_2 \rangle_G = \langle Q^{-1/2} D^2_G u(z) g_1, Q^{-1/2} g_2 \rangle_H,$$

one arrives at the following relations:

$$D_G u(x) = Q Du(x), \qquad D^2_G u(x) = Q D^2 u(x).$$

Moreover if (h_n) is an orthonormal basis in H then $g_m = Q^{1/2} h_m$, $m = 1, \dots$, is an orthonormal basis in G and

$$\begin{aligned}
\Delta_G u(x) &= \sum_{m=1}^{\infty} \langle D^2_G u(x) g_m, g_m \rangle_G && (3.4.3) \\
&= \sum_{m=1}^{\infty} \langle Q^{-1/2}(Q D^2 u(x)) Q^{1/2} h_m, Q^{-1/2} Q^{1/2} h_m \rangle_H \\
&= \sum_{m=1}^{\infty} \langle Q^{1/2} D^2 u(x) Q^{1/2} h_m, h_m \rangle_H.
\end{aligned}$$

Therefore

$$\Delta_G u(x) = \text{Tr}[Q^{1/2}D^2 u(x)Q^{1/2}], \qquad (3.4.4)$$

and for regular functions u the Gross Laplacian is identical, up to the constant 2, with the operator in the right hand side of the heat equation .

Note that if $G = Q^{1/2}(H)$ then

$$D_Q u(x) = Q^{-1/2}D_G u(x), \quad D_Q^2 u(x) = Q^{-1/2}D_G^2 u(x)Q^{1/2}.$$

3.5 The heat semigroup and its generator

We are here concerned with the heat semigroup

$$P_t\varphi(x) = \int_H \varphi(x+y)N_{tQ}(dy), \quad \varphi \in UC_b(H), \quad x \in H. \qquad (3.5.1)$$

Let us first prove that (P_t) is strongly continuous on $UC_b(H)$.

Proposition 3.5.1 *Let $Q \in L_1^+(H)$. Then the heat semigroup (P_t) is a strongly continuous semigroup on $UC_b(H)$.*

Proof. Let $\varphi \in UC_b(H)$. Since

$$P_t\varphi(x) = \int_H \varphi(x+\sqrt{t}\,z)N_Q(dz),$$

we have

$$|P_t\varphi(x) - \varphi(x)| = \left| \int_H [\varphi(x+\sqrt{t}\,z) - \varphi(x)]N_Q(dz) \right|$$

$$\leq \int_H \omega_\varphi(\sqrt{t}\,z)N_Q(dz),$$

where ω_φ is the uniform continuity modulus of φ. Therefore the conclusion follows letting t tend to 0. $\square$

As we said before, even when the space H is finite dimensional, P_t is not strongly continuous on $C_b(H)$. This is the reason why we are considering the heat semigroup on $UC_b(H)$.

Moreover, if $\dim H < \infty$ the heat semigroup is continuous on $(0, +\infty)$ in the operator norm and it is even analytic. It follows from the next theorem and basic results on semigroups theory, see e.g. E. B. Davies [104], A. Pazy

[180], that neither of those properties holds if $\dim H = \infty$. The following proposition, due to W. Desh and A. Rhandi [105], extends a previous result due to P. Guiotto [140]. The present proof is taken from J. van Neerven and J. Zabczyk [174].

Proposition 3.5.2 *If* $\dim H = \infty$ *and* $\mathrm{Ker}\, Q = \{0\}$ *then the heat semigroup is not continuous on* $(0, +\infty)$ *in the operator norm.*

Proof. If $t \neq s$ the operator $\frac{t-s}{t} I$ is not Hilbert-Schmidt and by the Feldman-Hajek theorem (see Example 1.3.8), the measures $N(x, tQ)$ and $N(x, sQ)$ are singular. Consequently

$$
\begin{aligned}
\|P_t - P_s\| &= \sup_{\substack{\varphi \in UC_b(H) \\ \|\varphi\|_0 \leq 1}} \|P_t\varphi - P_s\varphi\| \\
&= \sup_{\substack{\varphi \in UC_b(H) \\ \|\varphi\|_0 \leq 1}} \left| \int_H \varphi(y) N_{x,tQ}(dy) - \int_H \varphi(y) N_{x,tQ}(dy) \right| \\
&= \mathrm{var}(N_{x,tQ} - N_{x,sQ}) = 2,
\end{aligned}
$$

where "var" stands for the variation. $\square$

Since the semigroup (P_t) on $UC_b(H)$ is strongly continuous, it is determined by its infinitesimal generator Δ_Q. In particular if $\varphi \in D(\Delta_Q)$ then the equation

$$ D_t u = \Delta_Q u, \quad u(0) = \varphi, $$

has a unique solution in $UC_b(H)$ given by $u(t) = P_t\varphi$.

We study now some properties of the infinitesimal generator Δ_Q of (P_t), and its relations with the differential operator

$$ L\varphi = \frac{1}{2}\mathrm{Tr}[D_Q^2\varphi], \quad \varphi \in UC_b^2(H). $$

Let us recall, see e.g. A. Lunardi [162], that when $H = \mathbb{R}^d$ and $\det Q > 0$, then the domain of Δ_Q is given by

$$ D(\Delta_Q) = \left\{ \varphi \in W_{loc}^{2,p} : p \geq 1, \ \Delta\varphi \in UC_b(\mathbb{R}^d) \right\}. $$

Moreover we have that

$$ D(\Delta_Q) \subset UC_b^{2-\varepsilon}(\mathbb{R}^d), \quad \varepsilon \in [0, 1), \tag{3.5.2} $$

and only if $d = 1$ do we have that $D(\Delta_Q) = UC_b^2(\mathbb{R}^d)$.

We shall prove that for $\varphi \in UC_b^2(H)$ one has $\Delta_Q\varphi = L\varphi$. However, if $\dim H = \infty$, the space $UC_b^2(H)$ is not dense in $UC_b(H)$, see §2.2. Therefore we introduce a natural extension $\Delta_{0,Q}$ of L, which has a dense domain on $UC_b(H)$.

We set, that is,

$$\Delta_{0,Q}\varphi(x) = \frac{1}{2}\,\mathrm{Tr}[D_Q^2\varphi(x)], \quad x \in H, \ \varphi \in D(\Delta_{0,Q}), \tag{3.5.3}$$

where

$$D(\Delta_{0,Q}) = \{\varphi \in UC_Q^2(H) : \ D_Q^2\varphi \in UC_b(H, L_1(H))\}. \tag{3.5.4}$$

If (e_k, λ_k) is the eigensequence determined by Q, then

$$\Delta_{0,Q}\varphi(x) = \frac{1}{2}\sum_{k=1}^{\infty}\langle D_Q^2\varphi(x)e_k, e_k\rangle = \frac{1}{2}\sum_{k=1}^{\infty}\lambda_k D_k^2\varphi(x), \quad x \in H. \tag{3.5.5}$$

We are going to prove that $D(\Delta_{0,Q})$ is dense in $UC_b(H)$ and that Δ_Q is an extension of $\Delta_{0,Q}$.

The following result is due to E. Priola [189], we here present a proof given in J. Zabczyk [220].

Theorem 3.5.3 *Let $Q \in L_1^+(H)$ and let (P_t) be the heat semigroup defined by (3.5.1), and Δ_Q its infinitesimal generator. Then Δ_Q is an extension of $\Delta_{0,Q}$. Moreover $D(\Delta_{0,Q})$ and $D(\Delta_{0,Q}) \cap UC_b^1(H)$ are cores for Δ_Q.* ([6])

Proof. We will show that for $\varphi \in D(\Delta_{0,Q})$, $t > 0$ and $x \in H$, we have

$$P_t\varphi(x) = \varphi(x) + \int_0^t P_s(\Delta_{0,Q}\varphi)(x)ds. \tag{3.5.6}$$

Since $P_s(\Delta_{0,Q}\varphi) \to \Delta_{0,Q}\varphi$ as $s \to 0$ in $UC_b(H)$, this identity implies that $\varphi \in D(\Delta_Q)$ and $\Delta_{0,Q}\varphi = \Delta_Q\varphi$.

Define

$$R_n x = \sum_{k=1}^{n}\langle x, e_k\rangle e_k, \quad Q_n x = \sum_{k=1}^{n}\lambda_k\langle x, e_k\rangle e_k, \quad x \in H.$$

---

[6]Let (P_t) be a strongly continuous semigroup on a Banach space E, and let $L : D(L) \subset E \to E$ its infinitesimal generator. A subspace Y of $D(L)$ is said to be a *core* if it is dense in $D(L)$ endowed with the graph norm. A sufficient condition in order that a subspace Y is a core is that it is dense in E and invariant for P_t, $t \geq 0$, see [104]. The generator is uniquely determined on a core.

Let (P_t^n) be the heat semigroup corresponding to Q_n and Δ_Q^n its infinitesimal generator. By the finite dimensional theory, $D(\Delta_{0,Q}) \subset D(\Delta_Q^n)$ and

$$P_t^n \varphi(x) = \varphi(x) + \int_0^t P_s^n (\Delta_Q^n \varphi)(x) ds.$$

Moreover, for $\varphi \in UC_b(H)$,

$$P_t^n \varphi(x) = \int_H \varphi(x + R_n y) N_{tQ}(dy),$$

and consequently $P_t^n \varphi(x) \to P_t \varphi(x)$ as $n \to \infty$. Note also, that

$$\Delta_Q^n \varphi(x) = \frac{1}{2} \sum_{k=1}^{\infty} \langle D_{Q_n}^2 \varphi(x) e_k, e_k \rangle = \frac{1}{2} \sum_{k=1}^{n} \langle D_Q^2 \varphi(x) e_k, e_k \rangle$$

$$\to \frac{1}{2} \sum_{k=1}^{\infty} \langle D_Q^2 \varphi(x) e_k, e_k \rangle = \Delta_{0,Q} \varphi(x), \quad \text{as } n \to \infty.$$

In addition,

$$|\Delta_Q^n \varphi(x)| \le \frac{1}{2} \sum_{k=1}^{\infty} |\langle D_Q^2 \varphi(x) e_k, e_k \rangle| \le \frac{1}{2} \|D_Q^2 \varphi(x)\|_{L^1(H)}, \qquad (3.5.7)$$

as for self-adjoint operators S and arbitrary orthonormal basis (f_k),

$$\sum_{k=1}^{\infty} |\langle S f_k, f_k \rangle| \le \|S\|_{L_1(H)}$$

with the identity holding if (f_k) is determined by S. Thus the sequence $(\Delta_Q^n \varphi)$ is uniformly bounded. To establish (3.5.6) it is enough to prove that

$$\lim_{n \to \infty} \int_0^t P_s^n (\Delta_Q^n \varphi)(x) ds = \int_0^t P_s (\Delta_{0,Q} \varphi)(x) ds.$$

Since the sequence of functions $P_s^n (\Delta_Q^n \varphi)(x)$, $s \in [0, t]$, is bounded it is sufficient to verify that for $s \in [0, t]$,

$$\lim_{n \to \infty} P_s^n (\Delta_Q^n \varphi)(x) = P_s (\Delta_{0,Q} \varphi)(x).$$

However,

$$|P_s^n(\Delta_Q^n\varphi)(x) - P_s(\Delta_{0,Q}\varphi)(x)|$$

$$\leq |P_s^n(\Delta_Q^n\varphi)(x) - P_s(\Delta_Q^n\varphi)(x)| + |P_s(\Delta_Q^n\varphi)(x) - P_s(\Delta_{0,Q}\varphi)(x)|$$

$$\leq \left| \int_H [\Delta_Q^n\varphi(x + R_n y) - \Delta_Q^n\varphi(x + y)] N_{sQ}(dy) \right|$$

$$+ \left| \int_H [\Delta_Q^n\varphi(x + y) - \Delta_{0,Q}\varphi(x + y)] N_{sQ}(dy) \right| \leq I_1 + I_2.$$

The term I_2 converges to zero by the dominated convergence theorem. By an obvious generalization of (3.5.8), for arbitrary $x, z \in H$ and $n \in \mathbb{N}$,

$$|\Delta_Q^n\varphi(x) - \Delta_Q^n\varphi(x + z)| \leq \frac{1}{2} \|D_Q^2\varphi(x) - D_Q^2\varphi(x + z)\|_1.$$

Since $D_Q^2\varphi \in UC_b(H, L_1(H))$ by the same theorem, also $I_1 \to 0$, so that the proof of the first part of the theorem is complete.

We pass now to the proof of the second part. The resolvent of Δ_Q is given, from the Hille-Yosida theorem, by

$$R(\lambda, \Delta_Q)\varphi(x) = \int_0^{+\infty} e^{-\lambda t} P_t\varphi(x)dt, \ \lambda > 0.$$

We first show that if $\varphi \in UC_b^1(H)$, $t > 0$, and $\lambda > 0$ then

$$P_t\varphi \in D(\Delta_{0,Q}) \cap UC_b^1(H).$$

It follows from the statement and proof of Theorem 3.3.7 that $\|D_Q^2 P_t\varphi(x)\|_{L^1(H)} < \infty$, $x \in H$ and

$$\|D_Q^2 P_t\varphi(x) - D_Q^2 P_t\varphi(z)\|_{L^1(H)}$$

$$\leq \frac{1}{\sqrt{t}} (\mathrm{Tr}\ Q)^{1/2} \int_H |D\varphi(x + y) - D\varphi(z + y)|^2 N_{tQ}(dy).$$

So $P_t\varphi \in D(\Delta_{0,Q})$. Since

$$DP_t\varphi(x) = \int_H D\varphi(x + y) N_{tQ}(dy), \quad x \in H,$$

$P_t\varphi \in UC_b^1(H)$ as well. Consequently $D(\Delta_{0,Q}) \cap UC_b^1(H)$ is invariant for P_t. One shows in exactly the same way that $R(\lambda, \Delta_Q)\varphi \in D(\Delta_{0,Q}) \cap UC_b^1(H)$.

Let us prove now that $D(\Delta_{0,Q}) \cap UC_b^1(H)$ is dense in $UC_b(H)$. Take $\varphi \in D(\Delta_Q)$ and set $g = \varphi - \Delta_Q\varphi$. Since $UC_b^1(H)$ is dense in $UC_b(H)$ (see Theorem 2.2.1), there exists a sequence $(g_n) \subset UC_b^1(H)$ such that $g_n \to g$ in $UC_b(H)$. However,

$$\varphi_n = R(1, \Delta_Q)g_n \in D(\Delta_{0,Q}) \cap UC_b^1(H), \quad n \in \mathbb{N}.$$

Since

$$\varphi = R(1, \Delta_Q)g = \lim_{n\to\infty} R(1, \Delta_Q)g_n = \lim_{n\to\infty} \varphi_n,$$

the density of $D(\Delta_{0,Q}) \cap UC_b^1(H)$ in $D(\Delta_Q)$ and therefore also in $UC_b(H)$ follows. This implies that $D(\Delta_{0,Q}) \cap UC_b^1(H)$ is a core for $\Delta_{0,Q}$, see e.g. E. B. Davies [104]. Consequently the set $D(\Delta_{0,Q})$ is also a core. $\square$

The following result generalizes (3.5.3).

Proposition 3.5.4 *For any $\theta \in (0,1)$ the following inclusion holds:*

$$D(\Delta_Q) \subset UC_Q^{1+\theta}(H). \tag{3.5.8}$$

Proof. By Theorem 3.3.7 we have

$$\|P_t\varphi\|_0 \le \|\varphi\|_0, \quad \|D_Q^2 P_t\varphi\|_0 \le \frac{\sqrt{2}}{t}\|\varphi\|_0.$$

Therefore by Theorem 3.3.2 we have

$$\|D_Q^2 P_t\varphi\|_{1+\theta} \le 2^{\frac{1+\theta}{2}} t^{-\frac{1+\theta}{2}} \|\varphi\|_0.$$

Therefore

$$\|R(\lambda, \Delta_Q)\|_{1+\theta,Q} \le 2^{\frac{1+\theta}{2}}\Gamma(1-\theta)\lambda^\theta. \quad \square$$

Concerning the spectrum of Δ_Q we have the following result due to G. Metafune, A. Rhandi and R. Schnaubelt [173].

Proposition 3.5.5 *The spectrum $\sigma(\Delta_Q)$ of Δ_Q coincides with the half-plane* Re $\lambda \le 0$.

Proof. Let us identify H with ℓ^2, and define for each n and λ with negative real part

$$g_n(x) = e^{\lambda c_n} e^{\sum_{j=1}^n \lambda_j^{-1} x_j^2},$$

where

$$c_n = \left[\sum_{j=1}^n \lambda_j^{-1}\right]^{-1} \to 0,$$

as n tends to ∞. By a direct calculation we have

$$(\lambda g_n - \Delta_Q g_n)(x) = -4\lambda^2 c_n^2 \sum_{j=1}^{n} \lambda_j^{-1} x_j^2 \, e^{\lambda c_n \sum_{j=1}^{n} \lambda_j^{-1} x_j^2} = f_n(x),$$

$$\|f_n\|_0 = \frac{4|x|^2}{|\mathrm{Re}\,\lambda|} \, c_n e^{-1}.$$

Assume in contradiction that λ belongs to the resolvent set of Δ_Q, then $R(\lambda, \Delta_Q) f_n = g_n$. Since $\|g_n\|_0 = 1$, we have

$$\|R(\lambda, \Delta_Q)\| \geq \frac{\|R(\lambda, \Delta_Q) f_n\|_0}{\|f_n\|_0} \geq \frac{1}{\|f_n\|_0} \to +\infty,$$

a contradiction. Thus $\sigma(\Delta_Q) \supset \{\lambda \in \mathbb{C} : \mathrm{Re}\,\lambda \leq 0\}$. The proof of the result is complete. $\square$

We note that if H is finite dimensional and $Q > 0$, then the spectrum $\sigma(\Delta_Q)$ of Δ_Q is equal to the interval $(-\infty, 0]$.

Chapter 4

Poisson's equation

As before, H represents a separable Hilbert space, Q a symmetric nonnegative operator of trace class with Ker $Q = \{0\}$, (e_k) an orthonormal basis in H, and (λ_k) a sequence of positive numbers such that $Qe_k = \lambda_k e_k$, $k \in \mathbb{N}$. We consider the heat semigroup (P_t) defined by (3.5.1). We recall that (P_t) is strongly continuous in $UC_b(H)$ and that its infinitesimal generator Δ_Q is the closure of the operator $\Delta_{0,Q}$ defined by

$$
\begin{cases}
\Delta_{0,Q}\varphi(x) = \frac{1}{2} \operatorname{Tr}[D_Q^2\varphi(x)], & x \in H, \ \varphi \in D(\Delta_{0,Q}), \\
D(\Delta_{0,Q}) = \{\varphi \in UC_Q^2(H); \ D_Q^2\varphi \in UC_b(H; L_1(H))\},
\end{cases}
$$

see Proposition 3.5.1 and Theorem 3.5.3. This chapter is devoted to studying existence, uniqueness and regularity of solutions for elliptic equations related to Δ_Q and $\Delta_{0,Q}$: §4.1 concerns existence and uniqueness, §4.2 Schauder estimates , and §4.3 potential theory .

4.1 Existence and uniqueness results

We are here concerned with the elliptic equation on H,

$$
\lambda\varphi(x) - \frac{1}{2}\operatorname{Tr}[D_Q^2\varphi(x)] = g(x), \ x \in H, \tag{4.1.1}
$$

where $\lambda > 0$ and $g \in UC_b(H)$. The case $\lambda = 0$ will be studied in §4.3. Following tradition equation (4.1.1) will be called the *Poisson* equation.

A function φ is said to be a *strict solution* of (4.1.1) if it belongs to the domain of $\Delta_{0,Q}$ and fulfills (4.1.1).

It is well known that, even if H is d-dimensional (with $d > 1$), there does not exist in general a strict solution of (4.1.1). One has to look for weak or generalized solutions . By a *generalized solution* of equation (4.1.1) we mean a function φ that belongs to the domain of Δ_Q and fulfills

$$\lambda\varphi - \Delta_Q\varphi = g. \tag{4.1.2}$$

Proposition 4.1.1 *For any $\lambda > 0$ and $g \in UC_b(H)$, equation (4.1.1) has a unique generalized solution φ given by*

$$\varphi(x) = R(\lambda, \Delta_Q)g(x) = \int_0^{+\infty} e^{-\lambda t} P_t g(x)dt, \quad x \in H. \tag{4.1.3}$$

Moreover there exists a sequence $(g_n) \subset UC_b(H)$ converging to g in $UC_b(H)$ such that the equation

$$\lambda\varphi_n(x) - \frac{1}{2}\mathrm{Tr}[D_Q^2\varphi_n(x)] = g_n(x)$$

has a strict solution φ_n for any $n \in \mathbb{N}$ and $\varphi_n \to \varphi$ in $UC_b(H)$.

Proof. Since (P_t) is a strongly continuous semigroup in $UC_b(H)$, the first statement is an immediate consequence of the Hille-Yosida theorem. The last statement follows from the fact that $D(\Delta_{0,Q})$ is a core for Δ_Q, see Theorem 3.3.3. $\square$

Obviously the function $\varphi = R(\lambda, \Delta_Q)g$ is not a strict solution of equation (4.1.1) in general because

(i) φ does not belong to $C_Q^2(H)$ in general when the dimension of H is greater than 1,

(ii) even if $\varphi \in C_Q^2(H)$, in the case of infinite dimensional H, $D_Q^2\varphi$ is not necessarily of trace class.

In order to have a strict solution of (4.1.1), we need stronger conditions on g. A sufficient condition is given by the following result.

Theorem 4.1.2 *If $\lambda > 0$ and $g \in UC_b^1(H)$ then $\varphi = R(\lambda, \Delta_Q)g$ is a strict solution of (4.1.1).*

Proof. By Theorem 3.3.7 we have

$$\varphi(x) = \int_0^{+\infty} e^{-\lambda t}u(t, x)dt, \ x \in H,$$

where u is the solution of the problem

$$D_t u(t, x) = \frac{1}{2}\mathrm{Tr}[D_Q^2 u(t, x)], \ u(0, x) = g(x), \ t > 0, \ x \in H.$$

Since, by Theorem 3.3.7, $D_Q^2 u(t, x)$ is of trace class and

$$\|D_Q^2 u(t,x)\|_{L_1(H)} \le t^{-1/2} \, (\text{Tr } Q)^{1/2} \, \|Dg\|_0, \; x \in H,$$

one can easily show that

$$D_Q^2 \varphi(x) = \int_0^{+\infty} e^{-\lambda t} D_Q^2 u(t, x) dt, \; x \in H.$$

Moreover $D_Q^2 \varphi(x)$ is of trace class and

$$\text{Tr}[D_Q^2 \varphi(x)] = \int_0^{+\infty} e^{-\lambda t} \text{Tr}[D_Q^2 u(t, x)] dt, \; x \in H.$$

Consequently

$$\begin{aligned}
\frac{1}{2}\text{Tr}[D_Q^2 \varphi(x)] &= \frac{1}{2} \lim_{\substack{s \to 0 \\ T \to +\infty}} \int_s^T e^{-\lambda t} \text{Tr}[D_Q^2 u(t, x)] dt \\
&= \lim_{\substack{s \to 0 \\ T \to +\infty}} \int_s^T e^{-\lambda t} D_t u(t, x) dt \\
&= \lim_{\substack{s \to 0 \\ T \to +\infty}} \left[\left| e^{-\lambda t} u(t, x) \right|_s^T + \lambda \int_s^T e^{-\lambda t} u(t, x) dt \right] \\
&= -g(x) + \lambda u(x). \quad \square
\end{aligned}$$

4.2   Regularity of solutions

In this section we investigate the regularity of generalized solutions φ of (4.1.1). First we show that φ is always of class $C_Q^{1+\theta}(H)$, $\theta \in [0,1)$.

Proposition 4.2.1 *Let $\lambda > 0$ and $g \in UC_b(H)$, and let $\varphi = R(\lambda, \Delta_Q)g$ be a generalized solution of (4.1.1). Then $\varphi \in C_Q^{1+\theta}(H)$, for all $\theta \in [0,1)$, and there is $C_\theta > 0$ such that*

$$|\varphi|_{1+\theta,Q} \le C_\theta \lambda^{\frac{-(1-\theta)}{2}} \, \|g\|_0. \qquad (4.2.1)$$

Proof. Set $u(t, x) = P_t \varphi(x)$. By (3.3.5) it follows that

$$\int_0^\infty e^{-\lambda t} |D_Q u(t, x)| dt \le \sqrt{\frac{\pi}{\lambda}} \, \|\varphi\|_0, \; x \in H.$$

Consequently it is easy to see that $\varphi \in C^1_Q(H)$ and

$$\langle D_Q\varphi(x), h \rangle = \int_0^\infty e^{-\lambda t} \langle D_Q u(t,x), h \rangle dt, \ h, x \in H.$$

Moreover by the interpolatory estimate (3.3.2), there exists $C > 0$ such that

$$\|u(t,\cdot)\|_{1+\theta,Q} \le C \|u(t,\cdot)\|_0^{\frac{1-\theta}{2}} \|u(t,\cdot)\|_{2,Q}^{\frac{1+\theta}{2}},$$

which yields, thanks to (3.3.6),

$$\|u(t,\cdot)\|_{1+\theta,Q} \le C \left(\frac{\sqrt{2}}{t} \right)^{\frac{1+\theta}{2}} \|\varphi\|_0.$$

It follows that

$$\int_0^\infty e^{-\lambda t} \|u(t,\cdot)\|_{1+\theta,Q} dt \le C\sqrt{2}^{\frac{1+\theta}{2}} \int_0^\infty e^{-\lambda t} t^{-\frac{1+\theta}{2}} dt.$$

Consequently $\varphi \in C^{1+\theta}_Q(H)$ as required. $\square$

Proposition 4.2.2 *Let* $\lambda > 0$, $\theta \in (0,1)$, $g \in C^\theta_Q(H)$ *and* $\varphi = R(\lambda, \Delta_Q)g$. *Then* $\varphi \in UC^2_Q(H)$, *and*

$$\|D^2_Q\varphi(x)\| \le 2^{\frac{1-\theta}{2}} \Gamma(\theta/2) \lambda^{-\frac{\theta}{2}} \|g\|_{\theta,Q}, \tag{4.2.2}$$

where Γ *is the Euler function. Moreover* $D^2_Q\varphi(x) \in L_2(H)$ *and*

$$\|D^2_Q\varphi(x)\|_{L_2(H)} \le \Gamma(1-\theta) 2^{1-\theta} \lambda^{-\frac{\theta}{2}} \|g\|_{1,Q}. \tag{4.2.3}$$

Proof. By Proposition 3.3.6 we have

$$\|D^2_Q P_t g(x)\| \le t^{-1/2} \|g\|_{1,Q}, \ x \in H. \tag{4.2.4}$$

Moreover by (3.3.6), we have

$$\|D^2_Q P_t g(x)\| \le \sqrt{2}\, t^{-1} \|g\|_0, \ x \in H. \tag{4.2.5}$$

By a straightforward generalization of Theorem 2.3.3 we see that $C^\theta_Q(H)$ is an interpolation space between $UC_b(H)$ and $UC^1_Q(H)$, and

$$\left(UC_b(H), UC^1_Q(H) \right)_{\theta,\infty} = C^\theta_Q(H),$$

therefore by (4.2.4) and (4.2.5), it follows that

$$\|D_Q^2 P_t g(x)\| \le 2^{\frac{1-\theta}{2}} t^{-1+\frac{\theta}{2}} \|g\|_{\theta,Q}.$$

Now (4.2.2) follows by taking the Laplace transform. Finally, (4.2.3) follows from Theorems 3.3.5 and 3.3.7, by applying interpolation again. $\square$

We now establish regularity results in the form of Schauder estimates . We follow P. Cannarsa and G. Da Prato [31].

Theorem 4.2.3 *Let $\lambda > 0$, $\theta \in (0,1)$ and $g \in C_Q^\theta(H)$. Then*

$$\varphi = R(\lambda, \Delta_Q)g \in C_Q^{2+\theta}(H)$$

and there exists $C > 0$ such that

$$\|\varphi\|_{2+\theta,Q} \le C\|g\|_{\theta,Q}. \tag{4.2.6}$$

For the proof we need some information about the interpolation spaces between $D(\Delta_Q)$ and $UC_b(H)$. Let $D_{\Delta_Q}(\theta,\infty)$, $\theta \in (0,1)$, be the interpolation space $(UC_b(H), D(\Delta_Q))_{\theta,\infty}$. We recall, see Appendix A, that

$$D_{\Delta_Q}(\theta,\infty) = \{u \in UC_b(H) : [\varphi]_{\theta,\infty} < \infty\},$$

where

$$[\varphi]_{\theta,\infty} = \sup_{t>0} t^{-\theta} \|P_t\varphi - \varphi\|_0, \quad \varphi \in D_{\Delta_Q}(\theta,\infty).$$

Moreover

$$\|\varphi\|_{D_{\Delta_Q}(\theta,\infty)} = \|\varphi\|_0 + [\varphi]_{\theta,\infty}, \quad \varphi \in D_{\Delta_Q}(\theta,\infty).$$

The following result is due to A. Lunardi, see [163].

Proposition 4.2.4 *For all $\theta \in (0,1/2)$ the following inclusion holds:*

$$D_{\Delta_Q}(\theta,\infty) \subset C_Q^{2\theta}(H).$$

Proof. Let $\varphi \in D_{\Delta_Q}(\theta,\infty)$. Then from the obvious identity

$$\Delta_Q R(\lambda, \Delta_Q)\varphi = \lambda \int_0^{+\infty} e^{-\lambda t}(P_t\varphi - \varphi)dt = \lambda R(\lambda, \Delta_Q)\varphi - \varphi,$$

it follows that

$$\|\Delta_Q R(\lambda, \Delta_Q)\varphi\|_0 \le \Gamma(1+\theta)[\varphi]_{\theta,\infty}\lambda^{-\theta}, \quad \lambda > 0. \tag{4.2.7}$$

Let $x, y \in H$. By the inequality (4.2.7) and

$$\varphi(Q^{1/2}x) - \varphi(Q^{1/2}y) = -\Delta_Q R(\lambda, \Delta_Q)(\varphi(Q^{1/2}x) - \varphi(Q^{1/2}y))$$

$$+ \lambda R(\lambda, \Delta_Q)(\varphi(Q^{1/2}x) - \varphi(Q^{1/2}y)),$$

it follows that

$$|\varphi(Q^{1/2}x) - \varphi(Q^{1/2}y)| \leq 2\Gamma(1 + \theta)[\varphi]_{\theta,\infty} \lambda^{-\theta}$$

$$+ \|Q^{1/2}D(\lambda R(\lambda, \Delta_Q)\varphi)\|_0 |x - y|. \quad (4.2.8)$$

Let us estimate $\|Q^{1/2}D(\lambda R(\lambda, \Delta_Q)\varphi)\|_0$. Since

$$\lambda R(\lambda, \Delta_Q) - R(1, \Delta_Q) = \int_1^\lambda \frac{d}{d\mu}(\mu R(\mu, \Delta_Q))d\mu$$

$$= -\int_1^\lambda \Delta_Q R^2(\mu, \Delta_Q)d\mu,$$

it follows that

$$\lambda Q^{1/2}(DR(\lambda, \Delta_Q)\varphi) = Q^{1/2}DR(1, \Delta_Q)\varphi$$

$$- \int_1^\lambda Q^{1/2}D[R(\mu, \Delta_Q)\Delta_Q R(\mu, \Delta_Q)\varphi]d\mu.$$

Thus, using the estimates (4.2.1) and (4.2.2), we have

$$\|\lambda Q^{1/2}(DR(\lambda, \Delta_Q)\varphi)\|_0$$

$$\leq \sqrt{\pi} \left(\|\varphi\|_0 + \int_1^\lambda \mu^{-1/2} \|\Delta_Q R(\mu, \Delta_Q)\varphi\|_0 d\mu \right)$$

$$\leq \sqrt{\pi} \left[\|\varphi\|_0 + \frac{2\Gamma(1 + \theta)}{1 - 2\theta} [\varphi]_{\theta,\infty} \lambda^{1/2-\theta} \right]. \quad (4.2.9)$$

Substituting (4.2.9) in (4.2.8) yields

$$|\varphi(Q^{1/2}x) - \varphi(Q^{1/2}y)| \leq 2\Gamma(1 + \theta)\lambda^{-\theta}[\varphi]_{\theta,\infty}$$

$$+ \sqrt{\pi} \left[\|\varphi\|_0 + \frac{2\Gamma(1 + \theta)}{1 - 2\theta} [\varphi]_{\theta,\infty} \lambda^{1/2-\theta} \right] |x - y|.$$

It is enough to consider the case when $|x - y| \leq 1$. In this case we have

$$|\varphi(Q^{1/2}x) - \varphi(Q^{1/2}y)| \leq 2\Gamma(1 + \theta)\lambda^{-\theta}[\varphi]_{\theta,\infty}$$

$$+\sqrt{\pi}\left[\|\varphi\|_0|x - y|^{2\theta} + \frac{2\Gamma(1 + \theta)}{1 - 2\theta}[\varphi]_{\theta,\infty}\lambda^{1/2-\theta}|x - y|\right].$$

Setting $\lambda = |x - y|^{-2}$ the conclusion follows. $\square$

We are now ready to prove the main result of this section.

Proof of Theorem 4.2.3. By Proposition 4.2.2 we know that $\varphi \in UC_Q^2(H)$ and for any $\alpha, \beta \in H$ we have

$$\langle D_Q^2\varphi(x)\alpha, \beta\rangle = \int_0^\infty e^{-\lambda t}\langle D_Q^2 P_t g(x)\alpha, \beta\rangle dt, \quad x \in H.$$

If we set $h(x) = \langle D_Q^2\varphi(x)\alpha, \beta\rangle$, $x \in H$, then there exists a constant $c_\theta > 0$ such that $\|h\|_{D_{A_Q}(\theta/2,\infty)} \leq c_\theta|\alpha|\,|\beta|\,\|g\|_{\theta,Q}$. We have in fact

$$P_\xi h(x) - h(x)$$

$$= \int_0^{+\infty} e^{-\lambda t}P_\xi\langle D_Q^2 P_t g(x)\alpha, \beta\rangle dt - \int_0^{+\infty} e^{-\lambda t}\langle D_Q^2 P_t g(x)\alpha, \beta\rangle dt$$

$$= \int_0^{+\infty} e^{-\lambda t}\langle D_Q^2 P_{t+\xi} g(x)\alpha, \beta\rangle dt - \int_0^{+\infty} e^{-\lambda t}\langle D_Q^2 P_t g(x)\alpha, \beta\rangle dt$$

$$= (e^\xi - 1)\int_0^{+\infty} e^{-\lambda t}\langle D_Q^2 P_t g(x)\alpha, \beta\rangle dt - e^\xi\int_0^\xi e^{-\lambda t}\langle D_Q^2 P_t g(x)\alpha, \beta\rangle dt.$$

It follows by (4.2.2) that

$$|P_\xi h(x) - h(x)| \leq c\left[(e^\xi - 1)\int_0^{+\infty} e^{-\lambda t}t^{\theta/2-1}dt\right]|\alpha|\,|\beta|\,\|g\|_{\theta,Q}$$

$$+ \left[ce^\xi\int_0^\xi \frac{dt}{t^{1-\theta/2}}\right]|\alpha|\,|\beta|\,\|g\|_{\theta,Q},$$

where c is a suitable constant. Therefore there exists a constant $c_1 > 0$ such that if $\xi \in (0, 1)$

$$|P_\xi h - h|_0 \leq C_1\xi^{\theta/2}|\alpha|\,|\beta|\,\|g\|_{\theta,Q}, \quad \xi > 0,$$

and the conclusion follows. Thus, by Proposition 4.2.4, we have

$$h \in D_{\Delta_Q}(\theta/2, \infty) \subset C_Q^\theta(H),$$

which yields our thesis. $\square$

Remark 4.2.5 Notice that we are not able to prove that $D^2\varphi(x) \in L_1(H)$. However, one can show, see E. Priola and L. Zambotti [197], that (under the assumptions of Theorem 4.2.3) $D_Q^2\varphi(x) \in L_2(H)$ and there exists $C_{\alpha,\lambda}^1 > 0$ such that

$$\|D_Q^2\varphi(Q^{1/2}x) - D_Q^2\varphi(Q^{1/2}y)\|_{L^2(H)} \le C_{\alpha,\lambda}^1 |(x-y)|^\theta, \quad x,y \in H. \quad (4.2.10)$$

4.3 The equation $\Delta_Q u = g$

It is well known that if $H = \mathbb{R}^n$, $n \ge 3$, and $g \in UC_b(H)$ has a bounded support then there exists a unique function (up to an additive constant) $\varphi \in C_b^{1+\theta}(H), \theta \in (0,1)$, such that

$$-\frac{1}{2}\Delta\varphi = g, \quad (4.3.1)$$

in the distributional sense. Moreover,

$$\varphi(x) = \int_0^\infty P_t g(x)dt = C_n \int_{\mathbb{R}^n} \frac{g(y)}{|x-y|^{n-2}} \, dy, \quad x \in H, \quad (4.3.2)$$

where

$$C_n = \frac{1}{2\pi^{n/2}} \Gamma(n/2 - 1).$$

The function φ is called the *potential* of g.

If $n = 1, 2$ then equation (4.3.1) does not necessarily have a solution even if g is of class C^∞ with bounded support.

From now on we shall assume that H is an infinite dimensional space. Then we can still define, following L. Gross [138], the potential of any function $g \in UC_b(H)$ having a bounded support. For this we need a preliminary result on the asymptotic behavior in time of $P_t g, g \in UC_b(H)$.

Lemma 4.3.1 *Let $g \in UC_b(H)$ having bounded support K. Then for all $N \in \mathbb{N}$ there exists a constant $C_N = C_N(K)$ such that the following hold.*

(i) *The following estimates hold.*

$$|P_t g(x)| \leq C_N \, t^{-N/2} \|g\|_0, \quad x \in H, \qquad (4.3.3)$$
$$|D_Q P_t g(x)| \leq C_N \, t^{-(N+1)/2} \|g\|_0, \quad x \in H, \qquad (4.3.4)$$
$$\|D_Q^2 P_t g(x)\| \leq C_N \, t^{-(N+2)/2} \|g\|_0, \quad x \in H. \qquad (4.3.5)$$

(ii) *If* $g \in UC_b^1(H)$, *then*

$$\|D_Q^2 P_t g(x)\| \leq C_N \, t^{-(N+1)/2} \|g\|_{1,Q}. \qquad (4.3.6)$$

(iii) *If* $g \in C_Q^\theta(H)$, $\theta \in (0,1)$, *then*

$$\|D_Q^2 P_t g(x)\| \leq C_N 2^{\frac{1-\theta}{2}} t^{-(N+2-\theta)/2} \|g\|_{\theta,Q}. \qquad (4.3.7)$$

Proof. We first prove (4.3.3), following H. Kuo [149, Lemma 4.5]. Let us start from the identity

$$P_t g(x) = \int_H g(x+y) N_{tQ}(dy) = \int_K g(y) N_{x,tQ}(dy), \quad x \in H.$$

It follows that

$$|P_t g(x)| \leq \|g\|_0 \int_K N_{x,tQ}(dy), \quad x \in H.$$

Set $L = \sup_{z \in K}\{|\langle z, e_k \rangle| : k = 1, \ldots, N\}$ and $\Lambda = \{x \in H : |x_k| \leq L, \, k = 1, \ldots, N\}$, where $x_k = \langle x, e_k \rangle$. [1] We have clearly $K \subset \Lambda$. Since Λ is a cylindrical set, one finds easily

$$\int_\Lambda N_{x,tQ}(dy) = (2\pi t)^{-N/2} \prod_{k=1}^N \lambda_k^{-1/2} \prod_{k=1}^N \int_{-L}^L e^{-\frac{|y_k - x_k|^2}{2\lambda_k t}} \, dy_k. \qquad (4.3.8)$$

Thus

$$|P_t g(x)| \leq \|g\|_0 \int_\Lambda N_{x,tQ}(dy) \leq \|g\|_0 \frac{(2L)^N}{(2\pi)^{N/2}} t^{-N/2} \prod_{k=1}^N \lambda_k^{-1/2},$$

and (4.3.3) follows.

Let us prove (4.3.4). By (3.3.3), we have

$$\langle D_Q P_t g(x), h \rangle = \frac{1}{\sqrt{t}} \int_H \langle h, (tQ)^{-1/2} y \rangle g(x+y) N_{tQ}(dy), \quad h \in H.$$

---

[1] We recall that (e_k) is an orthonormal basis in H and (λ_k) a sequence of positive numbers such that $Q e_k = \lambda_k e_k$, $k \in \mathbb{N}$.

Consequently, by the Hölder inequality it follows that

$$|\langle D_Q P_t g(x), h\rangle|^2 \leq \frac{1}{t} \int_H |\langle h, (tQ)^{-1/2}y\rangle|^2 N_{tQ}(dy) \int_H g^2(x+y) N_{tQ}(dy),$$

$$\leq \frac{C}{t} |h|^2 \int_K g^2(y) N_{x,tQ}(dy), \ h \in H.$$

Therefore

$$|D_Q P_t g(x)|^2 \leq \frac{1}{t} \|g\|_0^2 \int_K N_{x,tQ}(dy).$$

Now (4.3.4) follows from (4.3.8). The proof of (4.3.5) is similar, using (3.3.4). Also (4.3.6) follows in the same way, starting from the formula

$$DP_t g(x) = \int_H Dg(x+y) N_{tQ}(dy) = \int_K Dg(y) N_{x,tQ}(dy).$$

Finally (4.3.7) follows by interpolation. $\square$

We are now ready to prove an existence result.

Proposition 4.3.2 *Let* $g \in UC_b^1(H)$ *with bounded support and let*

$$\varphi(x) = \int_0^{+\infty} P_t g(x) \, dt, \ x \in H. \tag{4.3.9}$$

Then u belongs to the domain of $\Delta_{0,Q}$ *and*

$$\frac{1}{2} \text{Tr}[D_Q^2 u(x)] = -g(x), \ x \in H. \tag{4.3.10}$$

Proof. The proof is similar to that of Theorem 4.1.2 and uses the same notations. From Theorem 3.3.7 it follows that

$$\|D_Q^2 P_t g(x)\|_{L_1(H)} \leq t^{-1/2} (\text{Tr } Q)^{1/2} \left[\int_H |Dg(x+y)|^2 N_{tQ}(dy) \right]^{1/2}.$$

Therefore if the support of g is included in $B_R = \{x \in H : |x| \leq R\}$, we have

$$\|D_Q^2 P_t g(x)\|_{L_1(H)} \leq t^{-1/2} (\text{Tr } Q)^{1/2} \|Dg\|_0 N_{x,tQ}(B_R).$$

By Lemma 4.3.1, for any $N \in \mathbb{N}$, there exists C_N such that

$$N_{x,tQ}(B(0,R)) \leq C_N t^{-N}.$$

Therefore

$$\|D_Q^2 P_t g(x)\|_{L_1(H)} \leq t^{-(N+1)/2} (\text{Tr } Q)^{1/2} \|Dg\|_0,$$

and consequently it is not difficult to show that $\varphi \in UC_Q^2(H)$,

$$D_Q^2\varphi(x) = \int_0^{+\infty} D_Q^2 P_t g(x)dt, \quad x \in H,$$

and

$$\mathrm{Tr}[D_Q^2\varphi(x)] = \int_0^{+\infty} \mathrm{Tr}[D_Q^2 P_t g(x)]dt, \quad x \in H.$$

Now, in order to prove that φ satisfies (4.3.10) we can proceed as in the proof of Theorem 4.1.2. $\square$

Theorem 4.3.3 *Let $g \in C_b^\theta(H)$, $\theta \in (0,1)$, have bounded support and let*

$$\varphi(x) = \int_0^{+\infty} P_t g(x)dt, \quad x \in H. \qquad (4.3.11)$$

Then

(i) $\varphi \in D(\Delta_Q)$ and $\Delta_Q\varphi = -g$,

(ii) $\varphi \in C_Q^{2+\theta}(H)$.

Proof. Let us prove (i). We have

$$P_s\varphi(x) - \varphi(x) = \int_0^{+\infty} P_{t+s}g(x)dt - \int_0^{+\infty} P_t g(x)dt$$

$$= -\int_0^s P_t g(x)dt.$$

Consequently there exists the limit $\lim_{s\to 0}\frac{1}{s}(P_s\varphi - \varphi) = -g$, and we have $\Delta_Q\varphi = -g$. So (i) is proved. To prove (ii) fix $\lambda > 0$. Then we have

$$\lambda\varphi - \Delta_Q\varphi = \lambda\varphi - g =: g_1.$$

Since $g \in C_Q^\theta(H)$ and $\varphi \in D(\Delta_Q) \subset C_Q^{1+\theta}(H)$ by Proposition 3.5.4, we have that $g_1 \in C_Q^\theta(H)$. Therefore we may apply Theorem 4.2.3 and conclude that $\varphi \in C_Q^{2+\theta}(H)$. $\square$

4.3.1 The Liouville theorem

We notice that constant functions are always solutions of the equation

$$\frac{1}{2}\mathrm{Tr}[D_Q^2\varphi(x)] = 0, \quad x \in H. \tag{4.3.12}$$

In this subsection we want to see whether they are the only solutions.

Note that for any strongly continuous semigroup (T_t) with generator A, an element $h \in D(A)$ satisfies $Ah = 0$ if and only if $T_t h = h$ for all $t \geq 0$. If $Aa = 0$, then a is called harmonic. We have therefore the following definition.

A Borel bounded function φ is said to be *harmonic* if

$$P_t\varphi(x) = \varphi(x), \quad x \in H, t \geq 0. \tag{4.3.13}$$

We can prove the following generalization of the classical *Liouville* theorem .

Theorem 4.3.4 *The only continuous harmonic functions are the constants.*

Proof. We will present here two different proofs.

First proof.
Let $\varphi \in C_b(H)$ be such that $P_t\varphi = \varphi$, $t \geq 0$. Then by Theorem 3.3.3 it follows that $\varphi \in C_Q^1(H)$ and

$$|D_Q\varphi(x)| \leq \frac{1}{\sqrt{t}}\,\|\varphi\|_0, \quad t \geq 0, x \in H.$$

Letting $t \to \infty$ we see that $D_Q\varphi(x) = 0$ for all $x \in H$. This implies that φ is constant in the directions $Q^{1/2}h$ for all $h \in H$. Since φ is continuous and $Q^{1/2}(H)$ is dense in H, it follows that φ is a constant as required.

Second proof.
We will prove first that if $a - b \in Q^{1/2}(H)$ then

$$\mathrm{Var}(N_{a,Q} - N_{b,Q}) \leq \left(e^{|Q^{-1/2}(a-b)|^2} - 1\right)^{1/2}. \tag{4.3.14}$$

By the Cameron-Martin formula we have

$$\frac{dN_{b,Q}}{dN_{a,Q}}(y) = e^{-\frac{1}{2}|Q^{-1/2}(a-b)|^2 + \langle Q^{-1/2}(y-a), Q^{-1/2}(b-a)\rangle}, \quad y \in H.$$

Moreover

$$\mathrm{Var}(N_{a,Q} - N_{b,Q}) = \int_H \left|1 - \frac{dN_{b,Q}}{dN_{a,Q}}(y)\right| dN_{a,Q}(dy)$$

$$\leq \left[\int_H \left|1 - \frac{dN_{b,Q}}{dN_{a,Q}}(y)\right|^2 N_{a,Q}(dy)\right]^{1/2}.$$

Since the law of $\langle Q^{-1/2}(y-a), Q^{-1/2}(b-a)\rangle$ is $N_{0,|Q^{-1/2}(b-a)|^2}$, therefore (4.3.14) follows.

Assume now that $\varphi \in C_b(H)$ is such that (4.3.13) holds. It follows from (4.3.14) that

$$|\varphi(a) - \varphi(b)| = |P_t\varphi(a) - P_t\varphi(b)| \le \|\varphi\|_0 \left(e^{\frac{1}{t}|Q^{-1/2}(a-b)|^2} - 1\right)^{1/2}.$$

Letting t tend to $+\infty$ gives $\varphi(a) = \varphi(b)$. As φ is continuous and $Q^{1/2}(H)$ is dense in H it follows that φ is constant. $\square$

If φ is not continuous the conclusion of Theorem 4.3.4 does not hold. We have in fact the following result proved by V. Goodman [132], with an elementary proof.

Proposition 4.3.5 *There exists a nonconstant harmonic function.*

Proof. We will first prove that there exists a Borel linear subspace H_0 of H different from H such that

$$N_{tQ}(H_0) = 1, \ t \ge 0. \tag{4.3.15}$$

Let us choose a nondecreasing sequence (α_k) of positive numbers such that $\alpha_k \uparrow +\infty$, and $\sum_{k=1}^{\infty} \alpha_k \lambda_k < +\infty$, and define

$$H_0 = \left\{x \in H : \sum_{k=1}^{\infty} \alpha_k |x_k|^2 < +\infty\right\}.$$

This is certainly possible. For instance choose an increasing sequence (N_m) of natural numbers such that

$$\sum_{k \ge N_m} \lambda_k \le 2^{-m}, \ m \in \mathbb{N},$$

and define

$$\alpha_n = \begin{cases} 0 \ \text{if} \ n \le N_1, \\ m \ \text{if} \ N_m \le n < N_{m+1}. \end{cases}$$

Then

$$\sum_{n=1}^{\infty} \alpha_n \lambda_n = \sum_{m=1}^{\infty} \sum_{N_m \le n < N_{m+1}}^{\infty} \alpha_n \lambda_n$$

$$= \sum_{m=1}^{\infty} m \sum_{N_m \le n < N_{m+1}}^{\infty} \lambda_n \le \sum_{m=1}^{\infty} m 2^{-m} < +\infty.$$

One can check easily that $Q^{1/2}(H) \subset H_0$. Moreover $N_{tQ}(H_0) = 1$ since

$$\int_H \sum_{k=1}^{\infty} \alpha_k |x_k|^2 N_{tQ}(dx) = t \sum_{k=1}^{\infty} \alpha_k \lambda_k < +\infty.$$

Therefore the set H_0 has the required properties.

Since for $x \in H_0$, $H_0 + x = H_0$, and for $x \notin H_0$, $(H_0 + x) \cap H_0 = \emptyset$, the function $\varphi = \chi_{H_0}$ is harmonic. $\square$

Chapter 5

Elliptic equations with variable coefficients

This chapter is devoted to studying the equation

$$\lambda\varphi(x) - \frac{1}{2}\text{Tr}[Q(x)D^2\varphi(x)] = g(x), \quad x \in H,$$

when the diffusion operator $Q(x) \in L(H)$ depends on x.

Here H is a separable Hilbert space, $g \in UC_b(H)$ and $\lambda > 0$. The idea is to exploit the results of Chapter 4, which concerns equations with constant coefficients, using perturbation results.

The first case, which we consider in §5.1, is when $Q(x)$ is a *small* perturbation to $Q(0)$. In §5.2 we allow "large" perturbation by using Schauder estimates .

5.1 Small perturbations

Let us first recall some classical results when dim $H = d < +\infty$. Consider the following equation:

$$\lambda\varphi(x) - \frac{1}{2}\text{Tr}[Q(x)D^2\varphi(x)] = g(x), \quad x \in H, \tag{5.1.1}$$

where $\lambda > 0$, $g \in UC_b(H)$, and $Q \in UC_b(H; H)$. Assume moreover that the ellipticity condition

$$\langle Q(x)\xi, \xi \rangle \geq \nu|\xi|^2, \ \forall x, \xi \in H,$$

holds for some $\nu > 0$. Now we set $Q(0) = Q$ and assume for simplicity that $Q = I$. Then we rewrite equation (5.1.1) in the form

$$\lambda\varphi(x) - \frac{1}{2}\Delta\varphi(x) - \frac{1}{2}\mathrm{Tr}[F(x)D^2\varphi(x)] = g(x), \ x \in H, \qquad (5.1.2)$$

where $F(x) = Q(x) - I$.

Moreover we set $\psi = \lambda\varphi - \frac{1}{2}\Delta\varphi$, so that equation (5.1.2) becomes

$$\psi - \gamma(\psi) = g, \qquad (5.1.3)$$

where

$$\gamma(\psi)(x) = \frac{1}{2}\mathrm{Tr}\left[F(x)D^2 R\left(\lambda, \frac{1}{2}\Delta\right)\psi(x)\right], \ x \in H. \qquad (5.1.4)$$

Notice that if $d > 1$ the mapping γ does not map $UC_b(H)$ into itself, since the domain of the Laplacian is not $UC_b^2(H)$, but for any $\theta \in (0, 1/2)$ it maps $C_b^\theta(H)$ into itself, provided $F \in C_b^\theta(H; H)$, in view of the Schauder estimates . Consequently if we assume that $F \in C_b^\theta(H; H)$ and that the norm $\|F\|_{UC_b^\theta(H;H)}$ is sufficiently small, we have that γ is a linear bounded application of $C_b^\theta(H; H)$ into itself with norm less than 1. By the Banach-Caccioppoli fixed point theorem it follows that equation (5.1.4) has a unique solution $\psi \in C_b^\theta(H)$. Consequently equation (5.1.1) has a unique solution $\varphi \in C_b^{2,\theta}(H)$.

Summarizing we can say that equation (5.1.1) has a unique solution provided $Q(x) - I$ is small in norm $C_b^\theta(H; H)$. It is well known that this condition (usually called the *Cordes condition*) implies that all the eigenvalues of $Q(x)$ are close to 1.

Let us try to repeat the above argument in infinite dimensions choosing $Q(x)$ of the following form: $Q(x) = Q^{1/2}(1 + F(x))Q^{1/2}$. We write equation (5.1.1) as

$$\lambda\varphi(x) - \Delta_Q\varphi(x) - \frac{1}{2}\mathrm{Tr}[F(x)D_Q^2\varphi(x)] = g(x), \ x \in H, \qquad (5.1.5)$$

where the operator Δ_Q is the infinitesimal generator of the heat semigroup defined by (3.5.3)-(3.5.4). Setting $\psi = \lambda\varphi - \frac{1}{2}\Delta\varphi$, (5.1.5) becomes

$$\psi - \gamma(\psi) = g, \qquad (5.1.6)$$

where

$$\gamma(\psi)(x) = \frac{1}{2}\mathrm{Tr}[F(x)D_Q^2 R(\lambda, \Delta_Q)\psi(x)], \ x \in H. \qquad (5.1.7)$$

Similarly to the finite dimensional case, we want to prove that γ maps $C_Q^\theta(H)$ into itself. But in this case it is not enough to assume that the norm $C_b^\theta(H; H)$ of F is small (for instance $F = \varepsilon I$, with ε small), because the operator $F(\cdot)D_Q^2 R(\lambda, \Delta_Q)\psi$ is not necessarily of trace class. Therefore we shall choose $F(x) \in L_1(H)$, $x \in H$, and such that its $L_1(H)$ norm is small.

Let us introduce some notation. We denote by $UC_b(H; L_1(H))$ the Banach space of all mappings $F : H \to L_1(H)$ that are uniformly continuous and bounded endowed with the norm

$$\|F\|_0 = \sup_{x \in H} \|F(x)\|_{L_1(H)}.$$

Moreover, for any $\theta \in (0, 1)$ we denote by $C_Q^\theta(H; L_1(H))$ the space of all mappings $F \in UC_b(H; L_1(H))$ such that

$$[F]_{\theta,Q} := \sup_{x,y \in H} \frac{\|F(x) - F(y)\|_{L_1(H)}}{|Q^{1/2}(x - y)|^\theta} < \infty.$$

$C_Q^\theta(H; L^1(H))$, endowed with the norm

$$\|F\|_{\theta,Q} = \|F\|_0 + [F]_{\theta,Q},$$

is a Banach space.

Theorem 5.1.1 *Let* $\theta \in (0, 1)$, $\lambda > 0$, $g \in C_Q^\theta(H)$ *and* $F \in C_Q^\theta(H; L_1(H))$ *be such that* $I + F(x) \in L^+(H)$ *for all* $x \in H$. *There exists* $\varepsilon_0 > 0$ *such that if* $\|F\|_{\theta,Q} \le \varepsilon_0$, *then there is a unique* $\varphi \in C_Q^{2+\theta}(H) \cap D(\Delta_Q)$ *such that*

$$\lambda\varphi(x) - \Delta_Q\varphi(x) - \frac{1}{2}\mathrm{Tr}[F(x)D_Q^2\varphi(x)] = g(x), \quad x \in H. \qquad (5.1.8)$$

Proof. As we have noticed before, equation (5.1.8) can be written in the form (5.1.6). By Theorem 4.2.3 one can easily check that $\gamma \in L(C_Q^\theta(H))$ and

$$\|\gamma\|_{L(UC_Q^\theta(H))} \le C_{\theta,\lambda}\|F\|_{\theta,Q}.$$

Therefore the conclusion follows again from the Banach-Caccioppoli fixed point theorem. $\square$

Remark 5.1.2 By a result due to L. Zambotti [223], it follows that, under the assumptions of Theorem 5.1.1, there exists a unique martingale solution to the differential stochastic equation

$$dX = \sqrt{Q^{1/2}(1 + F(X))Q^{1/2}}\ dW(t), \quad X(0) = x \in H,$$

where W is a cylindrical Wiener process taking values on H.

Note that the operator

$$L\varphi(x) = \frac{1}{2}\text{Tr}[(1 + F(x)D_Q^2\varphi(x)], \ x \in H,$$

is a natural candidate for the generator of the solution.

5.2 Large perturbations

Here we are looking for a solution $\varphi \in C_Q^{2+\theta}(H) \cap D(\Delta_Q)$ of equation (5.1.5) under the stronger assumption that $F \in C_b^\theta(H; L^1(H))$ but without assuming that the norm of F is small. We shall extend to infinite dimensions a well known procedure, see for instance A. Lunardi [162], based on the maximum principle, localization and the continuity method. We shall follow P. Cannarsa and G. Da Prato [31].

We need a last notation. We denote by $C_b^\theta(H, L^1(H))$ the Banach space of all mappings $F \in UC_b(H; L^1(H))$ such that

$$[F]_\theta := \sup_{x,y \in H} \frac{\|F(x) - F(y)\|_{L^1(H)}}{|x - y|^\theta} < \infty.$$

If $F \in C_Q^\theta(H, L^1(H))$, we set

$$\|F\|_\theta = \|F\|_0 + [F]_\theta.$$

The main result of this section is the following, see [31].

Theorem 5.2.1 *Let $\theta \in (0,1)$, $\lambda > 0$, $g \in C_Q^\theta(H)$, and $F \in C_b^\theta(H; L_1(H))$ be such that $I + F(x) \in L^+(H)$ for all $x \in H$. Then there exists a unique solution φ to the equation (5.1.5).*

The proof is split into three parts: uniqueness which is a consequence of the maximum principle (Proposition 5.2.2) , localization (Proposition 5.2.3) and a classical continuity argument.

Proposition 5.2.2 *Let $\varphi \in D(\Delta_Q) \cap C_Q^{2+\theta}(H)$ be a solution of (5.1.5). Then we have*

$$\|\varphi\|_0 \le \frac{1}{\lambda} \|g\|_0. \tag{5.2.1}$$

Proof. It is enough to prove the proposition when $\varphi \geq 0$. Recall that $\Delta_{0,Q}$ is the linear operator defined by (3.5.3)-(3.5.4).

Step 1. If $\varphi \in D(\Delta_{0,Q}) \cap C_Q^{2+\theta}(H)$ then (5.2.1) holds.

Let $\varphi \in D(\Delta_{0,Q}) \cap C_Q^{2+\theta}(H)$. By Lemma 3.2.6 for any $\varepsilon > 0$ there exists $\varphi_\varepsilon \in UC_b(H)$ such that

  (i) φ_ε attains a maximum in H,

  (ii) $\varphi - \varphi_\varepsilon \in D(\Delta_{0,Q})$ and $\|\Delta_{0,Q}(\varphi - \varphi_\varepsilon)\|_0 \leq \varepsilon$.

Then we have

$$\lambda \varphi_\varepsilon - \frac{1}{2}\mathrm{Tr}[(1 + F(x))D_Q^2\varphi_\varepsilon] = g + \lambda(\varphi_\varepsilon - \varphi)$$

$$-\frac{1}{2}\mathrm{Tr}[(1 + F(x))D_Q^2(\varphi_\varepsilon - \varphi)].$$

If x_ε is a maximum point of φ_ε, we have clearly

$$\mathrm{Tr}[(1 + F(x))\, D_Q^2\varphi_\varepsilon(x_\varepsilon)] = \sum_{k=1}^\infty \lambda_k D_k^2\varphi_\varepsilon(x_\varepsilon) \leq 0,$$

where (λ_k) are the eigenvalues of $Q^{1/2}(1 + F(x))Q^{1/2}$, and so

$$\frac{1}{2}\mathrm{Tr}[(1 + F(x))D_Q^2\varphi_\varepsilon(x_\varepsilon)] \leq 0.$$

Consequently $\lambda\|\varphi_\varepsilon\|_0 \leq \|g\|_0 + o(\varepsilon)$, and the conclusion follows letting ε tend to zero.

Step 2. General case.

Given $\varphi \in D(\Delta_Q) \cap C_Q^{2+\theta}(H)$ we set $\psi = \varphi - \Delta_Q\varphi$. Let $\varepsilon \in (0, \theta)$, and let $(\psi_n) \subset UC_b^1(H)$ be such that

$$\lim_{n\to\infty} \psi_n = \psi \text{ in } C_b^{\theta-\varepsilon}(H),$$

and set $\varphi_n = R(\lambda, \Delta_Q)\psi_n$. Then by Theorem 4.1.2 we have $\varphi_n \in D(\Delta_{0,Q})$. Moreover

$$\lim_{n\to\infty} \varphi_n = \varphi \text{ in } C_Q^{2+\theta-\varepsilon}(H),$$

by Theorem 4.2.3. Now

$$\lambda\varphi_n - \frac{1}{2}\mathrm{Tr}[(1+F(\cdot))D_Q^2\varphi_n] = g + \lambda(\varphi_n - \varphi)$$

$$-\frac{1}{2}\mathrm{Tr}[(1+F(\cdot))D_Q^2(\varphi_n - \varphi)].$$

By Step 1 we have

$$\|\varphi_n\|_0 \le \frac{1}{\lambda}(\|g\|_0 + o(n))$$

and the conclusion follows letting n tend to infinity. $\square$

Let us prove an a priori estimate by localization.

Proposition 5.2.3 *Let $\lambda > 0$ and $\theta \in (0,1)$. There exists a constant $C = C(\lambda, \theta) > 0$ such that if $g \in C_Q^\theta(H)$ and $\varphi \in C_Q^{2+\theta}(H)$ is a solution to equation (5.1.5), we have*

$$\|\varphi\|_{2+\theta,Q} \le C\|g\|_{\theta,Q}. \tag{5.2.2}$$

Proof. Let $\alpha \in C_b^\infty([0,+\infty); [0,+\infty))$ be such that $0 \le \alpha \le 1$, $\alpha(r) = 1$ if $0 \le r \le 1$, and $\alpha(r) = 0$ if $r \ge 2$. Set

$$\rho_{x,r}(z) = \alpha\left(\frac{|x-z|}{r}\right), \quad x, z \in H, \ r \in [0,1].$$

Then there exist $L_1, L_2 > 0$ such that

$$|D\rho_{x,r}(z)| \le \frac{L_1}{r}, \quad |D^2\rho_{x,r}(z)| \le \frac{L_2}{r^2}, \quad x, z \in H, \ r > 0.$$

Now for any $x \in H$ we set $\psi = \rho_{x,r}\varphi$ and look for the equation fulfilled by ψ. We have

$$D_Q^2\psi = \varphi D_Q^2\rho_{x,r} + \rho_{x,r}D_Q^2\varphi + 2D_Q\rho_{x,r} \otimes D_Q\varphi,$$

which implies

$$\mathrm{Tr}[(1+F(x))D_Q^2\psi] = \varphi\,\mathrm{Tr}[(1+F(x))D_Q^2\rho_{x,r}]$$

$$+\rho_{x,r}\,\mathrm{Tr}[(1+F(x))D_Q^2\varphi] + 2\langle(1+F(x))D_Q\rho_{x,r}, D_Q\varphi\rangle.$$

It follows that

$$\lambda\psi - \frac{1}{2}\mathrm{Tr}[(1+F(x_0))D_Q^2\psi] = g_1 + g_2, \tag{5.2.3}$$

where

$$g_1 = \frac{1}{2}\mathrm{Tr}[(F(x) - F(x_0))D_Q^2\psi],$$

and

$$g_2 = 2\langle(1 + F(x))D_Q\rho_{x,r}, D_Q\varphi\rangle + \varphi\,\mathrm{Tr}[(1 + F(x))D_Q^2\rho_{x,r}] + \rho_{x,r}g.$$

We first estimate g_1. In the proof we shall denote by C_k, $k \in \mathbb{N}$, different constants depending only on λ, Q, F. We recall that, by the maximum principle (Proposition 5.2.2), we have $\|\varphi\|_0 \leq C_1\|g\|_0$. Moreover

$$\|g_1\|_0 \leq C_2\|\psi\|_{2,Q}, \tag{5.2.4}$$

and, recalling that g_1 vanishes outside $B(x_0, 2r)$, it follows that

$$[g_1]_{\theta,Q} \leq \frac{1}{2}[F]_\theta\|\psi\|_{2,Q}$$

$$+ \frac{1}{2}\|\psi\|_{2+\theta,Q}\sup_{x\in B(x_0,2r)}\|F(x) - F(x_0)\|_{L_1(H)}.$$

Now, taking into account that $F \in C_b^\theta(H; L_1(H))$ we find

$$\|g_1\|_{\theta,Q} \leq C_3(r^\theta\|\psi\|_{2+\theta,Q} + \|\psi\|_{2,Q}). \tag{5.2.5}$$

Moreover, by the interpolatory estimate (3.3.2) and by (5.2.5) we find that

$$\|g_1\|_{\theta,Q} \leq C_4(r^\theta\|\psi\|_{2+\theta,Q} + K\|\psi\|_0^{\frac{2}{2+\theta}}\|\psi\|_{2+\theta,Q}^{\frac{\theta}{2+\theta}}).$$

From Young's inequality, ([1]) we get

$$\|g_1\|_{\theta,Q} \leq C_5(r^\theta\|\psi\|_{2+\theta,Q} + r^{-\theta^2/2}\|\psi\|_0).$$

Using the maximum principle again we end up with the estimate

$$\|g_1\|_{\theta,Q} \leq C_6(r^\theta\|\psi\|_{2+\theta,Q} + r^{-\theta^2/2}\|g\|_0). \tag{5.2.6}$$

As far as the estimate of g_2 is concerned, we have

$$\|g_2\|_{\theta,Q} \leq C_7\left(\frac{1}{r}\|\varphi\|_{1+\theta,Q} + \frac{1}{r^2+\theta}\|g\|_0 + \frac{1}{r^\theta}\|g\|_{\theta,Q}\right). \tag{5.2.7}$$

---

[1]If $x, y > 0$, $p, q > 1$, and $1/p + 1/q = 1$, we have $xy \leq \frac{x^p}{p} + \frac{y^q}{q}$.

In conclusion, from (5.2.5), (5.2.7), and the Schauder estimates we find

$$\|\psi\|_{2+\theta,Q} \le c\|g_1 + g_2\|_{\theta,Q}$$

$$\le C_8\left(r^\theta\|\psi\|_{2+\theta,Q} + r^{-\theta^2/2}\|g\|_0 + \frac{1}{r}\|\varphi\|_{1+\theta,Q} + \frac{1}{r^2+\theta}\|g\|_0 + \frac{1}{r^\theta}\|g\|_{\theta,Q}\right).$$

$$(5.2.8)$$

We fix now once and for all $r \in (0,1]$ such that $C_8 r^\theta < \frac{1}{2}$. From (5.2.8) we find

$$\|\psi\|_{2+\theta,Q} \le C_9(\|\varphi\|_{1+\theta,Q} + \|g\|_{\theta,Q}),$$

which implies

$$\|\varphi\|_{C_Q^{2+\theta}(B(x_0,r))} \le C_9(\|\varphi\|_{1+\theta,Q} + \|g\|_{\theta,Q}),$$

and, due to the arbitrariness of x_0,

$$\|\varphi\|_{2+\theta,Q} \le C_9(\|\varphi\|_{1+\theta,Q} + \|g\|_{\theta,Q}).$$

Using once again an interpolatory estimate, we obtain

$$\|\varphi\|_{2+\theta,Q} \le C_{10}(\|\varphi\|_0^{\frac{1+\theta}{2+\theta}} \|\varphi\|_{2+\theta,Q}^{\frac{1}{2+\theta}} + \|g\|_{\theta,Q}).$$

By Young's estimate, we find that for each $\varepsilon > 0$ there exists a constant $C(\varepsilon) > 0$ such that

$$\|\varphi\|_{2+\theta,Q} \le C_{11}(\varepsilon\|\varphi\|_{2+\theta,Q} + C(\varepsilon)\|g\|_{\theta,Q}).$$

Choosing finally ε such that $\varepsilon C_{11} < \frac{1}{2}$, we have

$$\|\varphi\|_{2+\theta,Q} \le C_{12}\|g\|_{\theta,Q}. \quad \square$$

Proof of Theorem 5.2.1. We use the classical continuity argument. Fixing $\lambda > 0$, we want to prove that for any $g \in C_Q^\theta(H)$, $\theta \in (0,1)$, there exists a unique solution φ to the equation (5.1.8).

For any $\alpha \in [0,1]$, we consider the equation

$$\lambda\varphi - \Delta_Q\varphi - \frac{\alpha}{2}\operatorname{Tr}[F(x)D_Q^2\varphi] = g, \qquad (5.2.9)$$

and we set

$$\Lambda = \{\alpha \in [0,1] : (5.2.9) \text{ has a unique solution in } C_Q^{2+\theta}(H)\}.$$

We now prove that the set Λ is open and closed in $[0,1]$, which will imply $\Lambda = [0,1]$. In fact that Λ is open in $[0,1]$ follows from Theorem 5.1.1. Let us prove that Λ is closed.

Let $(\alpha_n) \subset \Lambda$ be convergent to $\bar{\alpha}$, and set $\varphi_n = T_{\alpha_n} g$, so that

$$\lambda u \varphi_n - \Delta_Q \varphi_n - \frac{\alpha_n}{2} \operatorname{Tr}[F(x) D_Q^2 \varphi_n] = g. \tag{5.2.10}$$

Then we have

$$\lambda(\varphi_n - \varphi_m) \quad - \quad \Delta_Q(\varphi_n - \varphi_m) - \frac{\alpha_m}{2} \operatorname{Tr}[F(x) D_Q^2(\varphi_n - \varphi_m)]$$

$$= \quad \frac{\alpha_m - \alpha_n}{2} \operatorname{Tr}[F(x) D_Q^2 \varphi_m].$$

By Theorem 4.2.3 it follows that

$$\|\varphi_n - \varphi_m\|_{2+\theta,Q} \le \frac{C}{2} |\alpha_m - \alpha_n| \|F\|_{\theta} \|\varphi_n\|_{2+\theta,Q},$$

and also

$$\|\varphi_n - \varphi_m\|_{2+\theta,Q} \le \frac{C^2}{2} |\alpha_m - \alpha_n| \|F\|_{\theta} \|g\|_{\theta,Q}.$$

Consequently, there exists $\varphi \in C_Q^{2+Q}(H)$ such that $\varphi_n \to \varphi$ in $UC_Q^{2+Q}(H)$. Letting n tend to infinity in (5.2.10), we find that φ is a solution to (5.2.9) with $\alpha = \bar{\alpha}$ and the conclusion follows. $\square$

Chapter 6

Ornstein-Uhlenbeck equations

In this chapter we consider the following parabolic equation on a separable Hilbert space H: ([1])

$$\begin{cases} D_t u(t,x) & = \frac{1}{2}\mathrm{Tr}[QD^2 u(t,x)] + \langle Ax, Du(t,x) \rangle, & t > 0, x \in D(A), \\ u(0,x) & = \varphi(x), & x \in H, \ \varphi \in UC_b(H). \end{cases}$$

(6.0.1)

Here $A : D(A) \subset H \to H$ is the infinitesimal generator of a C_0 semigroup e^{tA}, $t \geq 0$, and $Q \in L^+(H)$.

Our basic assumption is

$$Q_t \in L_1(H), \quad t > 0, \tag{6.0.2}$$

where

$$Q_t x = \int_0^t e^{sA} Q e^{sA^*} x \, ds, \quad x \in H, \ t \geq 0. \tag{6.0.3}$$

This assumption is needed in order that the problem is well posed, see §6.1 below. Notice that, unlike the case $A = 0$ studied in Chapter 3, the choice $Q = I$ is allowed for a suitable A; it is enough that $\int_0^t \mathrm{Tr}[e^{sA} e^{sA^*}] \, ds < +\infty$.

Of the sections, §6.1 and §6.2 are devoted to existence and uniqueness of strict and classical solutions respectively, §6.3 to properties of the semigroup corresponding to the evolution equation (6.0.1), and §6.4 to the elliptic equation

$$\lambda \varphi(x) - \frac{1}{2}\mathrm{Tr}[QD^2\varphi(x)] + \langle Ax, D\varphi(x) \rangle = g(x), \quad t > 0, \ x \in D(A), \quad (6.0.4)$$

---

[1]We set everywhere $D_x = D$.

where $\lambda > 0$ and $g \in UC_b(H)$. Finally §6.5 and §6.6 are devoted to the following perturbed equations:

$$\begin{cases} D_t u(t,x) = \frac{1}{2}\mathrm{Tr}[QD^2 u(t,x)] + \langle Ax + F(x), Du(t,x) \rangle, \ t > 0, \ x \in D(A), \\ u(0,x) = \varphi(x), \quad x \in H, \end{cases}$$

$$(6.0.5)$$

and

$$\lambda \varphi(x) - \frac{1}{2}\mathrm{Tr}[QD^2\varphi(x)] + \langle Ax + F(x), D\varphi(x) \rangle = g(x), \quad t > 0, \ x \in D(A),$$

$$(6.0.6)$$

where $\varphi, g \in UC_b(H)$ and $F : H \to H$ is continuous and bounded.

6.1   Existence and uniqueness of strict solutions

We are here concerned with the parabolic equation (6.0.1) under the basic assumption (6.0.2).

In order to study (6.0.1) it is useful to consider an auxiliary problem, see Yu. Daleckij and S. V. Fomin [63] and P. Cannarsa and G. Da Prato [31]:

$$\begin{cases} D_t v(t,x) &= \frac{1}{2}\mathrm{Tr}[e^{tA}Qe^{tA^*}D^2 v(t,x)], \ t > 0, \quad x \in H, \\ v(0,x) &= \varphi(x), \quad x \in H, \end{cases} \qquad (6.1.1)$$

where $\varphi \in UC_b(H)$. As we shall see, the solutions of problems (6.0.1) and (6.1.1) are related by the formula

$$u(t,x) = v(t, e^{tA}x), \quad t \geq 0, \ x \in H. \qquad (6.1.2)$$

We will start by proving that, if φ is sufficiently regular and a natural additional condition is fulfilled, then there exists a unique strict solution of (6.0.1).

A function $u(t,x)$, $t \geq 0$, $x \in H$, is said to be a *strict solution* to equation (6.0.1) if

(i) u is continuous on $[0, +\infty) \times H$ and $u(0, \cdot) = \varphi$,

(ii) $u(t, \cdot) \in UC_b^2(H)$ for all $t \geq 0$, and $QD^2 u(t,x) \in L_1(H)$ for all $x \in H$ and $t \geq 0$,

(iii) for any $x \in D(A)$, $u(\cdot, x)$ is continuously differentiable on $[0, +\infty)$, and fulfills (6.0.1).

Similarly, a function $v(t, x)$, $t \geq 0$, $x \in H$, is said to be a *strict solution* to (6.1.1) if

(i) v is continuous on $[0, +\infty) \times H$ and $v(0, \cdot) = \varphi$,

(ii) $v(t, \cdot) \in UC_b^2(H)$ for all $t \geq 0$, and $e^{tA} Q e^{tA^*} D^2 u(t, x) \in L_1(H)$, for all $x \in H$ and $t > 0$,

(iii) for any $x \in H$, $u(\cdot, x)$ is continuously differentiable on $(0, +\infty)$, and fulfills (6.1.1) for $t > 0$.

Let us first consider the auxiliary problem (6.1.1). It is similar to problem (3.1.6). Thus we will proceed as in Chapter 3. If H is finite dimensional then one can easily solve problem (6.1.1). It is in fact a parabolic equation with coefficient depending on time but not on space. We find, after some straightforward calculations,

$$v(t, x) = \int_H \varphi(x + y) N_{Q_t}(dy), \quad x \in H, \ t \geq 0, \tag{6.1.3}$$

where Q_t is defined by (6.0.3). This formula justifies our basic assumption (6.0.2).

We say that $v(t, x)$, defined by (6.1.3), is a generalized solution to (6.1.1). The following result can be proved as Theorems 3.2.3 and 3.2.7.

Lemma 6.1.1 *If $\varphi \in UC_b^2(H)$, then problem (6.1.1) has a unique strict solution v given by formula (6.1.3).*

We now prove the result.

Theorem 6.1.2 *Assume that assumption (6.0.2) holds. Let $\varphi \in UC_b^2(H)$ be such that $Q D^2 \varphi \in UC_b(H; L_1(H))$. Then problem (6.0.1) has a unique strict solution u given by the formula*

$$u(t, x) = \int_H \varphi(e^{tA} x + y) N_{Q_t}(dy), \quad t \geq 0, \ x \in H. \tag{6.1.4}$$

Proof.
Existence.

We check that u, given by the formula (6.1.4), is a strict solution of (6.0.1). First it is clear that u has the required regularity properties. Thus we have only to show that, if $x \in D(A)$, then (6.0.1) is fulfilled; but this follows from a straightforward verification.

Uniqueness.

We assume here that u is a strict solution of (6.0.1), and prove that it is given by formula (6.1.4).

We proceed in two steps. First we assume that A is bounded. Then setting $v(t, x) = u(t, e^{-tA}x)$, we see that v is a solution of (6.1.1). Thus by the uniqueness v is given by (6.1.3) and consequently u by (6.1.4).

Consider now the case of a general operator A. We start from the obvious identity,

$$
\begin{cases}
D_t u(t, x) &= \frac{1}{2}\mathrm{Tr}[QD^2 u(t, x)] + \langle A_n x, Du(t, x)\rangle \\
&\quad + \langle (A - A_n)x, Du(t, x)\rangle, \\
u(0, x) &= \varphi(x), \ x \in H, \ \varphi \in UC_b(H),
\end{cases}
$$

where $A_n = nAR(n, A)$, $n \in \mathbb{N}$, are the Yosida approximations of A. It is not difficult to show that u is given by the variation of constants formula,

$$
u(t, x) = \int_H \varphi(e^{tA_n}x + y)N_{Q_t^n}(dy)
$$

$$
+ \int_0^t \int_H \langle (A - A_n)e^{(t-s)A_n}x, Du(e^{(t-s)A_n}x + y)\rangle N_{Q_{t-s}^n}(dy),
$$

where

$$
Q_t^n x = \int_0^t e^{sA_n}Qe^{sA_n^*}x \, ds, \ x \in H, \ t \geq 0, \ n \in \mathbb{N}.
$$

Now the conclusion follows letting n tend to infinity. $\square$

Remark 6.1.3 If $Q \in L_1^+(H)$ the conclusions of the theorem hold assuming only that $\varphi \in UC_b^2(H)$.

Remark 6.1.4 It is interesting to notice that (6.0.1) is the Kolmogorov equation corresponding to the linear stochastic equation

$$
dX = AXdt + Q^{1/2}dW(t), \quad X(0) = x, \tag{6.1.5}
$$

where W is a cylindrical Wiener process on H, see G. Da Prato and J. Zabczyk [101] and Chapter 7 below.

Let us recall that, in virtue of assumption (6.1.2), equation (6.1.5) has a unique mild solution given by

$$
X(t, x) = e^{tA}x + \int_0^t e^{(t-s)A} \, dW(s), \quad t \geq 0. \tag{6.1.6}
$$

The corresponding transition semigroup, called the *Ornstein-Uhlenbeck* semi-group, is defined as

$$R_t\varphi(x) = \int_H \varphi(e^{tA}x + y)N_{Q_t}(dy), \qquad x \in H, \ t \geq 0, \tag{6.1.7}$$

where Q_t is given by (6.0.3).

6.2 Classical solutions

We assume here that (6.0.2) holds and, given $\varphi \in B_b(H)$, we consider the generalized solution u,

$$u(t, x) = \int_H \varphi(e^{tA}x + y)N_{Q_t}(dy), \qquad x \in H, \ t \geq 0, \tag{6.2.1}$$

of equation (6.0.1).

It is convenient to write problem (6.0.1) in the following form:

$$\begin{cases} D_t u(t, x) &= \frac{1}{2}\text{Tr}[QD^2 u(t, x)] + \langle x, A^* Du(t, x)\rangle, \quad t > 0, \ x \in H, \\ u(0, x) &= \varphi(x), \quad x \in H, \end{cases} \tag{6.2.2}$$

where A^* is the adjoint operator of A.

As in Chapter 3, it is not true in general that a generalized solution $u(t, x)$ is Fréchet differentiable for $t > 0$, but it would be possible to prove differentiability in special directions as in §3.3. However, we limit ourselves to considering a class of equations for which generalized solutions are Fréchet differentiable for $t > 0$, for *each* bounded Borel initial function φ. ([2])

A function $u(t, x)$, $t \geq 0$, $x \in H$, is said to be a *classical solution* to (6.2.2) if

(i) u is continuous on $[0, +\infty) \times H$ and $u(0, \cdot) = \varphi$,

(ii) $u(t, \cdot) \in UC_b^2(H)$, for all $t > 0$, and $QD^2 u(t, x) \in L_1(H)$, for all $x \in H$, $t > 0$,

(iii) $Du(t, x) \in D(A^*)$ for all $x \in H$, $t > 0$,

(iv) for any $x \in H$ $u(\cdot, x)$ is continuously differentiable on $(0, +\infty)$, and fulfills (6.2.2).

---

[2]We denote by $B_b(H)$ the set of all Borel mappings from H into $\mathbb{R}$.

Classical solutions will exist under the following assumption, called the *controllability condition*, see J. Zabczyk [215]:

$$e^{tA}(H) \subset Q_t^{1/2}(H), \quad \text{for all } t > 0. \tag{6.2.3}$$

We will assume (6.2.3) from now on and we will define

$$\Lambda_t = Q_t^{-1/2} e^{tA}, \quad t > 0, \tag{6.2.4}$$

where $Q_t^{-1/2}$ denotes the pseudo-inverse of $Q_t^{1/2}$. It follows from the closed graph theorem that Λ_t is a bounded operator in H, for all $t > 0$.

Remark 6.2.1 Hypothesis (6.2.3) is equivalent to *null controllability*, see Appendix B, of the deterministic system

$$\xi' = A\xi + Q^{1/2}u, \quad \xi(0) = \xi_0. \tag{6.2.5}$$

If hypothesis (6.2.3) holds, then for any $t > 0$ there exists a constant $c_t > 0$ such that

$$Q_t \geq c_t e^{tA} e^{tA^*}. \tag{6.2.6}$$

Since Q_t is of trace class, this implies that $e^{tA} \in L_2(H)$, for all $t > 0$. Also we have

$$e^{sA}(H) \subset Q_t^{1/2}(H), \quad s < t. \tag{6.2.7}$$

For the proof of all these facts see Appendix B.

First we study regularity properties of the generalized solution , and then we show existence and uniqueness of a classical solution.

The controllability condition is closely related to the regularity of the generalized solution and of the strong Feller property for the transition semigroup (R_t), see G. Da Prato and J. Zabczyk [101].

Theorem 6.2.2 *Assume that (6.0.2) and (6.2.3) hold. Then for any $\varphi \in B_b(H)$ and any $t > 0$ we have $R_t\varphi \in UC_b^\infty(H)$.*

In particular for any $g, h, x \in H$, we have

$$\langle DR_t\varphi(x), g \rangle = \int_H \langle \Lambda_t g, Q_t^{-1/2} y \rangle \varphi(e^{tA} x + y) N_{Q_t}(dy), \tag{6.2.8}$$

and

$$\langle D^2 R_t\varphi(x)g, h\rangle$$

$$= \int_H [\langle \Lambda_t g, Q_t^{-1/2}y\rangle\langle\Lambda_t h, Q_t^{-1/2}y\rangle - \langle\Lambda_t g, \Lambda_t h\rangle]\varphi(e^{tA}x + y)N_{Q_t}(dy).$$

$$(6.2.9)$$

Moreover the following estimates hold:

$$|DR_t\varphi(x)| \leq \|\Lambda_t\|\,\|\varphi\|_0, \quad t > 0,\ x \in H, \qquad (6.2.10)$$

$$\|D^2 R_t\varphi(x)\| \leq \sqrt{2}\,\|\Lambda_t\|^2\,\|\varphi\|_0, \quad t > 0,\ x \in H. \qquad (6.2.11)$$

If conversely $R_t\varphi \in C_b(H)$ for any $\varphi \in B_b(H)$ and any $t > 0$, then assumption (6.2.3) holds.

The first statement of the theorem implies that (R_t) is *strong Feller*.
Proof of Theorem 6.2.2. Assume that (6.2.3) holds. Let $t > 0$, $\varphi \in B_b(H)$ and $x \in H$. Since $e^{tA}x \in Q_t^{1/2}(H)$, the measures $N_{e^{tA}x,Q_t}$ and N_{Q_t} are equivalent and the corresponding density $\rho_t(x,\cdot)$ is given by the Cameron-Martin formula (see Theorem 1.3.6):

$$\frac{dN_{e^{tA}x,Q_t}}{dN_{Q_t}}(y) = \rho_t(x,y), \quad y \in H,$$

where

$$\rho_t(x,y) = e^{\langle\Lambda_t x, Q_t^{-1/2}y\rangle - \frac{1}{2}|\Lambda_t x|^2}, \quad y \in H.$$

Therefore we have

$$R_t\varphi(x) = \int_H \varphi(y)e^{\langle\Lambda_t x, Q_t^{-1/2}y\rangle - \frac{1}{2}|\Lambda_t x|^2}N_{Q_t}(dy).$$

Now, proceding as in the proof of Theorem 3.3.3, we see that $R_t\varphi(x)$ is differentiable an arbitrary number of times. In particular, for $g, h \in H$, (6.2.8) and (6.2.9) follow. Let us prove (6.2.10). Using the Hölder inequality we find from (6.2.8) that

$$|\langle DR_t\varphi(x), g\rangle|^2 \leq \|\varphi\|_0^2 \int_H |\langle\Lambda_t g, Q_t^{-1/2}y\rangle|^2 N_{Q_t}(dy) = \|\varphi\|_0^2|\Lambda_t g|^2.$$

Therefore (6.2.10) follows; (6.2.11) can be seen similarly.

Let us prove the last statement. Assume in contradiction that for arbitrary $\varphi \in B_b(H)$, $R_t\varphi$ is a continuous function but nevertheless for some

$x_0 \in H$, $e^{tA}x_0 \notin Q_t^{1/2}(H)$. Then, for all $n \in \mathbb{N}$, the measures $N_{\frac{1}{n}e^{tA}x_0, Q_t}$ and N_{Q_t} are singular by the Cameron-Martin theorem. Consequently, for arbitrary $n \in \mathbb{N}$ there exists a Borel set $K_n \subset H$ such that

$$N_{\frac{1}{n}e^{tA}x_0, Q_t}(K_n) = 0, \quad N_{Q_t}(K_n) = 1,$$

and if $K = \bigcap_{n=1}^{\infty} K_n$, then

$$N_{\frac{1}{n}e^{tA}x_0, Q_t}(K) = 0, \quad N_{Q_t}(K) = 1.$$

If $\varphi = \chi_K$, then $R_t\varphi(\frac{x_0}{n}) = 0$, $n \in \mathbb{N}$ and $R_t\varphi(0) = 1$. Therefore the function $R_t\varphi$ is not continuous at 0. $\square$

Exercise 6.2.3 Prove that for all $n \in \mathbb{N}$ there exists a constant $C_n > 0$ such that

$$\|D^n R_t\varphi(x)\| \le C_n \|\Lambda_t\|^n \|\varphi\|_0, \; n \in \mathbb{N}, \; t \ge 0, \; x \in H. \tag{6.2.12}$$

Our aim now is to prove the following existence and uniqueness result.

Theorem 6.2.4 *Assume that* (6.0.2) *and* (6.2.3) *hold, that the operator* $\Lambda_t A$ *has a continuous extension to* H *and* $Q^{1/2}\Lambda_t \in L_4(H)$, $t > 0$. *Then problem* (6.2.2) *has a unique classical solution.*

Proof. Existence will follow from Propositions 6.2.5 and 6.2.8 below and uniqueness from Theorem 6.1.2 since the restriction of a classical solution to $[\varepsilon, +\infty) \times H$ is clearly a strict solution on $[\varepsilon, +\infty)$ for any $\varepsilon > 0$. $\square$

We show first that, under appropriate assumptions, $DR_t\varphi(x) \in D(A^*)$ for any $t > 0$ and any $\varphi \in B_b(H)$.

Proposition 6.2.5 *Assume that* (6.0.2) *and* (6.2.3) *hold and moreover that the operator* $\Lambda_t A$ *has a continuous extension* $\overline{\Lambda_t A}$ *to* H, $\forall t > 0$. *If* $\varphi \in B_b(H)$ *and* $t > 0$ *then* $DR_t\varphi(x) \in D(A^*)$ *for all* $x \in H$, *and it results that*

$$\langle A^* DR_t\varphi(x), h \rangle = \int_H \langle \overline{\Lambda_t A}\, h, Q_t^{-1/2}y \rangle \varphi(e^{tA}x + y) N_{Q_t}(dy), \; h \in H. \tag{6.2.13}$$

Moreover

$$\|A^* DR_t\varphi\|_0 \le \|\overline{\Lambda_t A}\| \, \|\varphi\|_0 \tag{6.2.14}$$

Proof. The conclusion follows easily from (6.2.8). $\square$

Remark 6.2.6 If e^{tA} is a differentiable semigroup then $\Lambda_t A$ is closable and its closure is given by

$$\overline{\Lambda_t A} = Q_t^{-1/2} e^{t/2A} A e^{t/2A}.$$

Thus $\overline{\Lambda_t A}$ is bounded in view of (6.2.7).

To show that, under a suitable assumption, $D^2 R_t \varphi(x)$ is of a trace class for any $t > 0$, we need a lemma due to J. Zabczyk, see [220].

Lemma 6.2.7 *Let* $\mu = N_R$ *be a Gaussian measure on* $(H, \mathcal{B}(H))$ *and let* $\psi \in B_b(H)$. *Define a linear bounded operator* G_ψ *in* H *by*

$$\langle G_\psi \alpha, \beta \rangle = \int_H [\langle \alpha, R^{-1/2} y \rangle \langle \beta, R^{-1/2} y \rangle - \langle \alpha, \beta \rangle] \psi(y) \mu(dy), \quad \alpha, \beta \in H. \tag{6.2.15}$$

Then $G \in L_2(H)$ *and*

$$\|G\|_{L_2(H)} \le 2\|\varphi\|_0. \tag{6.2.16}$$

Proof. Let $(e_k), (\lambda_k)$ be eigensequences relative to Q, and set $y_h = \langle y, e_h \rangle$, $y \in H$ and $h \in \mathbb{N}$. Then for any $h, k \in \mathbb{N}$ we have

$$\langle Ge_h, e_k \rangle = \begin{cases} \sqrt{2} \, \langle \varphi, \eta_h \rangle_{L^2(H,\mu)} & \text{if } h = k, \\ \\ \langle \varphi, \zeta_{h,k} \rangle_{L^2(H,\mu)} & \text{if } h \ne k, \end{cases}$$

where

$$\eta_h(y) = 2^{-1/2} \left[\lambda_h^{-1} y_h^2 - 1 \right], \quad \zeta_{h,k}(y) = (\lambda_h \lambda_k)^{-1/2} y_h y_k, \quad y \in H.$$

By a direct verification we can see that the system of functions on $L^2(H, \mu)$, η_h, $\zeta_{h,k}$, $h, k \in \mathbb{N}$, is orthonormal. Therefore, from the Parseval equality it follows that

$$\sum_{h,k=1}^{\infty} |\langle Ge_h, e_k \rangle|^2 \le 2 \sum_{h=1}^{\infty} |\langle \varphi, \eta_h \rangle_{L^2(H,\mu)}|^2 + \sum_{h,k=1}^{\infty} |\langle \varphi, \zeta_{h,k} \rangle_{L^2(H,\mu)}|^2$$

$$\le 2\|\varphi\|_0^2,$$

and the conclusion follows. $\square$

We are now in a position to prove, following J. Zabczyk [220], the result,

Proposition 6.2.8 *Assume that (6.0.2) and (6.2.3) hold and moreover that $Q^{1/2}\Lambda_t \in L_4(H)$, $t > 0$. If $\varphi \in B_b(H)$ and $t > 0$ then $QD^2R_t\varphi(x) \in L_1(H)$ and*

$$\mathrm{Tr}[QD^2R_t\varphi(x)] = (Q^{1/2}\Lambda_t)^*G_{\varphi(e^{tA}x+\cdot)}Q^{1/2}\Lambda_t, \tag{6.2.17}$$

where G is defined in Lemma 6.2.7. Moreover

$$\|QD^2R_t\varphi(x)\|_{L^1(H)} \leq 2\|Q^{1/2}\Lambda_t\|^2_{L_4(H)}\|\varphi\|_0, \ x \in H. \tag{6.2.18}$$

Proof. By (6.2.9) it follows that

$$\langle D^2R_t\varphi(x)Q^{1/2}h, Q^{1/2}g\rangle$$

$$= \int_H [\langle \Lambda_t Q^{1/2}g, Q_t^{-1/2}y\rangle\langle \Lambda_t Q^{1/2}h, Q_t^{-1/2}y\rangle - \langle \Lambda_t Q^{1/2}g, \Lambda_t Q^{1/2}h\rangle]$$

$$\times\varphi(e^{tA}x + y)N_{Q_t}(dy) = \langle G_{\varphi(e^{tA}x+\cdot)}\Lambda_t Q^{1/2}h, \Lambda_t Q^{1/2}g\rangle.$$

Therefore

$$Q^{1/2}D^2R_t\varphi(x)Q^{1/2} = Q^{1/2}\Lambda_t^*G_{\varphi(e^{tA}x+\cdot)}\Lambda_t Q^{1/2},$$

and the conclusion follows from Lemma 6.2.7. $\square$

The following estimates of the second derivative $D^2R_t\varphi$ will be useful in the sequel.

Proposition 6.2.9 *Assume that (6.0.2) and (6.2.3) hold. Then for any $\varphi \in UC_b^1(H)$, $t > 0$ and any $x, h, g \in H$ we have*

$$\langle DR_t\varphi(x), h\rangle = \int_H \langle D\varphi(e^{tA}x + y), e^{tA}h\rangle N_{Q_t}(dy), \tag{6.2.19}$$

and

$$\langle D^2R_t\varphi(x)h, g\rangle = \int_H \langle \Lambda_t g, Q_t^{-1/2}y\rangle\langle D\varphi(e^{tA}x + y), e^{tA}h\rangle N_{Q_t}(dy). \tag{6.2.20}$$

Moreover

$$\|D^2R_t\varphi(x)\| \leq \|e^{tA}\|\,\|\Lambda_t\|\,\|\varphi\|_1. \tag{6.2.21}$$

Finally, if $Q^{1/2}\Lambda_t \in L_4(H)$, for all $t > 0$, we have $QD^2R_t\varphi(x) \in L_1(H)$,

$$\mathrm{Tr}[QD^2R_t\varphi(x)] = \int_H \left\langle Q^{1/2}\Lambda_t^*Q_t^{-1/2}y, Q^{1/2}e^{tA^*}D\varphi(e^{tA}x + y)\right\rangle N_{Q_t}(dy),$$
$$\tag{6.2.22}$$

and the following estimate holds:

$$\|QD^2 R_t\varphi(x)\|_{L_1(H)} \leq \|\Lambda_t Q e^{tA^*}\|_{L_2(H)} \|\varphi\|_1, \quad x \in H. \tag{6.2.23}$$

Proof. Let $t > 0$ and $x \in H$. By (6.2.19) and Theorem 6.2.2 we easily find (6.2.20). Moreover we know by Proposition 6.2.8 that $QD^2 R_t\varphi(x) \in L_1(H)$. Now (6.2.22) follows from (6.2.20). We have in fact, denoting by (e_k) a complete orthonormal basis in H,

$$\text{Tr}[QD^2 R_t\varphi(x)]$$

$$= \sum_{k=1}^{\infty} \int_H \langle \Lambda_t Q^{1/2} e_k, Q_t^{-1/2} y \rangle \langle D\varphi(e^{tA}x + y), e^{tA} Q^{1/2} e_k \rangle N_{Q_t}(dy)$$

$$= \int_H \langle e^{tA} Q \Lambda_t^* Q_t^{-1/2} y, D\varphi(e^{tA}x + y) \rangle N_{Q_t}(dy).$$

Now, by the Hölder inequality, we have

$$\left| \text{Tr}[QD^2 R_t\varphi(x)] \right|^2 \leq \|\varphi\|_1^2 \int_H |e^{tA} Q \Lambda_t^* Q_t^{-1/2} y|^2 N_{Q_t}(dy)$$

$$= \|\varphi\|_1^2 \, \text{Tr}\left[e^{tA} Q \Lambda_t^* \Lambda_t Q e^{tA^*} \right],$$

and (6.2.23) follows. $\square$

Exercise 6.2.10 Prove that for all $n \in \mathbb{N}$ there exists a constant $C_{1,n} > 0$ such that

$$\|D^n R_t\varphi(x)\| \leq C_{1,n} \|e^{tA}\| \|\Lambda_t\|^{n-1} \|\varphi\|_1, \quad n \in \mathbb{N}, \, t \geq 0. \tag{6.2.24}$$

We end the section with an example.

Example 6.2.11 Let (e_n) be a complete and orthonormal basis in H, and assume that $Ae_n = -\alpha_n e_n, \quad Qe_n = \gamma_n e_n, \quad n \in \mathbb{N}$, where $\alpha_n, \gamma_n > 0$, $n \in \mathbb{N}$ and $\alpha_n \uparrow +\infty$. Then we have

$$Q_t e_n = \frac{\gamma_n}{2\alpha_n}(1 - e^{-2\alpha_n t}), \, t \geq 0, \, n \in \mathbb{N}.$$

Therefore the basic condition (6.0.2) is equivalent to

$$\sum_{k=1}^{\infty} \frac{\gamma_n}{\alpha_n} < +\infty, \, n \in \mathbb{N}. \tag{6.2.25}$$

Concerning the controllability condition (6.2.3), since

$$\Lambda_t^2 e_n = \frac{2\alpha_n}{\gamma_n(e^{2\alpha_n t} - 1)}, \quad t \ge 0, \ n \in \mathbb{N},$$

it is equivalent to

$$\|\Lambda_t\|^2 = \sup_{n \in \mathbb{N}} \frac{2\alpha_n}{\gamma_n(e^{2\alpha_n t} - 1)} < +\infty, \quad t \ge 0, \ n \in \mathbb{N}. \tag{6.2.26}$$

We come finally to conditions for the existence of a classical solution; they are fulfilled provided

$$\|\Lambda_t A\|^2 = \sup_{n \in \mathbb{N}} \frac{2\alpha_n^3}{\gamma_n(e^{2\alpha_n t} - 1)} < +\infty, \quad t \ge 0, \ n \in \mathbb{N}, \tag{6.2.27}$$

and, since

$$\Lambda_t Q^{1/2} e_n = \left[\frac{2\alpha_n}{e^{2\alpha_n t} - 1} \right]^{1/2} e_n, \quad t \ge 0, \ n \in \mathbb{N},$$

if

$$\|\Lambda_t Q^{1/2}\|_4^4 = 4 \sum_{n=1}^{\infty} \frac{\alpha_n^2}{(e^{2\alpha_n t} - 1)^2} < +\infty, \quad t \ge 0, \ n \in \mathbb{N}. \tag{6.2.28}$$

It is interesting to notice that the condition $\Lambda_t Q^{1/2} \in L_4(H)$ involves only the operator A and not the covariance operator Q.

Finally we consider the particular case where

$$\alpha_n = n^\alpha, \quad \gamma_n = n^{-\gamma}, \quad n \in \mathbb{N}, \tag{6.2.29}$$

where α, γ are positive numbers. In this case the basic condition (6.0.2) is equivalent to $\alpha + \gamma > 1$. In this case the controllability condition holds since

$$\|\Lambda_t\|^2 = \sup_{n \in \mathbb{N}} \frac{2n^{\alpha+\gamma}}{e^{2n^\alpha t} - 1} < +\infty, \quad t \ge 0, \ n \in \mathbb{N}.$$

It is easy to see that there exists $C > 0$ such that

$$\|\Lambda_t\| \le C t^{-\frac{\alpha+\gamma}{2\alpha}}, \quad t \ge 0. \tag{6.2.30}$$

Also conditions (6.2.27) and (6.2.28) are fulfilled and

$$\|\Lambda_t A\| \le C_1 \, t^{-\frac{3\alpha+\gamma}{2\alpha}}, \quad t \ge 0, \tag{6.2.31}$$

$$\|\Lambda_t Q^{1/2}\| \le C_2 \, t^{-\frac{2\alpha+1}{4\alpha}}, \quad t \ge 0, \tag{6.2.32}$$

for suitable constants C_1 and C_2.

6.3 The Ornstein-Uhlenbeck semigroup

We are here concerned with the family (R_t) of linear bounded operators on $UC_b(H)$ defined by $R_0 = I$ and

$$R_t\varphi(x) = \int_H \varphi(e^{tA}x + y)N_{Q_t}(dy), \quad x \in H,\ t > 0,\ \varphi \in UC_b(H). \quad (6.3.1)$$

As we shall see, (R_t) is a semigroup of linear bounded operators on $UC_b(H)$. However, it is not strongly continuous, unless $A = 0$, see S. Cerrai [35]. We have in fact

$$R_t\varphi_h(x) = \mathrm{Re}[e^{i\langle e^{tA}x,h\rangle}e^{-\frac{1}{2}(Q_t h,h)}], \ \text{if}\ \varphi_h(x) = \mathrm{Re}(e^{i\langle h,x\rangle})\quad x, h \in H,$$

and consequently the limit $\lim_{t\to 0} R_t\varphi_h(x) = \mathrm{Re}(e^{i\langle h,x\rangle})$ is not uniform on $x \in H$.

We are going to find a characterization of the maximal subspace of $UC_b(H)$ where (R_t) is strongly continuous. For this it is useful to introduce an auxiliary family $(G_t)_{t\geq 0}$ of linear operators on $UC_b(H)$ (which is not a semigroup),

$$G_t\varphi(x) = \int_H \varphi(y)N_{x,Q_t}(dy), \quad t \geq 0,\ \varphi \in UC_b(H).$$

Clearly we have

$$R_t\varphi(x) = (G_t\varphi)(e^{tA}x), \quad t \geq 0,\ \varphi \in UC_b(H).$$

Then we prove the following result.

Proposition 6.3.1 *Let $\varphi \in UC_b(H)$. Then the following statements are equivalent.*

(i) $\lim_{t\to 0} R_t\varphi = \varphi$ *in* $UC_b(H)$.

(ii) $\lim_{t\to 0} \varphi(e^{tA}x) = \varphi(x)$, *uniformly on* $x \in H$.

Proof. We show first that for any $\varphi \in UC_b(H)$ we have

$$\lim_{t\to 0} G_t\varphi = \varphi \quad \text{in}\ UC_b(H). \quad (6.3.2)$$

In fact since $\|G_t\varphi\|_0 \leq \|\varphi\|_0$, $\varphi \in UC_b(H)$, it is enough to prove (6.3.2) for $\varphi \in UC_b^1(H)$. In this case we have

$$|G_t\varphi(x) - \varphi(x)|^2 \leq \int_H |\varphi(x+y) - \varphi(x)|^2 N_{Q_t}(dy)$$

$$\leq \|\varphi\|_1^2 \int_H |y|^2 N_{Q_t}(dy) = \text{Tr}[Q_t],$$

and the conclusion follows letting t tend to 0.

We now prove that (i) $\Rightarrow$ (ii). We have in fact

$$|\varphi(e^{tA}x) - \varphi(x)| \leq |\varphi(e^{tA}x) - G_t\varphi(e^{tA}x) + |R_t\varphi(x) - \varphi(x)|.$$

It follows that

$$\lim_{t\to 0} |\varphi(e^{tA}x) - \varphi(x)| = \lim_{t\to 0} |R_t\varphi(x) - \varphi(x)| = 0.$$

The implication (ii)$\Rightarrow$(i) can be proved in a similar way. $\square$

The semigroup (R_t) belongs to a special class of semigroups of linear bounded operators on $UC_b(H)$, called π-*semigroups*, introduced and extensively studied by E. Priola [189]. In the following subsection we collect some results on π-semigroups needed in what follows.

6.3.1 π-Convergence

We first introduce the notion of π-convergence. It arises systematically as *bpc-convergence* in the book by S. N. Ethier and T. G. Kurz [112]. A sequence $(\varphi_n) \subset UC_b(H)$ is said to be π-*convergent* to a function $\varphi \in UC_b(H)$ (and we shall write $\varphi_n \xrightarrow{\pi} \varphi$) if for any $x \in H$ we have $\lim_{n\to\infty} \varphi_n(x) = \varphi(x)$ and if $\sup_{n\in\mathbb{N}} \|\varphi_n\|_0 < +\infty$.

A subset Λ of $UC_b(H)$ is said to be π-*dense* if for any $\varphi \in UC_b(H)$ there exists a sequence $(\varphi_n) \subset \Lambda$ such that $\varphi_n \xrightarrow{\pi} \varphi$.

Now we introduce the space $\mathcal{E}(H)$ of all *exponential* functions in H. $\mathcal{E}(H)$ is the linear subspace of $UC_b(H)$ spanned by the set of all real parts of the functions $(e^{i\langle x,h\rangle})_{h\in H}$. In some cases it will be useful to consider also the space $\mathcal{E}_A(H)$ of all real parts of the functions $(e^{i\langle x,h\rangle})_{h\in D(A^*)}$, where A^* is the adjoint of A.

Let us prove an approximation result.

Lemma 6.3.2 *For any* $\varphi \in UC_b(H)$ *there exists a multi-sequence* (φ_{n_1,n_2,n_3}) *in* $\mathcal{E}_A(H)$ *such that* *(i)* $\|\varphi_{n_1,n_2,n_3}\|_0 \leq \|\varphi\|_0$, *for all* $n_1, n_2, n_3 \in \mathbb{N}$,

(ii) we have

$$\lim_{n_1 \to \infty} \lim_{n_2 \to \infty} \lim_{n_3 \to \infty} \varphi_{n_1,n_2,n_3}(x) = \varphi(x), \ \forall x \in H.$$

Proof. Let $\varphi \in UC_b(H)$ and let (P_n) be a sequence of orthogonal finite dimensional projectors strongly convergent to the identity. Set

$$\varphi_{n_1}(x) = \varphi(P_{n_1}(x)), \ x \in H, \ n_1 \in \mathbb{N}.$$

Fix $n_1 \in \mathbb{N}$. Since the range of P_{n_1} is finite dimensional, it is not difficult to construct a sequence $(\varphi_{n_1,n_2}) \in \mathcal{E}(H)$ such that $\|\varphi_{n_1,n_2}\|_0 \leq \|\varphi\|_0$ and $\lim_{n_2 \to \infty} \varphi_{n_1,n_2}(x) = \varphi_{n_1}(x), \ \forall x \in H$. Now it is easy to check that the sequence

$$\varphi_{n_1,n_2,n_3}(x) = \varphi_{n_1,n_2}(n_3 R(n_3, A)x), \ x \in H,$$

fulfills the required conditions. $\square$

Remark 6.3.3 One can introduce locally convex topologies on $UC_b(H)$ for which the convergence of sequences is identical with π-convergence, see E. Priola [189]. Other locally convex topologies on $UC_b(H)$ are investigated in B. Goldys and B. M. Kocan [130] and F. Kühnemund [151].

6.3.2 Properties of the π-semigroup (R_t)

Proposition 6.3.4 *Let (R_t) be defined by (6.3.1) for $t > 0$ and $R_0 = I$.*
(i) (R_t) is a semigroup of linear bounded operators on $UC_b(H)$. Moreover

$$\|R_t\varphi\|_0 \leq \|\varphi\|_0, \quad \varphi \in UC_b(H), \ t \geq 0.$$

(ii) If $\varphi_n \xrightarrow{\pi} \varphi$ in $UC_b(H)$ then $R_t\varphi_n \xrightarrow{\pi} R_t\varphi$ in $UC_b(H)$ $\forall t \geq 0$.
(iii) For all $\varphi \in UC_b(H)$ and for all $x \in H$ the function $[0, +\infty) \to \mathbb{R}, \ t \to R_t\varphi(x)$ is continuous.

Proof. It is easy to see that $R_t \in L(UC_b(H))$ for all $t > 0$, and that $\|R_t\varphi\|_0 \leq \|\varphi\|_0$, for all $t \geq 0$. The semigroup law, $R_{t+s} = R_t R_s$, $t, s \geq 0$, follows from classical properties of Gaussian measures. Here is a different proof. In view of Lemma 6.3.2 it is enough to prove the result for $\varphi_h(x) = \text{Re}(e^{i\langle x,h\rangle})$, $h \in D(A^*)$. In this case we have

$$R_t\varphi_h(x) = \text{Re}\left[e^{i\langle e^{tA}x,h\rangle - \frac{1}{2}\langle Q_t h,h\rangle}\right], \quad t \geq 0, \ x \in H.$$

Consequently

$$R_s R_t \varphi_h(x) = \text{Re} \left[e^{i\langle e^{(t+s)A}x,h\rangle - \frac{1}{2}\langle (Q_t + e^{tA}Q_s e^{tA^*})h,h\rangle} \right], \quad t \geq 0, \ x \in H.$$

Since $Q_t + e^{tA}Q_s e^{tA^*} = Q_{t+s}$, (i) is proved. Part (ii) follows easily using the dominated convergence theorem. Finally, let us prove part (iii). We only verify continuity of $R_t \varphi(x)$ at 0. Let first $\varphi \in UC_b^1(H)$, then we have

$$R_t \varphi(x) - \varphi(x) = \int_H [\varphi(e^{tA}x + y) - \varphi(x)] N_{Q_t}(dy), \quad t > 0, \ x \in H,$$

and so

$$|R_t \varphi(x) - \varphi(x)| \leq \|\varphi\|_1 |e^{tA}x - x| + \int_H |y| N_{Q_t}(dy).$$

Since

$$\left[\int_H |y| N_{Q_t}(dy) \right]^2 \leq \int_H |y|^2 N_{Q_t}(dy) \ \text{Tr} \ Q_t,$$

(iii) also follows.

In the general case it is enough to approximate φ by a sequence $\varphi_n \to \varphi$ in $UC_b(H)$. $\square$

Remark 6.3.5 We could replace π-convergence by stronger notions of convergence requiring for instance uniform convergence on compact or on bounded sets, see S. Cerrai [35], E. Priola [189], B. Goldys and B. M. Kocan [130], and G. Tessitore and J. Zabczyk [209]. Also one could replace in several places $UC_b(H)$ with $C_b(H)$ but, for the sake of simplicity, we limit ourselves to π-convergence in $UC_b(H)$.

6.3.3   The infinitesimal generator

Following E. Priola [189], we define the *infinitesimal generator* L of (R_t) by

$$\begin{cases} D(L) = \{\varphi \in UC_b(H) : \exists \psi \in UC_b(H) : \Delta_h \varphi \xrightarrow{\pi} \psi \text{ as } h \to 0\}, \\ L\varphi = \psi, \end{cases}$$

where $\Delta_h = \frac{1}{h}(R_h - I)$, $h > 0$. It is easy to see that $D(L)$ is π-dense on $UC_b(H)$, and that if $\varphi \in D(L)$ then $R_t \varphi \in D(L), \forall\, t \geq 0$ and $LR_t \varphi = R_t L\varphi$. Moreover if $\varphi \in D(L)$, then $R_t \varphi(x)$ is differentiable $\forall\, t \geq 0$, and

$$\frac{d}{dt} R_t \varphi(x) = LR_t \varphi(x) = R_t L\varphi(x), \ x \in H.$$

The following result is due to E. Priola [189].

Proposition 6.3.6 *The resolvent set $\rho(L)$ of L contains $(0, +\infty)$ and its resolvent is given by*

$$R(\lambda, L)g(x) = \int_0^\infty e^{-\lambda t} R_t g(x) dt, \ g \in UC_b(H), \ \lambda > 0, \ x \in H. \quad (6.3.3)$$

Moreover if $\varphi_k \xrightarrow{\pi} \varphi$, we have $R(\lambda, L)\varphi_k \xrightarrow{\pi} R(\lambda, L)\varphi$, for any $\lambda > 0$.

Proof. Let $g \in UC_b(H)$ and set

$$\varphi(x) = \int_0^\infty e^{-\lambda t} R_t g(x) dt, \ x \in H.$$

The integral above is well defined for any $x \in H$ since $R_t g(x)$ is continuous on t and $|R_t g(x)| \leq \|g\|_0$. Let us prove that $\varphi \in UC_b(H)$. Assume first that $g \in UC_b^1(H)$, then by the very definition of $R_t g$ it follows that

$$|R_t g(x) - R_t g(y)| \leq \|g\|_1 |e^{tA}(x - y)|, \ x, y \in H,$$

so that $R_t g$ is Lipschitz continuous uniformly for $t \in [0, T]$. This implies that $\varphi \in UC_b(H)$. Now, since $UC_b^1(H)$ is dense in $UC_b(H)$ by Theorem 2.2.1, it follows easily that $\varphi \in UC_b(H)$ for any $g \in UC_b(H)$.

We prove now that $\varphi \in D(L)$ and is a solution to the equation

$$\lambda \varphi - L\varphi = g. \quad (6.3.4)$$

We have in fact

$$\Delta_h \varphi(x) = \frac{1}{h} \int_0^\infty e^{-\lambda t}(R_{t+h}g(x) - R_t g(x))dt$$

$$= \frac{1}{h}\left[(e^{\lambda h} - 1)\int_0^\infty e^{-\lambda t} R_t g(x) dt - e^{\lambda h}\int_0^h e^{-\lambda t} R_t g(x)dt\right].$$

Letting $h \to 0$, we see that the limit exists and (6.3.4) holds.

It remains to prove uniqueness for (6.3.4). Assume that $\varphi \in D(L)$ is such that $\lambda \varphi - L\varphi = 0$. Then we have

$$\lambda \int_0^\infty e^{-\lambda t} R_t \varphi(x) dt = \int_0^\infty e^{-\lambda t} R_t L\varphi(x) dt$$

$$= \int_0^\infty e^{-\lambda t} \frac{d}{dt} R_t g(x) dt = -\varphi(x) + \lambda \int_0^\infty e^{-\lambda t} R_t \varphi(x) dt.$$

This implies $\varphi(x) = 0$ as required. The last statement is easily proved. $\square$

6.4   Elliptic equations

We are here concerned with the elliptic equation

$$\lambda\varphi(x) - \frac{1}{2}\mathrm{Tr}[QD^2\varphi(x)] - \langle Ax, D\varphi(x)\rangle = g(x), \quad x \in D(A), \qquad (6.4.1)$$

where $\lambda > 0$ and $g \in UC_b(H)$.

We shall assume, besides (6.0.2) and (6.2.3), that there are $\kappa > 0$ and $\delta \in [1/2, 1)$ such that

$$\|\Lambda_t\| \leq \kappa t^{-\delta}, \quad t > 0, \qquad (6.4.2)$$

where Λ_t is defined by (6.2.4) and moreover (for the sake of simplicity) that $\|e^{tA}\| \leq M$, $t \geq 0$, for a suitable $M > 0$.

We say that φ, given by

$$\varphi(x) = R(\lambda, L)g(x) = \int_0^\infty e^{-\lambda t} R_t\varphi(x)dt, \ x \in H,$$

is a *generalized solution* to (6.4.1). A function φ is called a *strict solution* if

(i) $\varphi \in UC_b^2(H)$,

(ii) $QD^2\varphi(x) \in L_1(H)$ for any $x \in H$,

(iii) $D\varphi(x) \in D(A^*)$ for any $x \in H$,

(iv) φ fulfills (6.4.1).

We want to show now that a generalized solution φ enjoys additional regularity properties and that in some cases it is a strict solution. To this purpose the following lemma will be useful.

Lemma 6.4.1 *Let $0 \leq \beta < \alpha$. Then there exists $C_{\alpha,\beta} > 0$ such that*

$$\|R_t g\|_\alpha \leq C_{\alpha,\beta} t^{-\delta(\alpha-\beta)}\|g\|_\beta, \ g \in UC_b^\beta(H). \qquad (6.4.3)$$

Proof. Let first $\alpha \in (0,1)$ and $\beta = 0$. Then, recalling (6.2.12), we have

$$|DR_t\varphi(x)| \leq C_1\|\Lambda_t\|\|\varphi\|_0, \quad x \in H, \ t > 0. \qquad (6.4.4)$$

Since $|R_t\varphi(x)| \leq \|\varphi\|_0$, by an elementary interpolatory argument we find

$$[R_t\varphi]_\alpha \leq C_1^\alpha\|\Lambda_t\|^\alpha\|\varphi\|_0, \quad t > 0,$$

and (6.4.3) follows in this case recalling (6.4.2).

Let now $\alpha = 1$, $\beta \in (0,1)$. Then by (6.2.24) we have

$$|DR_t\varphi(x)| \leq C_{1,1}M\|\Lambda_t\|\|\varphi\|_1, \quad x \in H. \tag{6.4.5}$$

Due to Theorem 2.3.3 $\left(UC_b(H), UC_b^1(H)\right)_{\beta,\infty} = C_b^\beta(H)$, and then from (6.4.4) and (6.4.5) we obtain $|DR_t\varphi(x)| \leq C_1^{1-\beta}C_{1,1}^\beta M^\beta\|\Lambda_t\|^{1-\beta}\|g\|_\beta$, and also in this case (6.4.3) follows. The general case can be treated in a similar way. $\square$

Now we can prove the following result.

Proposition 6.4.2 *Let* $\lambda > 0$ *and* $g \in UC_b(H)$, *and let* $\varphi = R(\lambda, L)g$ *be the generalized solution to* (6.4.1).

(i) $\varphi \in C_b^\beta(H)$ *for any* $\beta \in [0, 1/\delta)$, *and*

$$\|\varphi\|_\beta \leq C_{\beta,0}\Gamma(1 - \delta\beta)\lambda^{\delta\beta-1}\|g\|_0. \tag{6.4.6}$$

(ii) If $g \in C_b^\theta(H)$ *for some* $\theta \in (0,1)$, *and* $\beta < \theta + \frac{1}{\delta}$, *then* $\varphi \in UC_b^\beta(H)$, *and*

$$\|\varphi\|_\beta \leq C_{\beta,\theta}\Gamma(1 - (\beta - \theta)\delta)\lambda^{(\beta-\theta)\delta-1}\|g\|_\theta, \quad x \in H. \tag{6.4.7}$$

(iii) If $g \in C_b^\theta(H)$ *for some* $\theta \in (0,1)$, *and in addition*

$$\int_0^\infty e^{-\lambda t}\|\Lambda_t Q^{1/2}\|_{L_4(H)}^{2(1-\theta)}\|\Lambda_t Q e^{tA^*}\|_{L_2(H)}^\theta dt < +\infty, \tag{6.4.8}$$

then $QD^2\varphi(x) \in L_1(H)$ *for any* $x \in H$, *and* φ *is a strict solution.*

Proof. Let us prove (i). Taking into account (6.4.3) we have

$$\|\varphi\|_\beta \leq \int_0^\infty e^{-\lambda t}\|R_t g\|_\beta dt \leq C_{\beta,0}\int_0^\infty e^{-\lambda t}t^{-\delta\beta}dt\|g\|_0$$

$$- C_{\beta,0}\Gamma(1 - \delta\beta)\lambda^{\delta\beta-1}\|g\|_0,$$

and (i) follows. Part (ii) is proved similarly. Let us show (iii). By (6.2.18) and (6.2.23) we have respectively that

$$\|QD^2R_t g(x)\|_{L_1(H)} \leq 2\|Q^{1/2}\Lambda_t\|_{L_4(H)}^2\|g\|_0,$$

and

$$\|QD^2R_t g(x)\|_{L_1(H)} \leq \|\Lambda_t Q e^{tA^*}\|_{L_2(H)}\|g\|_1.$$

By interpolation we have

$$\|QD^2 R_t g(x)\|_{L_1(H)} \leq 2^{1-\theta} \|Q^{1/2} \Lambda_t\|_{L_4(H)}^{2(1-\theta)} \|\Lambda_t Q e^{tA^*}\|_{L_2(H)}^{2\theta} \|g\|_\theta,$$

which yields the conclusion. $\square$

The following result is an immediate consequence of Proposition 6.4.2.

Corollary 6.4.3 *We have*

$$D(L) \subset C_\theta^\beta(H), \quad \beta \in [0, 1/\delta). \tag{6.4.9}$$

In particular, due to Example B.3.2, if $Q = I$

$$D(L) \subset C_\theta^\beta(H), \quad \beta \in [0, 2). \tag{6.4.10}$$

Example 6.4.4 We go back to Example 6.2.11. We set as there

$$Ae_n = -n^\alpha e_n, \quad Qe_n = n^{-\gamma} e_n, \quad n \in \mathbb{N},$$

with $\alpha + \gamma > 1$ and $\gamma < \alpha$. Therefore (6.4.2) holds with $\delta = \frac{\alpha+\gamma}{2\alpha} < 1$. Also (i) implies that if $g \in UC_b(H)$ then $\varphi \in UC_b^\beta(H)$ with $\beta < \frac{2\alpha}{\alpha+\gamma}$. Since $\frac{2\alpha}{\alpha+\gamma} > 1$, we have that $\varphi \in UC_b^1(H)$. However, $\beta < 2$ and we cannot conclude that $D^2\varphi$ exists.

Moreover by (ii) it follows that if $g \in UC_b^\theta(H)$ then $\varphi \in UC_b^\beta(H)$ with $\beta < \theta + \frac{2\alpha}{\alpha+\gamma}$. Thus if θ is such that $\theta + \frac{2\alpha}{\alpha+\gamma} > 2$, we can conclude that $D^2\varphi$ exists. In the particular case when $Q = 1$ we have $\delta = 1/2$, see Appendix B. In this case for any $\theta > 0$ we have $\varphi \in UC_b^2(H)$.

Let finally discuss part (iii) of Proposition 6.4.2.

We recall that

$$\|\Lambda_t Q^{1/2}\|_{L_4(H)} \leq C_1 t^{-\frac{3\alpha+\gamma}{2\alpha}}, \quad t > 0. \tag{6.4.11}$$

Moreover

$$\|\Lambda_t Q e^{tA^*}\|_{L^2(H)}^2 = \sum_{n=1}^\infty \frac{2n^{\alpha-\gamma}}{1 - e^{-2n^\alpha t}}.$$

Thus there exists $C_2 > 0$ such that

$$\|\Lambda_t Q e^{tA^*}\|_{L^2(H)}^2 \leq C_2 t^{-\frac{\alpha+1-\gamma}{2\alpha}}, \quad t > 0. \tag{6.4.12}$$

Then by (6.4.11) and (6.4.12) there exists $C_3 > 0$ such that we have

$$\|\Lambda_t Q^{1/2}\|_{L_4(H)}^{2(1-\theta)} \|\Lambda_t Q e^{tA^*}\|_{L^2(H)}^\theta \leq C_3 t^{-\frac{1}{2\alpha}[\theta(1-2\alpha-2\gamma)+3\alpha+\gamma]}.$$

Thus if

$$\theta > \frac{\alpha + \gamma}{2\alpha + 2\gamma - 1}, \tag{6.4.13}$$

part (iii) of Proposition 6.4.2 is fulfilled and φ is a classical solution.

In the particular case were $Q = I$ and $\alpha = 2$, (6.4.13) is equivalent to $\theta > 2/3$.

6.4.1 Schauder estimates

In this subsection we want to prove Schauder estimates for the solution of (6.4.1). If H is finite dimensional and $\det Q > 0$, Schauder estimates were proved by G. Da Prato and A. Lunardi [94]. Here we are going to extend this result to the infinite dimensional situation. In this case it is natural to assume that Q has a bounded inverse. We shall take for simplicity $Q = I$ and A of negative type. In this case there exists $K > 0$ such that, see Appendix B,

$$\|\Lambda_t\| \leq K t^{-1/2}, \ t > 0, \tag{6.4.14}$$

and assumption (6.2.3) holds. The following result was proved in P. Cannarsa and G. Da Prato [32].

Theorem 6.4.5 *Assume that* (6.0.2) *holds, A is of negative type, and that $Q = I$. Let $\theta \in (0, 1)$, $g \in C_b^\theta(H)$ and $\lambda > 0$. Then the function $\varphi = R(\lambda, L)g$ belongs to $C_b^{2+\theta}(H)$.*

Proof. The proof is based on the following interpolatory result, due to A. Lunardi [163], see Proposition 2.3.8:

$$\left(C_b^\alpha(H), C_b^{2+\alpha}(H)\right)_{1-\frac{\alpha-\theta}{2}, \infty} \subset C_b^{2+\theta}(H),$$

for all $\alpha \in (0, 1)$ and $\theta \in (0, \alpha)$. Thus, to prove the theorem it will be enough to show that for some $\alpha \in (\theta, 1)$, we have

$$\varphi \in \left(C_b^\alpha(H), C_b^{2+\alpha}(H)\right)_{1-\frac{\alpha-\theta}{2}, \infty}. \tag{6.4.15}$$

We first note that, in view of (6.4.3), we have

$$\|R_t \varphi\|_\alpha \leq C_{\alpha,\theta} \|\Lambda_t\|^{\alpha-\theta} \|\varphi\|_\theta \leq C_{\alpha,\theta} \, t^{\frac{\alpha-\theta}{2}} \|\varphi\|_\theta. \tag{6.4.16}$$

Now in order to prove (6.4.15) we set $\varphi(x) = a(t,x) + b(t,x)$, where

$$a(t,x) = \int_0^t e^{-\lambda s} R_s g(x) ds, \quad b(t,x) = \int_t^{+\infty} e^{-\lambda s} R_s g(x) ds.$$

Then from (6.4.16) it follows that

$$
\begin{aligned}
\|a(\cdot,t)\|_\alpha &\le C_{\alpha,\theta}^1 \int_0^t e^{-\lambda s} s^{-\frac{\alpha-\theta}{2}} ds \, \|g\|_\theta \\
&= C_{\alpha,\theta}^1 t^{1-\frac{\alpha-\theta}{2}} \int_0^1 e^{-\lambda t\sigma} \sigma^{-\frac{\alpha-\theta}{2}} d\sigma \, \|g\|_\theta \\
&\le \frac{C_{\alpha,\theta}^1}{1-\frac{\alpha-\theta}{2}} t^{1-\frac{\alpha-\theta}{2}} \|g\|_\theta,
\end{aligned}
$$

and

$$
\begin{aligned}
\|b(\cdot,t)\|_{2+\alpha} &\le C_{\alpha,\theta}^1 \int_t^{+\infty} e^{-\lambda s} s^{-\frac{\alpha-\theta}{2}-1} ds \, \|g\|_\theta \\
&= C_{\alpha,\theta}^1 t^{-\frac{\alpha-\theta}{2}} \int_1^{+\infty} e^{-\lambda t\sigma} \sigma^{-\frac{\alpha-\theta}{2}-1} d\sigma \, \|g\|_\theta \\
&\le 2\frac{C_{\alpha,\theta}^1}{\alpha-\theta} t^{\frac{\theta-\alpha}{2}} \|g\|_\theta.
\end{aligned}
$$

This implies (6.4.15). □

By Theorem 6.4.5 and Proposition 6.4.2 we find the result.

Corollary 6.4.6 *Assume that $\theta \in (0,1)$, $g \in C_b^\theta(H)$, $\lambda > 0$ and in addition that (6.4.8) holds. Then the generalized solution φ to (6.4.1) has the following properties.*

(i) $\varphi \in C_b^{2+\theta}(H)$ and $D^2\varphi(x) \in L_1(H)$, for any $x \in H$.

(ii) $\mathrm{Tr}[D^2\varphi(\cdot)] \in UC_b(H)$.

(iii) $\langle \cdot, A^* D\varphi \rangle \in UC_b(H)$.

Moreover φ is a strict solution of (6.4.1).

Remark 6.4.7 Let us consider the restriction (R_t^θ) of the semigroup (R_t) to $C_b^\theta(H)$, $\theta \in (0,1)$. Then (R_t^θ) is a semigroup on $C_b^\theta(H)$ with the same continuity properties as (R_t). Its infinitesimal generator L^θ is the part of L in $C_b^\theta(H)$:

$$D(L^\theta) = \{\varphi \in D(L) \cap C_b^\theta(H) : \ L\varphi \in C_b^\theta(H)\}.$$

Corollary 6.4.6 allows us to characterize, under suitable assumptions, the domain of L^θ. We recall that in the finite dimensional case, see [94],

$$D(L^\theta) = \{\varphi \in C_b^{2+\theta}(H) : \ \langle A\cdot, D\varphi \rangle \in C_b(H)\}.$$

Under the hypotheses of Corollary 6.4.6 we have the following characterization of $D(L^\theta)$:

$$D(L^\theta) = \Big\{\varphi \in UC_b^{2,\theta}(H) : \ D^2\varphi(x) \in L_1(H), \forall\, x \in H,$$

$$\mathrm{Tr}[D^2\varphi(x)] \in UC_b(H), \ \langle \cdot, A^*D\varphi \rangle \in UC_b(H)\Big\}.$$
$$(6.4.17)$$

6.4.2 The Liouville theorem

A bounded Borel function φ is said to be *harmonic* for the Ornstein-Uhlenbeck semigroup (R_t) if $R_t\varphi = \varphi$ for all $t > 0$.

Let us remark that if the controllability condition (6.2.3) holds then a harmonic function φ is necessarily continuous; in fact by Theorem 6.2.2 $\varphi \in C_b^\infty(H)$.

Let us prove the following simple result.

Proposition 6.4.8 *Assume that (6.2.3) holds and $\Lambda_t \to 0$ strongly as $t \to +\infty$. Then, harmonic functions for (R_t) are constant.*

Proof. It follows from (4.3.14) that

$$|\varphi(a) - \varphi(b)| = |R_t\varphi(a) - R_t\varphi(b)| \le \|\varphi\|_0 \left(e^{|\Lambda_t(a-b)|^2} - 1\right)^{1/2} \to 0$$

as $t \to \infty$, and therefore φ is constant. $\square$

By Theorem B.3.4 we have the following more analytical characterization.

Proposition 6.4.9 *Assume that (6.2.3) holds. If the only nonnegative solution P of the equation*

$$PA + A^*P - PBB^*P = 0$$

is 0, then harmonic functions for (R_t) are constant.

6.5   Perturbation results for parabolic equations

We are here concerned with the problem

$$
\begin{cases}
D_t u(t,x) &= \tfrac{1}{2}\mathrm{Tr}\left[QD^2 u(t,x)\right] + \langle Ax + F(x), Du(t,x)\rangle, \\
& t \geq 0, \ x \in D(A). \\
u(0,x) &= \varphi(x), \quad x \in H,
\end{cases}
\tag{6.5.1}
$$

under assumptions (6.0.2), (6.2.3) and (6.4.2) and with $F \in C_b(H;H)$. We shall write this problem as

$$
\begin{cases}
D_t u(t,\cdot) = Lu(t,\cdot) + \langle F(x), Du(t,\cdot)\rangle, & t \geq 0, \ x \in H, \\
u(0,\cdot) = \varphi.
\end{cases}
\tag{6.5.2}
$$

A *strict solution* to problem (6.5.2) is a continuous function $u : [0,\infty) \times H \to H$, differentiable in t, such that $u(t,\cdot) \in D(L)$ for all $t > 0$, and equations (6.5.2) hold.

Notice that this definition is meaningful since, in view of Proposition 6.4.2(i), if $\varphi \in D(L)$, then $\varphi \in UC_b^1(H)$ and so the term $\langle F(\cdot), Du(t,\cdot)\rangle$ in (6.5.2) is well defined and belongs to $UC_b(H)$.

It is easy to check that if u is a strict solution of (6.5.2) then it is a solution of the following integral equation:

$$
u(t,\cdot) = R_t\varphi + \int_0^t R_{t-s}(\langle F(\cdot), Du(s,\cdot)\rangle)ds, \quad t \geq 0.
\tag{6.5.3}
$$

A solution of (6.5.3) is called a *mild solution*.

The following result is proved in G. Da Prato and J. Zabczyk [102].

Proposition 6.5.1 *Assume that assumptions* (6.0.2), (6.2.3) *and* (6.4.2) *hold and that* $F \in C_b(H;H)$. *Then for any* $\varphi \in B_b(H)$ *there is a unique mild solution of equation* (6.5.3).

Proof. We fix $T > 0$ and write equation (6.5.3) as $u = f + \gamma(v)$, where

$$
f(t,x) \quad = \quad R_t\varphi(x), \ t \in [0,T], \quad x \in H,
$$

$$
\gamma(u)(t,\cdot) \quad = \quad \int_0^t R_{t-s}(\langle F(\cdot), Du(s,\cdot)\rangle)ds, \quad t \in [0,T].
$$

Then we solve the equation (6.5.3) on the space Z_T consisting of the set of all functions $u : [0,T] \times H \to \mathbb{R}$ such that

(i) $u \in B_b([0,T] \times H)$,

(ii) for all $t > 0$, $u(t, \cdot) \in UC_b^1(H)$,

(iii) $\sup\limits_{t \in (0,T]} t^\alpha \|u(t, \cdot)\|_1 < +\infty$,

(iv) for all $x \in H$, $Du(\cdot, x)$ is measurable.

Z_T, endowed with the norm

$$\|u\|_{Z_T} := \|u\|_0 + \sup\limits_{t \in (0,T]} t^\alpha \|u(t, \cdot)\|_1,$$

is a Banach space.

We first notice that f belongs to Z_T in view of Theorems 6.2.2 and 6.2.4. Then if we show that γ is a contraction on Z_T, provided T is sufficiently small, the conclusion will follow by a standard argument. We have in fact

$$|\gamma(u)(t, x)| \leq \|F\|_0 \int_0^t \|Du(s, \cdot)\|_0 ds \leq \frac{t^{1-\alpha}}{1-\alpha} \|u\|_{Z_T},$$

and, since

$$D\gamma(u)(t, x) = \int_0^t D[R_{t-s}(\langle F(\cdot), Du(s, \cdot) \rangle)] ds,$$

we have

$$t^\alpha |D\gamma(u)(t, x)| \leq t^\alpha \int_0^t |D[R_{t-s}(\langle F(\cdot), D_x u(s, \cdot) \rangle)]| ds$$

$$\leq t^\alpha C \|F\|_0 \int_0^t (t - s)^{-\alpha} s^{-\alpha} ds \|u\|_{Z_T}$$

$$= t^{1-\alpha} C \|F\|_0 \int_0^1 (1 - s)^{-\alpha} s^{-\alpha} ds \|u\|_{Z_T}.$$

Thus γ is a contraction, provided T is sufficiently small, and the proposition is proved. $\square$

Remark 6.5.2 Under the assumptions of Proposition 6.5.1 it follows, by a result due to D. Gątarek and B. Goldys [125], that there exists a unique martingale solution to the differential stochastic equation

$$dX = (AX + F(X))dt + Q^{1/2}dW(t), \; X(0) = x \in H,$$

where W is a cylindrical Wiener process taking values on H.

6.6   Perturbation results for elliptic equations

In this section we assume that (6.0.2) holds, that $Q = I$, A is of negative type, and $F \in UC_b(H; H)$. Therefore (6.4.14) holds.

We are going to show, following [82], that the linear operator

$$N\varphi = L\varphi + \langle F, D\varphi \rangle, \; \varphi \in D(L),$$

is m-dissipative on $UC_b(H)$.

Lemma 6.6.1 *For any $\lambda > \lambda_0 := \pi K^2 \|F\|_0^2$ and any $f \in UC_b(H)$, there is a unique solution $\varphi \in D(L) \cap C_b^1(H)$ of the equation*

$$\lambda\varphi - L\varphi - \langle F(x), D\varphi \rangle = f. \tag{6.6.1}$$

Proof. Let $\lambda > \lambda_0$. Then, setting $\psi = \lambda\varphi - L\varphi$, equation (6.6.1) becomes

$$\psi - T_\lambda\psi = f, \tag{6.6.2}$$

where T_λ is defined by

$$T_\lambda\psi(x) = \langle F(x), DR(\lambda, L)\psi(x) \rangle, \; \psi \in UC_b(H), \; x \in H. \tag{6.6.3}$$

Recalling Proposition 6.4.2 and Corollary 6.4.3, we see that

$$\|T_\lambda\psi\|_0 \le K\sqrt{\frac{\pi}{\lambda}} \, \|F\|_0 \, \|\psi\|_0.$$

Therefore if $\lambda > \lambda_0$ equation (6.6.1) has a unique solution φ. $\square$

Now we are going to show that N is m-dissipative on $UC_b(H)$. For this we first consider the case when F is in addition of class C^1.

Lemma 6.6.2 *Let $F : H \to H$ be bounded and of class C^1. Then N is m-dissipative. Moreover if $\varphi \in UC_b(H)$ is nonnegative then $R(\lambda, N)\varphi$ is nonnegative for all $\lambda > 0$.*

Proof. We have already seen in Lemma 6.6.1 that for any $\lambda > \lambda_0$ the range of $\lambda - N$ is $UC_b(H)$. Then it is sufficient to show that L is dissipative. It is convenient to introduce for any $h > 0$ an operator N_h approximating N,

$$N_h\varphi = L\varphi + \Delta_h\varphi, \; \varphi \in D(L),$$

where

$$\Delta_h\varphi(x) = \frac{1}{h}\left(\varphi(\eta(h, x)) - \varphi(x)\right), \tag{6.6.4}$$

and η is the solution to

$$\eta_t(t, x) = F(\eta(t, x)), \quad \eta(0, x) = x \in H. \tag{6.6.5}$$

Clearly for any $\varphi \in UC_b^1(H)$ we have

$$\lim_{h \to 0} \Delta_h \varphi = \langle F, D\varphi \rangle, \text{ in } UC_b(H). \tag{6.6.6}$$

Now, given $\lambda > 0$ and $f \in C_b(H)$, we consider the equation

$$\lambda \varphi_h - L\varphi_h - \Delta_h \varphi_h = f. \tag{6.6.7}$$

Equation (6.6.7) can be solved as before by a standard fixed point argument depending on the parameter h, and

$$\lim_{h \to 0} \varphi_h = \varphi \text{ in } C_b(H). \tag{6.6.8}$$

Now by (6.6.7) we find

$$\left(\lambda + \frac{1}{h}\right) \varphi_h - L\varphi_h = f + \frac{1}{h} \varphi(\eta(h, x)). \tag{6.6.9}$$

It follows that

$$\|\varphi_h\|_0 \le \frac{1}{\lambda + \frac{1}{h}} \left(\|f\|_0 + \frac{1}{h} \|\varphi_h\|_0\right),$$

which yields $\|\varphi_h\|_0 \le \frac{1}{\lambda} \|f\|_0$. Consequently, letting h tend to 0 gives

$$\|\varphi\|_0 \le \frac{1}{\lambda} \|f\|_0.$$

Therefore N is m-dissipative as required.

It remains to prove the last statement. Let $\varphi \in UC_b(H)$ be nonnegative, $h > 0$ and let φ_h be the solution of (6.6.7). It is clear by (6.6.9) that φ_h is nonnegative, so that the conclusion follows. $\square$

Finally we consider the general case. We shall denote by $C_b^{0,1}(H; H)$ the subspace of $UC_b(H; H)$ of all Lipschitz continuous functions from H into H. We shall need the following result due to F. A. Valentine [213].

Proposition 6.6.3 $C_b^{0,1}(H; H)$ *is dense in* $UC_b(H; H)$.

Proposition 6.6.4 *Let* $F : H \to H$ *be continuous and bounded. Then* N *is* m-*dissipative. Moreover if* $\varphi \in UC_b(H)$ *is nonnegative then* $R(\lambda, N)\varphi$ *is nonnegative for all* $\lambda > 0$.

Proof. By Proposition 6.6.3 there exists a sequence $(F_n) \subset C_b^{0,1}(H; H)$ such that $F_n \to F$ in $UC_b(H; H)$. Given $\lambda \geq \lambda_0 = \pi K^2 \|F\|_0^2$ and $f \in UC_b(H)$, consider the equation

$$\lambda \varphi_n - N_n \varphi_n = \lambda \varphi_n - L \varphi_n - \langle F_n(x), D\varphi_n \rangle = f, \qquad (6.6.10)$$

which can be solved as in Lemma 6.6.1 by successive approximations. Moreover, due to the dissipativity of $\lambda - N_n$, we have

$$\lim_{n \to \infty} \varphi_n = \varphi, \text{ in } UC_b(H; H),$$

where $\varphi = R(\lambda, N)f$. By Lemma 6.6.2 it follows that $|\varphi_n(x)| \leq \frac{1}{\lambda} \|f\|_0$, $x \in H$. Therefore

$$|\varphi(x)| \leq \frac{1}{\lambda} \|f\|_0, \qquad \forall x \in H,$$

and consequently N is m-dissipative.

Finally, the last statement follows from Lemma 6.6.2. □

The proof of the following result is straightforward, it is left to the reader as an exercise.

Proposition 6.6.5 *Assume, besides* (6.0.2)*, that* $Q = I$ *and* $F \in UC_b^1(H; H)$*. Then for any* $\lambda > 0$ *and any* $f \in UC_b^1(H)$ *there is a unique solution* $\varphi \in C_b^2(H)$ *of the equation*

$$\lambda \varphi(x) - L\varphi(x) - \langle F(x), D\varphi(x) \rangle = f(x), \qquad x \in H. \qquad (6.6.11)$$

Chapter 7

General parabolic equations

The present chapter is concerned with the following Kolmogorov equation in a separable Hilbert space H (norm $|\cdot|$, inner product $\langle\cdot,\cdot\rangle$):

$$
\begin{cases}
D_t u(t,x) = \frac{1}{2}\operatorname{Tr}[(G(x))^* D^2 u(t,x)(G(x))] \\
\qquad\qquad +\langle Ax + F(x), Du(t,x)\rangle, \ t \geq 0, \ x \in D(A), \qquad (7.0.1)\\
u(0,x) = \varphi(x), \ x \in H.
\end{cases}
$$

Here $A : D(A) \subset H \to H$ is the generator of a C_0 semigroup e^{tA} in H and $F : H \to H$, $G : H \to L(V,H)$ are at least Lipschitz continuous, and V is another separable Hilbert space. We will make rather strong assumptions on the coefficients and the initial function φ. The results of this chapter should be regarded as a starting point for a study of the Kolmogorov equations with less regular data. For reaction-diffusion equations see S. Cerrai [43].

A function $u(t,x)$, $t \geq 0$, $x \in H$, is said to be a *strict solution* to equation (7.0.1) if

(i) u is continuous on $[0,+\infty) \times H$, and $u(0,\cdot) = \varphi$,

(ii) $u(t,\cdot) \in UC_b^2(H)$ for all $t \geq 0$,

(iii) for any $x \in D(A)$, $u(\cdot,x)$ is continuously differentiable on $[0,+\infty)$, and fulfills (7.0.1).

We shall define a *generalized* solution to (7.0.1) by the formula

$$
u(t,x) = P_t\varphi(x) = \mathbb{E}\left[\varphi(X(t,x))\right], \quad t \geq 0, x \in H, \ \varphi \in C_b(H),
$$

127

where the process $X(t, x)$ $t \geq 0$, $x \in H$, is the solution of the differential stochastic equation,

$$\begin{cases} dX(t) = (AX(t) + F(X(t)))dt + G(X(t))dW(t), \\ X(0) = x \in H, \end{cases} \tag{7.0.2}$$

where W is a Q-Wiener process with values in a Hilbert space U such that $V = Q^{\frac{1}{2}}(U)$, see [101].

We start by recalling in §7.1 some results on implicit function theorems on which our approach is based. Basic definitions of Q-Wiener processes and of the stochastic integration theory, together with their fundamental properties, are recalled in §7.2. Then we shall prove results on existence of solutions to (7.0.2) and on their dependence on initial data in §7.3. Then §7.4 is devoted to the regular dependence of the generalized solutions to (7.0.1) on the initial condition. Existence and uniqueness of the strict solutions to (7.0.1) are the subjects of §7.5 and §7.6 respectively. Finally §7.7 presents an extension, based on a version of the Bismut-Elworthy-Xe formula, of the regularity results, to the case of only bounded initial functions φ.

We follow basically J. Zabczyk [220].

7.1 Implicit function theorems

We gather here abstract theorems on implicit functions needed in what follows. They are versions of similar results from [101], [102] and [220]. Some improvements can be found in C. Knoche and K. Frieler [?].

Let Λ be an open subset of a Banach space E and H a transformation from $\Lambda \times E$ into E. We shall assume that

$$\|H(\lambda, x) - H(\lambda, y)\|_E \leq \alpha \|x - y\|_E, \quad \text{for all } \lambda \in \Lambda, \ x, y \in E, \tag{7.1.1}$$

and consider the equation

$$x = H(\lambda, x), \quad (\lambda, x) \in \Lambda \times E. \tag{7.1.2}$$

Theorem 7.1.1 *Assume that the transformation H satisfies (7.1.1) with $\alpha \in [0, 1)$. Then for arbitrary $\lambda \in \Lambda$ there exists exactly one solution $x = \varphi(\lambda)$ of the equation (7.1.2). If in addition H is continuous with respect to the first variable then the function φ is continuous on Λ.*

If G is a mapping from E into F then its directional derivative at x and in direction y will be denoted by $\partial_x G(x; y)$. If the directional derivative is a linear and continuous mapping of y then this mapping will be called the *Gateaux derivative* of G at the point x and denoted by $\partial G(x)$. This convention naturally extends to higher and partial derivatives. In particular the nth Gateaux derivative $\partial^n G(x)$ at the point x is a continuous n-linear transformation from E into F. If the Gateaux derivative is continuous with respect to x for all fixed directions then we say that the Gateaux derivative is strongly continuous.

Theorem 7.1.2 *Assume that H is continuous with respect to the first variable and satisfies (7.1.1) with $\alpha \in [0,1)$. If, in addition, at any $(\lambda, x) \in \Lambda \times E$, there exist strongly continuous directional Gateaux derivatives $\partial_\lambda H(\lambda, x)$ and $\partial_x H(\lambda, x)$ then for arbitrary $\lambda \in \Lambda$ there exists the strongly continuous Gateaux derivative $\partial_\lambda \varphi$, and for any $\lambda, \mu \in \Lambda$,*

$$\partial_\lambda \varphi(\lambda; \mu) = [I - \partial_x H(\lambda, \varphi(\lambda))]^{-1} \partial_\lambda H(\lambda, \varphi(\lambda); \mu). \tag{7.1.3}$$

Theorem 7.1.3 *Assume that E_0 is a Banach space continuously embedded into E and that (7.1.1) holds in both E and E_0 with the same constant $\alpha \in [0,1)$. Let moreover the assumptions of Theorem 7.1.2 be satisfied in E and E_0. If, in addition, there exist strongly continuous second partial Gateaux derivatives from $\Lambda \times E_0$ into E, $\partial_\lambda \partial_x H$, $\partial_x \partial_\lambda H$ and $\partial_\lambda^2 H, \partial_x^2 H$, then there exists the strongly continuous second partial Gateaux derivative of $\partial_\lambda^2 \varphi$ and at any point $\lambda \in \Lambda$ and in any directions $\mu_0, \nu_0 \in \Lambda$,*

$$\partial_\lambda^2 \varphi(\lambda_0; \mu_0, \nu_0) = [I - \partial_x H(\lambda_0, \varphi(\lambda_0))]^{-1}$$

$$\times \Big[\partial_x^2 H(\lambda_0, \varphi(\lambda_0); \partial_\lambda \varphi(\lambda_0; \mu_0), \partial_\lambda \varphi(\lambda_0; \nu_0)$$

$$+ \partial_\lambda \partial_x H(\lambda_0, \varphi(\lambda_0); \partial_\lambda \varphi(\lambda_0; \mu_0), \nu_0) + \partial_x \partial_\lambda H(\lambda_0, \varphi(\lambda_0); \mu_0, \partial_\lambda \varphi(\lambda_0; \nu_0))$$

$$+ \partial_\lambda^2 H(\lambda_0, \varphi(\lambda_0); \mu_0, \nu_0) \Big].$$

$$\tag{7.1.4}$$

Remark 7.1.4 Note that by Theorems 7.1.1 and 7.1.2, the functions $\varphi(\lambda)$, $\partial_\lambda \varphi(\lambda; \mu)$, $\lambda, \mu \in \Lambda$, are continuous as E_0-valued functions. Therefore, by the assumptions and Theorem 7.1.3, the formula (7.1.4) defines a continuous function of all three variables.

Assume now that we have a sequence of mappings H_n, $n \in \mathbb{N}$ and a mapping H such that

$$\|H(\lambda, x) - H(\lambda, y)\|_E \leq \alpha \|x - y\|, \quad x, y \in E, \lambda \in \Lambda, \tag{7.1.5}$$

$$\|H_n(\lambda, x) - H_n(\lambda, y)\|_E \leq \alpha \|x - y\|, \quad x, y \in E, \lambda \in \Lambda. \tag{7.1.6}$$

If $\alpha \in [0, 1)$ then for each $\lambda \in \Lambda$ the equations

$$x = H(\lambda, x), \tag{7.1.7}$$
$$x = H_n(\lambda, x) \tag{7.1.8}$$

have unique solutions denoted by $\varphi(\lambda)$ and $\varphi_n(\lambda)$ respectively. Thus

$$\varphi(\lambda) = H(\lambda, \varphi(\lambda)), \quad \varphi_n(\lambda) = H_n(\lambda, \varphi_n(\lambda)), \quad \lambda \in \Lambda, \; n \in \mathbb{N}. \tag{7.1.9}$$

Theorem 7.1.5 *Assume that functions H, H_n, $n \in \mathbb{N}$, satisfy (7.1.5) and (7.1.6) with $\alpha \in [0, 1)$.*
  (i) If $H_n(\lambda, x) \to H(\lambda, x)$, for all $(\lambda, x) \in \Lambda \times E$, then $\varphi_n(\lambda) \to \varphi(\lambda)$, for all $\lambda \in \Lambda$
  (ii) If H, H_n, $n \in \mathbb{N}$ are continuous and $H_n \to H$ uniformly on compact sets then $\varphi_n \to \varphi$ uniformly on compact sets.

The convergence result from Theorem 7.1.5 can be extended to directional derivatives of φ as well.

Theorem 7.1.6 *Assume that mappings H, H_n, $n \in \mathbb{N}$, satisfy the conditions of Theorem 7.1.2. If*

$$H_n(\lambda, x) \to H(\lambda, x), \quad \partial_x H_n(\lambda, x; y) \to \partial_x H(\lambda, x; y),$$

$$\partial_\lambda H_n(\lambda, x; \mu) \to \partial_\lambda H(\lambda, x; \mu) \quad as \; n \to +\infty$$

uniformly in (λ, x) on compact sets and uniformly in y and μ from bounded sets, then

$$\partial_\lambda \varphi_n(\lambda; \mu) \to \partial_\lambda \varphi(\lambda; \mu) \quad as \; n \to +\infty \tag{7.1.10}$$

uniformly in λ from compact sets and μ from bounded sets.

Theorem 7.1.7 *Assume that mappings H, H_n, $n \in \mathbb{N}$, satisfy the conditions of Theorem 7.1.3 with the same space E_0. Assume that the conditions of Theorem 7.1.6 are satisfied in both E and E_0. If, in addition*

$$\begin{aligned}
\partial_\lambda \partial_x H_n(\lambda, x; y; \mu) &\to \partial_\lambda \partial_x H(\lambda, x; y; \mu), & as \; n \to +\infty, \\
\partial_x \partial_\lambda H_n(\lambda, x; \mu, y) &\to \partial_\lambda \partial_x H(\lambda, x; \mu, y), & as \; n \to +\infty, \\
\partial_\lambda^2 H_n(\lambda, x; \mu, \nu) &\to \partial_\lambda^2 H(\lambda, x; \mu, \nu), & as \; n \to +\infty, \\
\partial_x^2 H_n(\lambda, x; y, z) &\to \partial_x^2 H(\lambda, x; y, z), & as \; n \to +\infty,
\end{aligned}$$

uniformly in (λ, x) *from compact subsets of* $\Lambda \times E_0$ *and uniformly in* (μ, ν, y, z) *from bounded sets of* $\Lambda \times \Lambda \times E \times E$, *then*

$$\partial_\lambda^2 \varphi_n(\lambda; \mu, \nu) \to \partial_\lambda^2 \varphi(\lambda; \mu, \nu), \qquad as\ n \to +\infty, \tag{7.1.11}$$

uniformly in λ *from compact subsets of* Λ *and uniformly in* μ, ν *from bounded subsets of* $\Lambda \times \Lambda$.

7.2 Wiener processes and stochastic equations

7.2.1 Infinite dimensional Wiener processes

Let $(\Omega, \mathcal{F}, \mathbb{P})$ be a probability space with a given increasing family of σ-fields $\mathcal{F}_t \subset \mathcal{F}$, $t \geq 0$, and let $Q \in L^1_+(U)$. A family $W(t)$, $t \geq 0$, of U-valued random variables is called a *Q-Wiener process*, if and only if

(i) $W(0) = 0$ and $\mathcal{L}(W(t) - W(s)) = N_{(t-s)Q}$, $t \geq s$,

(ii) $W(t_1)$, $W(t_2) - W(t_1), \ldots, W(t_n) - W(t_{n-1})$ are independent random variables, $0 \leq t_1 < t_2 < \cdots < t_n$, $n \in \mathbb{N}$,

(iii) for almost all $\omega \in \Omega$, $W(t, \omega)$, $t \geq 0$, is a continuous function.

In particular we have

$$\mathbb{E}\langle W(t), a\rangle_U \langle W(s), b\rangle_U = t \wedge s \langle Qa, b\rangle_U, \qquad a, b \in U.$$

Let (e_k) be the sequence of all eigenvectors of Q corresponding to the sequence of eigenvalues (γ_k). If $\gamma_k > 0$ then

$$\beta_k(t) = (\gamma_k)^{-\frac{1}{2}} \langle W(t), e_k\rangle_U, \quad t \geq 0,$$

are one dimensional standard Wiener processses $(Q = 1)$, mutually independent. It is clear that

$$W(t) = \sum_{k=1}^{\infty} \sqrt{\gamma_k} \beta_k(t) e_k, \qquad t \geq 0. \tag{7.2.1}$$

Conversely, if (e_k) is a complete orthonormal system in U, (γ_k) is a summable sequence of nonnegative numbers and (β_k) are independent, standard Wiener processes then the formula (7.2.1) defines a U-valued Wiener process.

7.2.2   Stochastic integration

There is a natural class of operator valued processes which can be stochastically integrated with respect to a U-valued Q-Wiener process W. Denote, as before, $V = Q^{\frac{1}{2}}(U)$ the image of U by the operator $Q^{1/2}$ equipped with the following scalar product: $\langle a, b \rangle_V = \langle Q^{-1/2}a, Q^{-1/2}b \rangle_U$, where $Q^{-1/2}$ denotes the pseudo-inverse of $Q^{1/2}$. The space V is a separable Hilbert space.

Denote by $L_{HS}(V, H)$ the Hilbert space of all Hilbert-Schmidt operators from V into H equipped with the Hilbert-Schmidt norm. The space $L_{HS}(V, H)$ is again a separable Hilbert space which, from now on, will be denoted by $\mathcal{H}$. Note that operators belonging to $L_{HS}(V, H)$ are not, in general, defined on the whole space U and that different Wiener processes may lead to the same space $\mathcal{H}$.

An $\mathcal{H}$-valued process Ψ is called *measurable* on $[0, t]$ if it is measurable when treated as a transformation from the set $\Omega \times [0, t]$, equipped with the σ-field $\mathcal{F}_t \times \mathcal{B}([0, t])$, into $(\mathcal{H}, \mathcal{B}(\mathcal{H}))$ If the process Ψ is measurable on all intervals $[0, t], t \geq 0$, then it is called *progressively measurable*. A progressively measurable process Ψ such that

$$\mathbb{P} \left(\int_0^t \|\Psi(s)\|_{\mathcal{H}}^2 \, ds < +\infty, \quad t \geq 0 \right) = 1 \qquad (7.2.2)$$

is called *stochastically integrable*. For stochastically integrable processes the *stochastic Itô integral*

$$\int_0^t \Psi(s) \, dW(s), \quad t \geq 0, \qquad (7.2.3)$$

is well defined, see [101]. Moreover:

$$\mathbb{E} \left(\left| \int_0^t \Psi(s) \, dW(s) \right|_H^2 \right) \leq \mathbb{E} \left(\int_0^t \|\Psi(s)\|_{\mathcal{H}}^2 \, ds \right). \qquad (7.2.4)$$

If the right hand side of (7.2.4) is finite then (7.2.4) becomes an identity. In addition the following Burkholder-Davis-Gundy inequality holds . For arbitrary $p > 0$ there exists a constant $c_p > 0$ such that

$$\mathbb{E} \left(\sup_{s \leq t} \left| \int_0^s \Psi(u) \, dW(u) \right|_H^p \right) \leq c_p \mathbb{E} \left(\int_0^t \|\Psi(s)\|_{\mathcal{H}}^2 \, ds \right)^{p/2}, \quad t \geq 0. \quad (7.2.5)$$

Assume that Ψ is an $\mathcal{H}$-valued process stochastically integrable on $[0, T]$, ψ is an H-valued progressively measurable process, with trajectories Bochner integrable on $[0, T]$, $\mathbb{P}$-a.s., and $Z(0)$ is an $\mathcal{F}_0$-measurable H-valued random variable. Then the process

$$Z(t) = Z(0) + \int_0^t \psi(s)\, ds + \int_0^t \Psi(s)\, dW(s), \qquad t \in (0, T), \qquad (7.2.6)$$

is continuous and well defined. Let $F : [0, T] \times H \to \mathbb{R}$ be a function uniformly continuous on bounded subsets of $[0, T] \times H$ together with its partial derivatives F_t, F_x, F_{xx}. The following, *Itô's formula*, holds.

Theorem 7.2.1 *Under the above assumptions on the function F and the process Z, $\mathbb{P}$-a.s., for all $t \in [0, T]$,*

$$F(t, Z(t)) \;=\; F(0, Z(0)) + \int_0^t \langle F_x(s, Z(s)), \Psi(s)\, dW(s)\rangle_H$$

$$+ \int_0^t \Big[F_t(s, Z(s)) + \langle F_x(s, Z(s)), \psi(s)\rangle_H$$

$$+ \frac{1}{2}\mathrm{Tr}[(\Psi(s)Q^{1/2})^* F_{xx}(s, Z(s))(\Psi(s)Q^{1/2})] \Big] ds.$$

$$(7.2.7)$$

7.3 Dependence of the solutions to stochastic equations on initial data

It is convenient to write (7.0.2) in the *mild* form

$$X(t) = e^{tA}x + \int_0^t e^{(t-s)A}F(X(s))ds + \int_0^t e^{(t-s)A}G(X(s))dW(s). \quad (7.3.1)$$

We will analyze this equation using a functional analytic approach, based on an implicit function theorem, see [102] and [220], and on properties of deterministic and stochastic convolutions.

7.3.1 Convolution and evaluation maps

Denote by $H^p([0, T])$, $p \geq 1$, the space of all progressively measurable H-valued processses ψ defined on $[0, T]$, $T > 0$, equipped with the norm

$$\|\psi\|_{H^p([0,T])} = \sup_{t \in [0,T]} \left(\mathbb{E}|\psi(t)|_H^p\right)^{1/p}.$$

Similarly denote by $\mathcal{H}^p([0,T])$ the space of all progressively measurable $\mathcal{H}$-valued processes Ψ, defined on $[0,T]$, equipped with the norm

$$\|\Psi\|_{\mathcal{H}^p([0,T])} = \sup_{t\in[0,T]} \left(\mathbb{E}\|\Psi(t)\|_{\mathcal{H}}^p\right)^{1/p}.$$

We introduce also spaces $H^{p,p}([0,T])$ and $\mathcal{H}^{p,p}([0,T])$ defined similarly to $H^p([0,T])$, $\mathcal{H}^p([0,T])$ but with different norms:

$$\|\psi\|_{H^{p,p}([0,T])} = \left(\mathbb{E}\int_o^T |\psi(t)|_H^p dt\right)^{1/p}, \qquad (7.3.2)$$

$$\|\Psi\|_{\mathcal{H}^{p,p}([0,T])} = \left(\mathbb{E}\int_0^T \|\Psi(t)\|_{\mathcal{H}}^p dt\right)^{1/p}.$$

The normed spaces $H^p([0,T])$, $\mathcal{H}^p([0,T])$, $H^{p,p}([0,T])$ and $\mathcal{H}^{p,p}([0,T])$are Banach spaces denoted shortly by H^p, $\mathcal{H}^p$ and $H^{p,p}$, $\mathcal{H}^{p,p}$.

In some cases, it will be convenient to consider spaces $H_c^p([0,T])$ and $\mathcal{H}_c^p([0,T])$ consisting of those processes ψ and Ψ which are stochastically continuous in $t \in [0,T]$. More precisely $\psi \in H_c^p([0,T])$ if $\psi \in H^p([0,T])$ and for each $t_0 \in ([0,T])$,

$$\lim_{t\to t_0} \mathbb{E}|\psi(t) - \psi(t_0)|_H^p = 0.$$

In a similar way $\Psi \in \mathcal{H}_c^p([0,T])$ if $\Psi \in \mathcal{H}^p([0,T])$ and for each $t_0 \in [0,T]$,

$$\lim_{t\to[0,T]} \mathbb{E}\|\Psi(t) - \Psi(t_0)\|_{\mathcal{H}}^p = 0.$$

The spaces $H_c^p([0,T])$, $\mathcal{H}_c^p([0,T])$ are linear closed subspaces of $H^p[0,T]$, and $\mathcal{H}^p([0,T])$ respectively and therefore are Banach spaces as well.

We will need the following equivalent norms, see [14]:

$$\|\psi\|_{p,\lambda,T} = \sup_{t\in[0,T]} e^{-\lambda t} \left(\mathbb{E}|\psi(t)|_H^p\right)^{1/p},$$

and

$$\|\Psi\|_{p,\lambda,T} = \sup_{t\in[0,T]} e^{-\lambda t} \left(\mathbb{E}\|\Psi(t)\|_{\mathcal{H}}^p\right)^{1/p},$$

where $\lambda \in \mathbb{R}$.

Define now for $\psi \in H^p([0,T])$, $\Psi \in \mathcal{H}^p([0,T])$ two *convolution* type mappings

$$T_0(\psi)(t) = \int_0^t e^{(t-s)A}\psi(s)ds, \qquad t \in [0,T], \qquad (7.3.3)$$

and

$$T_1(\Psi)(t) = \int_0^t e^{(t-s)A}\Psi(s)dW(s), \quad t \in [0,T]. \tag{7.3.4}$$

We have the following easy consequence of the Burkholder-Davis-Gundy inequality , see [101].

Proposition 7.3.1 *For arbitrary $p \geq 2$ and $T > 0$, the formulae (7.3.3) and (7.3.4) define linear operators from $H^p([0,T])$ and $H^{p,p}([0,T])$ into $H^p([0,T])$ and from $\mathcal{H}^p([0,T])$ and $\mathcal{H}^{p,p}([0,T])$ into $H^p([0,T])$ respectively. In addition, for arbitrary $\alpha \in (0,1)$ there exists $\lambda > 0$ such that*

$$\|T_0(\psi)\|_{p,\lambda,T} \leq \alpha\|\psi\|_{p,\lambda,T}. \tag{7.3.5}$$

and

$$\|T_1(\Psi)\|_{p,\lambda,T} \leq \alpha\|\Psi\|_{p,\lambda,T}. \tag{7.3.6}$$

Proof. The proof is straightforward. We will show only how to select $\lambda > 0$ to fulfill (7.3.6). By the very definition:

$$\begin{aligned}
\mathbb{E}|T_1(\Psi)(t)|_H^p &= \mathbb{E}\left|\int_0^t e^{(t-s)A}\Psi(s)dW(s)\right|_H^p \\
&\leq M_T^p c_p \mathbb{E}\left(\int_0^t \|\Psi(s)\|_{\mathcal{H}}^2 ds\right)^{p/2} \\
&\leq M_T^p c_p \mathbb{E}\left(\int_0^t \|\Psi(s)\|_{\mathcal{H}}^p ds\right) \\
&\leq M_T^p c_p \left(\int_0^t e^{\lambda tp}\mathbb{E}(e^{-\lambda s}\|\Psi(s)\|)_{\mathcal{H}}^p ds\right) \\
&\leq M_T^p c_p \left(\int_0^t e^{\lambda tp}ds\right)\|\Psi\|_{p,\lambda,T}^p \\
&\leq M_T^p c_p \frac{e^{\lambda tp}-1}{\lambda p}\|\Psi\|_{p,\lambda,T}^p.
\end{aligned}$$

Therefore,

$$\|T_1(\Psi)\|_{p,\lambda,T}^p \leq \frac{1}{\lambda}M_T^p c_p\frac{e^{\lambda tp}-1}{p}\|\Psi\|_{p,\lambda,T}^p$$

and it is enough to choose λ such that

$$\frac{1}{\lambda}\, M_T^p\, c_p \frac{e^{\lambda t p} - 1}{p} \leq \alpha. \quad \square$$

The mappings $F : H \to H$, $G : H \to \mathcal{H}$ induce mappings $\mathcal{F}$ and $\mathcal{G}$ on stochastic processes according to the formulae

$$\mathcal{F}(\psi)(t) = F(\psi(t)), \quad \psi \in H^p([0,T]), \ t \in [0,T],$$

and

$$\mathcal{G}(\psi)(t) = G(\psi(t)), \quad \psi \in H^p([0,T]), \ t \in [0,T].$$

They are often called *evaluation maps*. The following proposition follows from the definitions of the appropriate norms.

Proposition 7.3.2 *If $F : H \to H$ and $G : H \to \mathcal{H}$ are Lipschitz continuous mappings then the mappings $\mathcal{F} : H^p([0,T]) \to H^{p,p}([0,T])$, $\mathcal{G} : H^p([0,T]) \to \mathcal{H}^{p,p}([0,T])$ are Lipschitz continuous as well.*

We have also the following crucial result.

Proposition 7.3.3 *Assume that the Lipschitz continuous mappings $F : H \to H$, $G : H \to \mathcal{H}$ have directional Gateaux derivatives*

$$\partial^k F(x)(y_1, \ldots, y_k), \quad \partial^k G(x)(y_1, \ldots, y_k), \quad x \in H, \ y_1, \ldots, y_k \in H,$$

for $k = 1, \ldots, n$, continuous in all variables and such that for $k = 1, \ldots, n$,

$$\sup_{\substack{x \in H \\ \|y_i\| \leq 1 \\ i=1,\ldots,k}} \|\partial^k F(x)(y_1, \ldots, y_k)\|_H < +\infty,$$

and

$$\sup_{\substack{x \in H \\ \|y_i\| \leq 1 \\ i=1,\ldots,k}} \|\partial^k G(x)(y_1, \ldots, y_k)\|_{\mathcal{H}} < +\infty,$$

Then the transformations $\mathcal{F} : H^p([0,T]) \to H^{p,p}([0,T])$ and $\mathcal{G} : H^p([0,T]) \to \mathcal{H}^{p,p}([0,T])$ have directional Gateaux derivatives

$$\partial^k \mathcal{F}(X)(Y_1, \ldots, Y_k), \quad \partial^k \mathcal{G}(X)(Y_1, \ldots, Y_k), \quad k = 0, 1, \ldots, n,$$

for each $X \in H^p([0,T])$ and $Y_1, \ldots, Y_k \in H^{kp}([0,T])$. The derivatives are continuous from $H^p \times \underbrace{H^{kp} \times \cdots \times H^{kp}}_{k \ times}$ into $H^{p,p}$ and $\mathcal{H}^{p,p}$ respectively and

$$\sup_{\substack{x \in H^p \\ \|Y_i\|_{H^{kp}} \leq 1 \\ i=1,\ldots,k}} \|\partial^k \mathcal{F}(x)(y_1, \ldots, y_k)\|_{H^{p,p}} < +\infty,$$

and

$$\sup_{\substack{x \in H^p \\ \|Y_i\|_{H^{kp}} \leq 1 \\ i=1,\ldots,k}} \|\partial^k \mathcal{G}(x)(y_1,\ldots,y_k)\|_{\mathcal{H}^{p,p}} < +\infty,$$

Moreover

$$\partial^k \mathcal{F}(X)(Y_1,\ldots,Y_k)(t) = \partial^k F(X(t))(Y_1(t),\ldots,Y_k(t)), \quad t \in [0,T],$$

$$\partial^k \mathcal{G}(X)(Y_1,\ldots,Y_k)(t) = \partial^k G(X(t))(Y_1(t),\ldots,Y_k(t)), \quad t \in [0,T].$$

Proof. We will consider for instance the operator $\mathcal{F}$. The theorem is true if $n = 1$. If it is true for some k then for arbitrary $X \in H^p$, $Y_1,\ldots,Y_{k+1} \in H^{(k+1)p}$ and $\sigma > 0$,

$$\frac{1}{\sigma}[\partial^k \mathcal{F}(X(t) + \sigma Y_{k+1}(t))(Y_1(t),\ldots,Y_k(t)) - \partial^k \mathcal{F}(X;Y_1,\ldots,Y_k)(t)]$$

$$= \frac{1}{\sigma}[\partial^k F(X(t) + \sigma Y_{k+1}(t))(Y_1(t),\ldots,Y_k(t))$$

$$-\partial^k F(X(t))(Y_1(t),\ldots,Y_k(t))]$$

$$= \int_0^1 \partial^{k+1} F(X(t) + \sigma s Y_{k+1}(t))(Y_1(t),\ldots,Y_k(t),Y_{k+1}(t))ds.$$

Therefore,

$$\left\| \frac{1}{\sigma}[\partial^k \mathcal{F}(X + \sigma Y_{k+1})(Y_1,\ldots,Y_k) - \partial^k \mathcal{F}(X)(Y_1,\ldots,Y_k)] \right.$$

$$\left. -\partial^{k+1} F(X(\cdot))(Y_1(\cdot),\ldots,Y_k(\cdot),Y_{k+1}(\cdot)) \right\|_{H^{p,p}}^p$$

$$\leq \mathbb{E} \int_0^T \int_0^1 \left[|\partial^{k+1} F(X(t) + \sigma s Y_{k+1}(t))(Y_1(t),\ldots,Y_{k+1}(t)) \right.$$

$$\left. -\partial^{k+1} F(X(t))(Y_1(t),\ldots,Y_{k+1}(t))|^p \right] ds dt$$

$$\leq \mathbb{E} \int_0^T \int_0^1 \left[|\partial^{k+1} F(X(t) + \sigma s Y_{k+1}(t)) - \partial^{k+1} F(X(t))|^p \right.$$

$$\left. \times |Y_1(t)|^p \ldots |Y_{k+1}(t)|^p \right] ds dt. \quad \square$$

Remark 7.3.4 Note that in the case $H = L^2(D)$, where $D \subset \mathbb{R}^d$ is an open set, and in the case where F and G are Nemytski operators, that is

$$f(x)(\xi) = f(\xi, x(\xi)), \quad g(x)(\xi) = g(\xi, x(\xi)), \quad \xi \in D,$$

for some functions $f : D \times \mathbb{R} \to \mathbb{R}$ and $g : D \times \mathbb{R} \to \mathbb{R}$, then the assumptions of Proposition 7.3.2 are not satisfied. To this purpose we refer to S. Cerrai [43, Chapters 4 and 6].

7.3.2   Solutions of stochastic equations

We go back to the equation (7.0.2) and investigate existence of solutions and the character of their dependence on initial conditions $x \in H$.

For each $x \in H$ and $Y \in H^p([0,T])$ define mappings $\mathcal{K}$, $\mathcal{K}_0$ and $\mathcal{K}_1$,

$$\mathcal{K}(x,Y)(t) = e^{tA}x + \int_0^t e^{(t-s)A}F(Y(s))ds + \int_0^t e^{(t-s)A}G(Y(s))dW(s),$$

$$\mathcal{K}_0(Y)(t) = \int_0^t e^{(t-s)A}F(Y(s))ds,$$

$$\mathcal{K}_1(Y)(t) = \int_0^t e^{(t-s)A}G(Y(s))dW(s),$$

from $H \times H^p([0,T])$ into $H^p([0,T])$.

Equation(7.0.2) is equivalent to (7.3.1) which in turn can be written as a fixed point problem:

$$X = \mathcal{K}(x, X). \tag{7.3.7}$$

We have the following existence result.

Theorem 7.3.5 *Assume that $F : H \to H$ and $G : H \to \mathcal{H}$ are Lipschitz continuous. Then for each $p \geq 2$, $T > 0$ and $x \in H$ the equation (7.3.1) has a unique solution $X(\cdot, x)$ in $H^p([0,T])$. Moreover the mapping $x \to X(\cdot, x)$ from H into $H^p([0,T])$ is Lipschitz continuous.*

Proof.   According to the definitions one has to show that the equation (7.3.7) has a unique solution in $H^p([0,T])$. Its right hand side is the sum of a linear mapping in the variable x and two transformations $\mathcal{K}_0$ and $\mathcal{K}_1$. Note that the transformations $\mathcal{K}_0$ and $\mathcal{K}_1$ are compositions of linear, integral mappings $\mathcal{T}_0$ and $\mathcal{T}_1$ with evaluation maps $\mathcal{F}$ and $\mathcal{G}$. By Proposition 7.3.2 we can apply Theorem 7.1.1. In this way we find existence of a solution $X(\cdot, x)$ which depends continuously on x. $\square$

Theorem 7.3.6 *In addition to the assumptions of Theorem 7.3.5 assume that $F \in C_b^k(H,H)$ and $G \in C_b^k(H,\mathcal{H})$. Then for each $p \geq 2$ and $T > 0$ the mapping $x \to X(\cdot,x)$, from H into $H^p([0,T])$ is k times Gateaux differentiable in x with bounded and strongly continuous derivatives up to order k. Moreover for any $y, z \in H$ the processes $\eta^y(t) = \partial_x X(t,x;y)$, $\zeta^{y,z}(t) = \partial_x^2 X(t,x;y,z)$, $t \in [0,T]$, are the unique solutions of the following equations:*

$$
\begin{aligned}
d\eta^y &= (A\eta^y + DF(X(t))\eta^y)dt + DG(X(t))\eta^y dW(t), \\
\eta^y(0) &= h,
\end{aligned}
\tag{7.3.8}
$$

$$
\begin{aligned}
d\zeta^{y,z} &= (A\zeta^{y,z} + DF(X(t))\zeta^{y,z})dt + DG(X(t))\zeta^{y,z}dW(t) \\
&\quad + D^2F(X(t))(\eta^y,\eta^z)dt + D^2G(X(t))(\eta^y,\eta^z)dW(t), \\
\zeta^y(0) &= 0.
\end{aligned}
\tag{7.3.9}
$$

Proof. To prove (7.3.9)-(7.3.10) we apply Theorem 7.1.3. Its assumptions are satisfied by Proposition 7.3.3. □

7.4 Space and time regularity of the generalized solutions

We can finally establish basic regularity properties of the generalized solution u to (7.0.1) given by

$$
u(t,x) = P_t\varphi(x) = \mathbb{E}\left[\varphi(X(t,x))\right], \quad t \geq 0, x \in H, \ \varphi \in C_b(H).
$$

Note that for a fixed $t \geq 0$ the function $x \to u(t,x)$ can be regarded as a composition of the mappings from H into $L^p(\Omega,H)$ and from $L^p(\Omega,H)$ into $L^q(\Omega,\mathbb{R})$, given by $x \to X(t,x)$, $\xi \to \varphi(\xi)$, and of the linear, integration operator $\eta \to \mathbb{E}(\eta)$ from $L^q(\Omega,\mathbb{R})$ into $\mathbb{R}$. The first mapping can be obtained from $x \to X(\cdot,x)$ by fixing the time argument t. So its Gateaux differentiability will be a consequence of Theorem 7.3.6. To obtain Fréchet differentiability the following result will be used.

Proposition 7.4.1 *If a mapping Ψ from a Banach space E_1 into a Banach space E_2 has all Gateaux derivatives $\partial^l \Psi$ uniformly bounded up to order k, then Ψ has all Fréchet derivatives $D^l \Psi$ continuous and bounded up to order $k-1$ identical with $\partial^l \Psi$.*

Proof. We will show only that ∂^{k-1} is norm continuous. Let us fix vectors $x, y, h_1, \dots, h_{k-1} \in E_1$. Then

$$|\partial^{k-1}\Psi(y; h_1, \dots, h_{k-1}) - \partial^{k-1}\Psi(x; h_1, \dots, h_{k-1})|$$

$$\leq \int_0^1 |\partial^k \Psi(x + \sigma(y - x); h_1, \dots, h_{k-1}, y - x)| d\sigma$$

$$\leq ||\partial^k \Psi||_0 \, |h_1| \dots |h_{k-1}| \, |x - y|.$$

Consequently

$$|\partial^{k-1}\Psi(y) - \partial^{k-1}\Psi(x)| \leq ||\partial^k \Psi||_0 \, |x - y|, \quad x, y \in E_1.$$

This proves the result. $\square$

Let $\varphi : H \to \mathbb{R}$ be a given function. If ξ is an H-valued random variable defined on $(\Omega, \mathcal{F}, \mathbb{P})$ then $\Phi(\xi)$ given by the evaluation formula

$$\Phi(\xi)(\omega) = \varphi(\xi(\omega)), \quad \omega \in \Omega,$$

is a real valued random variable provided φ is a Borel function.

The following proposition can be proved in a similar way to the previous general results on differentiability.

Proposition 7.4.2 *Assume that φ belongs to $C_b^k(H)$.*

(i) If $r \geq 1$ and $kr \leq p$ then the transformation Φ from $L^p(\Omega, H)$ into $L^r(\Omega, R)$ has strongly continuous and bounded Gateaux derivatives up to order k.

(ii) If $r \geq 1$ and $kr < p$ then the transformation Φ from $L^p(\Omega, H)$ into $L^r(\Omega, R)$ has strongly continuous and bounded Fréchet derivatives up to order k.

Proof. We assume that $kr < p$ and prove only the norm continuity in $X \in L^p(\Omega, H)$ of the k-linear transformation

$$\partial^k \Phi(X; Y_1, \dots, Y_k)(\omega) = D^k \varphi(X(\omega); Y_1(\omega), \dots, Y_k(\omega)), \quad \omega \in \Omega,$$

with values in $L^r(\Omega, R)$ and defined for $Y_1, \ldots, Y_k$ in $L^p(\Omega, H)$. Note that for arbitrary X, Z and $Y_1, \ldots, Y_k \in L^p(\Omega, H)$,

$$\int_\Omega |\partial^k \varphi(X(\omega); Y_1(\omega), \ldots, Y_k(\omega)) - \partial^k \varphi(Z(\omega); Y_1(\omega), \ldots, Y_k(\omega))|^r \mathbb{P}(d\omega)$$

$$\leq \int_\Omega |\partial^k \varphi(X(\omega)) - \partial^k \varphi(Z(\omega))|^r \, |Y_1(\omega)|^r \ldots |Y_k(\omega)|^r \mathbb{P}(d\omega)$$

$$\leq \left(\int_\Omega |\partial^k \varphi(X(\omega)) - \partial^k \varphi(Z(\omega))|^{\frac{pr}{p-kr}} \mathbb{P}(d\omega) \right)^{\frac{p-kr}{p}} \prod_{j=1}^k \left(\int_\Omega |Y_j(\omega)|^p \mathbb{P}(d\omega) \right)^{\frac{r}{p}}.$$

The continuity of $\partial^k \varphi$ implies, in a standard way, the required norm continuity of the transformation. $\square$

As a consequence of the previous results we have a theorem on differentiability of the generalized solution.

Theorem 7.4.3 *Assume that* $\varphi \in C_b^2(H, \mathbb{R})$, $F \in C_b^3(H, H)$ *and* $G \in C_b^3(H, \mathcal{H})$. *Then for arbitrary* $t \geq 0$, $u(t, \cdot) \in C_b^2(H, \mathbb{R})$.

Proof. Taking into account all the preparatory results it is enough to apply Theorem 7.3.6. $\square$

We will show now that the generalized solution u is continuous with respect to time and that the same is true for its first and second space derivatives. To do so we consider the equation in the subspaces $H_c^p([0, T])$ and $\mathcal{H}_c^p([0, T])$ of $H^p([0, T])$ and $\mathcal{H}^p([0, T])$ introduced at the beginning of the section. We have the following proposition.

Proposition 7.4.4 *The formulae (7.3.3) and (7.3.4) define continuous linear mappings from* $H^p([0, T])$ *and from* $\mathcal{H}^p([0, T])$ *into* $H_c^p([0, T])$ *respectively.*

Proof. We prove the result for T_1 only; the proof for T_0 is similar. Let $0 \le s < t \le T$ and $h = t - s$. Then

$$\mathbb{E}|T_1\Psi(s) - T_1\Psi(t)|^p$$

$$\le 2^{p-1}\mathbb{E}\left|\int_0^s e^{(s-\sigma)A}\Psi(\sigma)\,dW(\sigma) - \int_0^s e^{(s-\sigma)A}e^{hA}\Psi(\sigma)\,dW(\sigma)\right|^p$$

$$+2^{p-1}\mathbb{E}\left|\int_s^{s+h} e^{(s+h-\sigma)A}\Psi(\sigma)\,dW(\sigma)\right|^p$$

$$\le 2^{p-1}M_T^p s^{1-2/p}\left(\mathbb{E}\int_0^s \|(e^{hA}-I)\Psi(\sigma)\|_{\mathcal{H}}^p\,d\sigma\right)$$

$$+2^{p-1}M_T^p h^{1-2/p}\mathbb{E}\left(\int_0^h \|\Psi(\sigma)\|_{\mathcal{H}}^p\,d\sigma\right) = I_1 + I_2.$$

Since $\|(e^{hA}-I)\Psi(\sigma)\|_{\mathcal{H}}^p \le (M_T+1)^p\|\Psi(\sigma)\|_{\mathcal{H}}^p$, $\|(e^{hA}-I)\Psi(\sigma)\|_{\mathcal{H}}^p \to 0$ as $h \to 0$, $\mathbb{P}$-almost surely, and $\mathbb{E}\int_0^T \|\Psi(\sigma)\|_{\mathcal{H}}^p\,d\sigma < +\infty$, the first integral I_1 converges to 0, as $h \to 0$, (uniformly in $t, s \in [0,T]$). It is clear that also $I_2 \to 0$ as $h \to 0$ and the result follows. $\square$

Taking into account Proposition 7.4.4 and the proofs of Theorems 7.3.5 and 7.3.6 we arrive at the following time regularity result

Theorem 7.4.5 *Under the assumptions of Theorem 7.4.3 the generalized solution $u(t,x)$, $t \in [0,T]$, $x \in H$, is continuous with respect to both variables together with its first and second x-derivatives.*

7.5 Existence

Theorem 7.5.1 *Assume that $F \in C_b^3(H;H)$, $G \in C_b^3(H;\mathcal{H})$ and $\varphi \in C_b^2(H)$. Then the generalized solution u is a strict solution of the Kolmogorov equation (7.0.1).*

Proof. We assume first that the generator A is linear, bounded and continuous on H. In this case the process $X(t) = X(t,x)$, $t \ge 0$, $x \in H$, is a strong solution of the corresponding evolution equation and therefore

$$X(t,x) = x + \int_0^t [AX(s,x) + F(X(s,x))]ds + \int_0^t G(X(s,x))dW(s).$$

Applying the Itô formula to the process $\varphi(X(t)) = \varphi(X(t,x))$, $t \geq 0$, we have

$$d\varphi(X(t)) = \langle D\varphi(X(t)), dX(t) \rangle + \frac{1}{2} \mathrm{Tr}[G^*(X(t))D^2\varphi(X(t))G(X(t))]dt.$$

Consequently,

$$
\begin{aligned}
u(t,x) &= \mathbb{E}[\varphi(X(t,x))] \\
&= \varphi(x) + \mathbb{E}\left[\int_0^t \langle AX(s) + F(X(s)), D\varphi(X(s))\rangle ds\right] \\
&\quad + \frac{1}{2}\mathbb{E}\left[\int_0^t \mathrm{Tr}[G^*(X(s))D^2\varphi(X(s))G(X(s))]ds\right].
\end{aligned}
$$

By the dominated convergence theorem

$$
\begin{aligned}
D_t^+ u(0,x) &= \lim_{t\downarrow 0} \frac{u(t,x) - \varphi(x)}{t} \\
&= \frac{1}{2}\mathbb{E}\left[\mathrm{Tr}[G(X(0))^*D^2\varphi(X(0))G(X(0))]\right] \\
&\quad + \mathbb{E}\left[\langle AX(0) + F(X(0)), D\varphi(X(0))\rangle\right].
\end{aligned}
$$

Taking into account that $X(0) = x$, $u(0,x) = \varphi(x)$ we can write that

$$
\begin{aligned}
D_t^+ u(0,x) = \lim_{t\downarrow 0} \frac{u(t,x) - \varphi(x)}{t} &= \frac{1}{2}\mathrm{Tr}[G(x)^*D^2u(0,x)G(x)] \\
&\quad + \langle Ax + F(x), Du(0,x)\rangle.
\end{aligned}
$$

Let us fix now $s > 0$. Since

$$u(t+s,x) = P_{t+s}\varphi(x) = P_t(u(s,\cdot))(x),$$

therefore, applying the previous argument with φ replaced by $u(s,\cdot)$, we obtain that

$$
\begin{aligned}
D_t^+ u(s,x) = \lim_{t\downarrow 0} \frac{u(t+s,x) - u(s,x)}{t} &= \frac{1}{2}\mathrm{Tr}[G(x)^*D^2u(s,x)G(x)] \\
&\quad + \langle Ax + F(x), Du(s,x)\rangle.
\end{aligned}
$$

However, the right hand side of the above identity is a continuous function on $[0, T] \times H$ and consequently, by Lemma 3.2.4, $u(\cdot, x)$ is continuously differentiable in t and the required result follows.

To treat the case of an arbitrary generator A we consider a sequence (u_n) where $u_n(t, x) = \mathbb{E}\varphi[X_n(t, x))]$ and X_n is a solution of

$$X_n(t, x) = x + \int_0^t [A_n X(s, x) + F(X(s, x))]ds + \int_0^t G(X(s, x))dW(s),$$

with $A_n = nA(nI - A)^{-1}$ being the Yosida approximations of A. By the first part of the proof, for $s \geq 0$ and $x \in H$,

$$D_t^+ u_n(s, x) = \frac{1}{2}\text{Tr}[G(x)^* D^2 u_n(s, x)G(x)] + \langle Ax + F(x), Du_n(s, x)\rangle.$$

$$(7.5.1)$$

Let us assume that $x \in D(A)$, so that $A_n x \to Ax$. Define a sequence of mappings $(\mathcal{K}_n)$ acting from $H_c^p([0, T])$ into $H_c^p([0, T])$ as

$$\mathcal{K}_n(x, Y)(t) = e^{tA_n}x + \int_0^t e^{(t-s)A_n} F(Y(s))\, ds + \int_0^t e^{(t-s)A_n} G(Y(s))\, dW(s),$$

for every $Y \in H_c^p([0, T])$, $t \in [0, T]$.

Since the space $H_c^p([0, T])$ is separable it is easy to check that the sequence $(\mathcal{K}_n)$ converges strongly to $\mathcal{K}$ given by

$$\mathcal{K}(x, Y)(t) = e^{tA}x + \int_0^t e^{(t-s)A} F(Y(s))\, ds + \int_0^t e^{(t-s)A} G(Y(s))\, dW(s),$$

and the corresponding Lipschitz constants remain bounded. By Theorem 7.1.5 and Theorem 7.1.6, applied to $\mathcal{K}_n$, $\mathcal{K}$ and the solutions X_n and X, we have strong convergence in H_c^p of X_n and the Gateaux derivatives ∂X_n, $\partial^2 X_n$ to X and to ∂X, $\partial^2 X$ respectively. Taking into account that $\varphi \in C_b^2([0, T], H)$ we deduce the uniformly bounded convergence of $u_n(s, x)$ and the Gateaux derivatives $\partial u_n(s, x)$, $\partial^2 u_n(s, x)$, to $u(s, x)$, $\partial u(s, x)$ and $\partial^2 u(s, x)$, respectively. So we can pass, in the equation (7.5.1), to the limit, as $n \to \infty$. This implies that the function u is a strict solution of the Kolmogorov equation. □

7.6 Uniqueness

The problem of uniqueness of the solutions to the Kolmogorov equation is here discussed. Itô's formula and regularization schemes for stochastic evolution equations are used.

We first give another proof of the uniqueness for the infinite dimensional heat equation , considered in Chapter 3, then we shall consider the general case.

7.6.1 Uniqueness for the heat equation

We shall prove the following result.

Theorem 7.6.1 *Assume that the functions φ, u, $D_t u$, Du, $D^2 u$ are uniformly continuous on closed bounded subsets of $(0, +\infty) \times H$ and that the equation*

$$\begin{cases} D_t u(t, x) = \frac{1}{2}\mathrm{Tr}[D^2_Q u(t, x)], & t > 0, \ x \in H, \\ \lim_{t \to 0} u(t, x) = \varphi(x), & x \in H, \end{cases} \qquad (7.6.1)$$

holds. Moreover, assume that u is continuous on $[0, +\infty) \times H$ and for arbitrary $T > 0$ there exists $M > 0$ such that

$$|u(t, x)| + |Du(t, x)| \le Me^{M|x|}, \qquad (t, x) \in [0, T] \times H.$$

Then

$$u(t, x) = \mathbb{E}[\varphi(x + W(t))], \qquad t \ge 0, \ x \in E.$$

Proof. Fix $t > t_0 > 0$, $x \in H$ and define

$$\psi(s) = u(t - s, x + W(s)), \qquad s \in [0, t_0].$$

Due to the assumptions imposed on u one can apply Itô's formula to ψ and obtain

$$\psi(t_0) = \psi(0) + \int_0^{t_0} \left[\frac{1}{2} \mathrm{Tr}[D^2_Q u(t - s, x + W(s)) - Du(t - s, x + W(s))] \right] ds$$

$$+ \int_0^{t_0} \langle Du(t - s, x + W(s), dW(s) \rangle. \qquad (7.6.2)$$

But u satisfies the heat equation and therefore,

$$u(t - t_0, x + W(t_0)) = u(t, x) + \int_0^{t_0} \langle Du(t - s, x + W(s)), dW(s) \rangle. \qquad (7.6.3)$$

However,

$$\int_0^{t_0} |Du(t-s, x+W(s))|^2 ds \leq M^2 e^{2|x|} \int_0^{t_0} ds < +\infty, \mathbb{P}\text{-a.s.}$$

Thus, applying the expectation operator to both sides of (7.6.3), one arrives at $\mathbb{E}[u(t-t_0, x+W(t_0))] = u(t,x)$. Since

$$|u(t-t_0, x+W(t_0))| \leq M e^{|x|} e^{\sup_{s \leq t} |W(s)|}$$

and, by Fernique's theorem, see e.g. [101], $\mathbb{E}(e^{\sup_{s \leq t} \|W(s)\|}) < +\infty$, the Lebesgue theorem implies

$$u(t,x) = \lim_{t_0 \downarrow 0} \mathbb{E}[u(t-t_0, x+W(t_0))] = \mathbb{E}[\varphi(x+W(t))]. \quad \square$$

7.6.2 Uniqueness in the general case

We are here concerned with the Kolmogorov equations (7.0.1). As before, A denotes the infinitesimal generator of a C_0 semigroup e^{tA}, $t \geq 0$, on H, Q is a self-adjoint bounded nonnegative operator on H and $F : H \to H$ and $G : H \to \mathcal{H}$ are Lipschitz continuous mappings. Equation (7.0.1) corresponds to the stochastic equation (7.0.2), which has a unique solution $X(\cdot, x)$ in $H^p([0,T])$, for any $T > 0$ and $p \geq 2$.

Our aim is to prove the following result.

Theorem 7.6.2 *Assume that F and G are Lipschitz continuous mappings and $u(t,x)$, $t \geq 0$, $x \in H$, is a bounded continuous function such that for each $t > 0$, the first and second space derivatives $Du(t,x)$ and $D^2 u(t,x)$, $x \in H$, exist and are bounded on $(0, +\infty) \times H$ and uniformly continuous on bounded subsets of $[\varepsilon, +\infty) \times H$, for any $\varepsilon > 0$. If u satisfies (7.0.1) then*

$$u(t,x) = \mathbb{E}[\varphi(X(t,x))], \qquad t \geq 0, \ x \in H, \tag{7.6.4}$$

where X is the solution to (7.0.2).

For the proof we will need some preparatory work. We cannot repeat the proof from the previous subsection for two reasons. To apply the Itô formula to the process $u(t-s, X(s,x))$, $s \in [0,t_0]$, $t_0 < t$, the function $u(t,x)$, $t > 0$, $x \in H$, should have continuous first time derivative, which exists, in general, only if $x \in D(A)$. Moreover the solution X of (7.0.2) is not given in the integral form

$$X(t) = x + \int_0^t (AX(s) + F(X(s)))ds + \int_0^t G(X(s))dW(s), \qquad t \geq 0, \tag{7.6.5}$$

required by the formulation of the Itô formula, but it satisfies a convolution type version of the stochastic equation. To overcome these difficulties we will approximate both the function u and the process X in such a way that Itô's formula will be applicable and by passing to the limit in the formulae we will arrive at (7.6.4).

We first need the concept of strong solutions of the stochastic equations. If a solution to (7.0.2) taking values in $D(A)$ is such that

$$\int_0^t |AX(s)|\, ds < +\infty, \text{ for all } t \geq 0, \text{ P-a.s.,}$$

and (7.6.5) holds, then X is called a *strong solution* to (7.0.2).

We fix $n > \omega$ ([1]) and set

$$J_n = n(n-A)^{-1}, \quad F_n(x) = J_n F(x), \quad G_n(x) = J_n G(x), \quad x \in H.$$

It is easy to see that if F and G are Lipschitz continuous mappings with respect to the spaces H and $\mathcal{H}$, then F_n and G_n are Lipschitz with respect to $\widehat{H} = D(A)$ and $\widehat{\mathcal{H}} = L_{HS}(V, \widehat{H})$. Consequently, for each initial $J_n x$, $x \in H$, there exists a unique solution X_n of the equation

$$\begin{cases} dX_n(t) = (\widehat{A}X_n(t) + F_n(X_n(t)))dt + G_n(X_n(t))dW(t), \\ X_n(0) = J_n x \in H, \end{cases} \qquad (7.6.6)$$

where $\widehat{A}$ is the restriction of A to $\widehat{H}$. The operator $\widehat{A}$ generates the same semigroup e^{tA}, $t \geq 0$, but restricted to $D(A)$.

Lemma 7.6.3 *If F and G are Lipschitz continuous mappings from H into H and H into $\mathcal{H}$ respectively then, for arbitrary $n > \omega$, $p \geq 2$, $T > 0$, the equation (7.6.6) has a unique mild solution X_n in $\widehat{H}$ which is a strong solution of (7.6.6) with $\widehat{A}$ replaced by A. Moreover if $n \to +\infty$, $X_n \to X$ in $H_c^p([0,T]; H)$.*

Proof. It has been already shown that the solution X_n exists. Since X_n is a solution to (7.6.6) it is also a solution to (7.0.2) with F, G and x replaced by F_n, G_n and $J_n x$. By the very definition, for all $t > 0$, $\int_0^t |AX_n(s)|\, ds < +\infty$, P-a.s., and this easily implies that X_n is the required strong solution. Define a sequence $(\mathcal{K}_n)$ of mappings acting from $H_c^p([0,T])$ into $H_c^p([0,T])$ by the formula

$$\mathcal{K}_n(x, Y)(t) = e^{tA}x + \int_0^t e^{(t-s)A} F_n(Y(s))\, ds + \int_0^t e^{(t-s)A} G_n(Y(s))\, dW(s),$$

---

[1]We recall that $\|e^{tA}\| \leq M_T e^{\omega t}$, $t \in [0,T]$.

where $Y \in H_c^p([0,T])$, $t \in [0,T]$. Since the space $H_c^p([0,T])$ is separable it is easy to check that the sequence $(\mathcal{K}_n)$ converges strongly to $\mathcal{K}$,

$$\mathcal{K}(x,Y)(t) = e^{tA}x + \int_0^t e^{(t-s)A}F(Y(s))\,ds + \int_0^t e^{(t-s)A}G(Y(s))\,dW(s),$$

and the corresponding Lipschitz constants remain bounded. The convergence of the solutions X_n to X, in $H_c^p([0,T])$, is therefore a consequence of Theorem 7.1.5. $\square$

We pass now to the proof of Theorem 7.6.2.

Proof. Assume that u satisfies equation (7.0.1) and has the properties formulated in Theorem 7.6.1. Fix any $t_0 \in (0,t)$, $n > 0$ and define $\psi(s) = u_n(t-s, X_n(s))$, where $u_n(s,x) = u(s, J_nx)$, $s \in [0,t_0]$. The assumptions of the Itô formula are satisfied and

$$d\psi(s) = [-D_t u_n(t-s, X_n(s)) + \mathcal{L}_n u_n(t-s, \cdot)(X_n(s))]\,ds$$

$$+\langle Du_n(t-s, X_n(s)), G_n(X_n(s))\,dW(s)\rangle, \qquad s \in [0,t_0].$$

$$(7.6.7)$$

Here

$$\mathcal{L}_n\varphi(x) = \frac{1}{2}\,\mathrm{Tr}[G_n(x))^*D^2\varphi(x)(G_n(x)]$$

$$+\langle Ax + F_n(x), D\varphi(x)\rangle, \qquad x \in D(A),$$

is the value of the operator determined by the process X_n, on the function φ. Note that

$$D_t u_n(s,x) = D_t u(s, J_nx), \qquad Du_n(s,x) = J_n^* Du(s, J_nx),$$

$$D^2 u_n(s,x) = J_n^* D^2 u(s, J_nx)J_n, \qquad s > 0,\ x \in H.$$

Therefore

$$\mathcal{L}_n u_n(s,\cdot)(x) = \frac{1}{2}\,\mathrm{Tr}[(J_nG_n(x))^*D^2u(s, J_nx)](J_nG_n(x))$$

$$+\langle AJ_nx + J_nF_n(x), Du(s, J_nx)\rangle,$$

and

$$-D_t u_n(s, x) + \mathcal{L}_n u_n(s, \cdot)(x) = -D_t u(s, J_n x) + \mathcal{L}_n u_n(s, \cdot)(x)$$

$$= \mathcal{L}_n u_n(s, \cdot)(x) - \mathcal{L}_n u_n(s, \cdot)(J_n x)$$

$$= \frac{1}{2} \text{Tr} \Big[[(J_n G_n(x) Q^{1/2})(J_n G_n(x) Q^{1/2})^*$$

$$- (G(J_n x) Q^{1/2})(G(J_n x) Q^{1/2})^*] D_x^2 u(s, J_n x) \Big]$$

$$+ \langle J_n F_n(x) - F(J_n x), Du(s, J_n x) \rangle, \ s > 0, \ x \in H.$$

Since $\mathbb{E}(\psi(t_0)) = \mathbb{E}(u_n(t - t_0, X_n(t_0))) = \mathbb{E}(u(t - t_0, J_n X_n(t_0))), \mathbb{E}(\psi(0)) = u(t, J_n x)$, one gets from (7.6.7) that

$$\mathbb{E}[u(t - t_0, J_n X_n(t_0)]$$

$$= u(t, J_n x) + \frac{1}{2} \mathbb{E} \int_0^{t_0} \text{Tr} \Big[[(J_n G_n(X_n(s)))(J_n G_n(X_n(s)))^*$$

$$- (G(J_n X_n(s)))(G(J_n X_n(s)))^*] D_x^2 u(t - s, J_n X_n(s)) \Big] ds$$

$$+ \mathbb{E} \int_0^{t_0} \langle J_n F_n(X_n(s)) - F(J_n X_n(s)), D_x u(t - s, J_n X_n(s)) \rangle ds$$

$$= u(t, J_n x) + \frac{1}{2} I_n^1(t_0) + I_n^2(t_0). \tag{7.6.8}$$

Taking into account that the process X_n has continuous trajectories one gets, by the Lebesgue dominated convergence theorem, that

$$\mathbb{E}(u(t, J_n X_n(t))) = u(t, J_n x) + \frac{1}{2} I_n^1(t) + I_n^2(t).$$

It is therefore enough to show that $I_n^1(t) \to 0$, $I_n^2(t) \to 0$ as $n \to +\infty$. We will prove for instance that $\lim_{n \to +\infty} I_n^1(t) = 0$. Note that if B and C are Hilbert-Schmidt operators then

$$\begin{aligned} |\text{Tr } BB^* - \text{Tr } CC^*| &\leq \|BB^* - CC^*\|_1 \tag{7.6.9} \\ &\leq \|(B - C)B^*\|_1 + \|C(B - C)^*\|_1 \\ &\leq \|B - C\|_{HS}(\|B\|_{HS} + \|C\|_{HS}) \end{aligned}$$

Let $M = \sup\{\|D_x^2 u(s,x)\| : \ s > 0, x \in H\}$. By (7.6.9)

$$I_n^1(t) \leq M \left(\mathbb{E} \int_0^t \|J_n G_n(X_n(s)) - G(J_n X_n(s))\|_{HS} \, ds \right)^{1/2}$$

$$\times \left(\mathbb{E} \int_0^t \left[\|J_n G_n(X_n(s)) Q^{1/2}\|_{HS} + \|G(J_n X_n(s))\|_{HS} \right] ds \right)^{1/2}.$$

Taking into account that G is Lipschitz form H into $\mathcal{H}$ and that $X_n \to X$ as $n \to +\infty$, one sees that $I_n^1(t) \to 0$ as $n \to +\infty$. $\square$

7.7 Strong Feller property

To prove that the generalized solution u is regular and satisfies the Kolmogorov equation we assumed, in the previous sections, that the initial function φ was in the space $C_b^2(H)$. In this section we show that if the diffusion operator G is nondegenerate then the generalized solution might be very regular for initial functions φ which are only bounded. Results of this type require a new technique which we will describe now. We restrict our considerations only to the regularity questions. The question of solvability of the corresponding Kolmogorov equation will not be answered here.

As before we are concerned with the stochastic equation

$$\begin{cases} dX(t) = (AX(t) + F(X(t)))dt + G(X(t))dW(t), \\ X(0) = x \in H, \end{cases} \tag{7.7.1}$$

but we assume that the space V is identical with H. Let (P_t) be the corresponding transition semigroup

$$u(t,x) = P_t \varphi(x) = \mathbb{E}\left[\varphi(X(t,x))\right], \ t \geq 0, x \in H, \ \varphi \in C_b(H).$$

The transition semigroup (P_t) is said to be *strong Feller* if for arbitrary $t > 0$ and an arbitrary function $\varphi \in B_b(H)$, $P_t \varphi \in C_b(H)$. Necessary and sufficient conditions for the strong Feller property in the case $F = 0$ and $G = I$ were given in Chapter 6. In this section we consider much more general data F and G and follow S. Peszat and J. Zabczyk [184].

Theorem 7.7.1 *Assume that $F : H \to H$, $G : H \to \mathcal{H}$ are Lipschitz continuous and that for all $x \in H$, $G(x)$ is an invertible mapping such that for some $K > 0$ $\|G^{-1}(x)\| \leq K$, $x \in H$. Assume in addition that there exists $\alpha > 0$ such that for all $t > 0$, $e^{tA} \in L_{HS}(H,H)$ and*

$$\int_0^t s^{-\alpha} \|e^{sA}\|_{HS}^2 ds < +\infty.$$

Then for arbitrary $T > 0$ there exists a constant $C_T > 0$ such that for all $\varphi \in B_b(H)$ and all $t \in [0, T]$,

$$|P_t\varphi(x) - P_t\varphi(y)| \leq \frac{C_T}{\sqrt{t}} \|\varphi\|_0 |x - y|, \quad x, y \in H. \tag{7.7.2}$$

Proof. We start by proving (7.7.2) in the case of smooth F, G and φ.

Lemma 7.7.2 *Assume that the mappings F, G and the function φ have uniformly continuous and bounded derivatives up to the second order. Then for each $t > 0$, $P_t\varphi \in UC_b^2(H)$ and*

$$\varphi(X(t, x)) = P_t\varphi(x)$$
$$+ \int_0^t \langle DP_{t-s}\varphi(X(t, x)), G(X(s, x)) \, dW(s) \rangle, \quad \mathbb{P}\text{-a.s.} \tag{7.7.3}$$

Proof. Let (e_n) be a complete orthonormal system in H. For each n, let $X_n = X_n(\cdot, x)$ be the solution to the equation

$$\begin{cases} dX_n(t) = (A_n X_n(t) + F(X_n(t)))dt + G(X_n(t))Q_n^{1/2}dW(t), \\ X_n(0) = x \in H, \end{cases}$$

where $A_n = nA(n - A)^{-1}$ is the Yosida approximation of A and Q_n is the orthogonal projection of H onto the subspace generated by $\{e_1, \ldots, e_n\}$, $n \in \mathbb{N}$.
    It can be shown that the function

$$u_n(t, x) = \mathbb{E}[\varphi(X_n(t, x))], (t, x) \in [0, +\infty) \times H,$$

is the classical solution of the Kolmogorov equation

$$\begin{cases} D_t u_n(t, x) = \frac{1}{2}\text{Tr}[(G(x)Q_n^{1/2})^* D^2 u_n(t, x)(G(x)Q_n^{1/2})] \\ \qquad\qquad + \langle A_n x + F(x), Du_n(t, x) \rangle, \ t \geq 0, \ x \in D(A), \\ u_n(0, x) = \varphi(x), \ x \in H. \end{cases}$$

Applying Itô's formula to the process $\psi(s) = u_n(t - s, X_n(s, x))$, $s \in [0, t]$, $x \in H$, one obtains $\mathbb{P}$-a.s.,

$$\varphi(X_n(t, x)) = u_n(t, x) + \int_0^t \langle Du_n(t - s, X_n(s, x)), G(X_n(x, s)) Q_n^{1/2}dW(s) \rangle. \tag{7.7.4}$$

Note that
$$Du(s, x) = \mathbb{E}[(DX(s, x))^* D\varphi(X(s, x))]$$
and
$$Du_n(s, x) = \mathbb{E}[(DX_n(s, x))^* D\varphi(X_n(s, x))]. \tag{7.7.5}$$

Applying Theorems 7.3.5, 7.3.6, and taking into account that the approximation procedure affects only the semigroup and the Wiener process, one can pass to the limit in (7.7.4) and (7.7.5) and arrive at (7.7.3). $\square$

We derive now the so called Bismut-Elworthy-Xe formula , see [184].

Lemma 7.7.3 *Under the assumptions of Lemma 7.7.2, the directional derivative* $\langle DP_t\varphi(x), h \rangle$ *is given by the formula*

$$\langle DP_t\varphi(x), h \rangle = \frac{1}{t} \mathbb{E}\left[\varphi(X(t, x)) \int_0^t \langle G^{-1}(X(s, x))(DX(s, x)h), dW(s) \rangle\right]. \tag{7.7.6}$$

Proof. Fix $h \in H$ and define $u(t, x) = P_t\varphi(x)$, $t \geq 0$, $x \in H$. Multiplying both sides of (7.7.3) by

$$\int_0^t \langle G^{-1}(X(s, x))[DX(s, x)h], dW(s) \rangle,$$

and taking expectations one gets

$$\mathbb{E}\left(\varphi(X(t, x)) \int_0^t \langle G^{-1}(X(s, x))(DX(s, x)h), dW(s) \rangle\right)$$
$$= \mathbb{E}\left[\int_0^t \langle G^*(X(s, x))DP_{t-s}\varphi(X(s, x)), G^{-1}(X(s, x))(DX(s, x)h) \rangle \, ds\right]$$
$$= \mathbb{E}\left[\int_0^t \langle DP_{t-s}\varphi(X(s, x)), DX(s, x)h \rangle ds\right]$$
$$= \int_0^t \langle D[\mathbb{E}(P_{t-s}\varphi(X(s, x)))], h \rangle ds$$
$$= \int_0^t \langle D(P_s P_{t-s}\varphi)(x), h \rangle ds = t\langle DP_t\varphi(x), h \rangle. \quad \square$$

Lemma 7.7.4 *Under the assumptions of Lemma 7.7.3, the estimate* (7.7.2) *holds true.*

Proof. Fix $T > 0$, then by Lemma 7.7.3,

$$\langle DP_t\varphi(x), h\rangle|^2 \;\leq\; \frac{1}{t^2}\,\|\varphi\|_0^2\,\mathbb{E}\left[\int_0^t |G^{-1}(X(s,x))\cdot DX(s,x)h|^2 ds\right]$$

$$\leq\; \frac{K^2}{t^2}\,\|\varphi\|_0^2\,\mathbb{E}\left[\int_0^t |DX(s,x)h|^2\,ds\right], \qquad (7.7.7)$$

and we need an estimate on the process $Y(t) = DX(t,x)h$, $t \geq 0$. By Theorem 7.3.5, the process Y is the mild solution of the equation

$$\begin{cases} dY(t) = (AY(t) + DF(X(t))\,Y(t))dt + DG(X(t))\,Y(t)\,dW(t), \\ Y(0) = h. \end{cases}$$

$$(7.7.8)$$

Equation (7.7.8) is linear, with random coefficients and equivalent to the integral equation

$$\begin{aligned} Y(t) \;=\;& e^{tA}h + \int_0^t e^{(t-s)A}DF(X(s))\,Y(s)ds \\ &+ \int_0^t e^{(t-s)A}DG(X(s))\,Y(s)dW(s). \end{aligned}$$

By our assumptions, for arbitrary $T > 0$, there exists $M > 0$, such that

$$|DF(x)y| + \|e^{\sigma A}DG(x)y\|_{HS} \leq M|y|, \quad x, y \in H, \ \sigma \in [0,T].$$

Applying the contraction mapping principle in $\mathcal{H}^2([0,T])$ one easily obtains that (7.7.8) has a unique solution satisfying

$$\mathbb{E}|Y(t)|^2 \leq C|h|^2, \qquad t \in [0,T], \qquad (7.7.9)$$

if T is sufficiently small. Reiterating the procedure one gets (7.7.9) for arbitrary $T > 0$, $h \in H$. Applying (7.7.9) to (7.7.7) one has that

$$|\langle DP_t\varphi(x), h\rangle|^2 \leq \frac{K^2}{t}\,C\,\|\varphi\|_0^2\,|h|^2, \qquad t \in [0,T].$$

Let x, y be arbitrary elements in H. Then by the mean value theorem, for $\sigma(y) \in [0,T]$

$$P_t\varphi(x) - P_t\varphi(y) = \langle DP_t\varphi(x + \sigma(y)(x-y)), x - y\rangle.$$

Therefore

$$|P_t\varphi(x) - P_t\varphi(y)| \le \frac{K\sqrt{C}}{\sqrt{t}}\|\varphi\|_0 |x - y|,$$

as required. $\square$

Lemma 7.7.5 *If* (7.7.2) *holds for arbitrary* $\varphi \in UC_b^2(H)$, *then it holds for all* $\varphi \in B_b(H)$.

Proof. If $\varphi \in UC_b(H)$, then there exists a sequence (φ_n) of functions from $UC_b^2(H)$ such that $\lim_n \varphi_n(x) = \varphi(x)$, $\|\varphi_n\|_0 \le \|\varphi\|_0$, and therefore (7.7.2) holds for all $\varphi \in UC_b(H)$. From elementary properties of measures on metric spaces one has the following estimate of the variation of the signed measure $P(t, x, \cdot) - P(t, y, \cdot)$: (^{2})

$$\mathrm{Var}\,(P(t, x, \cdot) - P(t, y, \cdot)) \quad = \quad \sup_{\substack{\varphi \in UC_b(H) \\ \|\varphi\|_0 \le 1}} |P_t\varphi(x) - P_t\varphi(y)|$$

$$\le \quad \frac{C_T}{\sqrt{t}}|x - y|, \quad x, y \in H.$$

But then for $\varphi \in B_b(H)$

$$\begin{aligned} |P_t\varphi(x) - P_t\varphi(y)| &\le \left| \int_H \varphi(z)[P(t, x, dz) - P(t, y, dz)] \right| \\ &\le \|\varphi\|_0 \,\mathrm{Var}(P(t, x, \cdot) - P(t, y, \cdot)) \\ &\le \|\varphi\|_0 \frac{C_T}{\sqrt{t}}|x - y|. \quad \square \end{aligned}$$

From Lemma 7.7.5 it follows that in order to prove the theorem one can assume that $\varphi \in UC_b^2(H)$. Now we show that it is possible to eliminate the hypothesis of twice order differentiability of the coefficients F and G. Actually it is possible to construct, see [184], approximations F_n of F and G_n of G with the following properties:

(i) F_n, G_n, $n \in \mathbb{N}$, are twice Fréchet differentiable with bounded and continuous derivatives,

(ii) F_n, G_n satisfy the Lipschitz condition uniformly with respect to $n \in \mathbb{N}$,

(iii) the operators G_n, $n \in \mathbb{N}$, are invertible and

$$\varlimsup_{n \to +\infty} \sup_{x \in H} \|G_n^{-1}(x)\| \le \sup_{x \in H} \|G^{-1}(x)\|,$$

$^2 P(t, x, \cdot)$ is the law of $X(t, x)$.

(iv) $\lim\limits_{n\to+\infty} |F_n(x) - F(x)| = 0, \qquad \lim\limits_{n\to+\infty} \|G_n(x) - G(x)\| = 0, \qquad x \in H.$

To complete the proof of the theorem set

$$P_t^n \varphi(x) = \mathbb{E}[\varphi(X_n(t,x))], \ \varphi \in C_b(H),$$

where X_n is the solution to problem (7.7.1) corresponding to coefficients F_n and G_n. Fix $T > 0$, $\varphi \in UC_b^2(H)$. By Lemma 7.7.4

$$|P_t^n \varphi(x) - P_t^n \varphi(y)| \le \frac{C_T}{\sqrt{t}} \, \|\varphi\|_0 \, |x - y|, \qquad x, y \in H, \ t \in [0, T].$$

However, for fixed $t > 0$ and $x \in H$ there exists a subsequence $(X_{n_k}(t,x))$ such that $X_{n_k}(t,x) \to X(t,x)$, $\mathbb{P}$-a.s. as $k \to +\infty$. Since φ is bounded and continuous function we have

$$P_t^{n_k} \varphi(x) = \mathbb{E}(\varphi(X_{n_k}(t,x))) \to \mathbb{E}(\varphi(X(t,x))) = P_t \varphi(x),$$

as $k \to +\infty$, and therefore

$$|P_t \varphi(x) - P_t(y)| \le \frac{C_T}{\sqrt{t}} \|\varphi\|_0 |x - y|, \qquad x, y \in H, \ t \in [0, T],$$

as required. $\square$

Chapter 8

Parabolic equations in open sets

The heat equation with a linear first order term is considered in an open subset $\mathcal{O}$ of a separable Hilbert space H. Boundary conditions are of the Dirichlet type, see §8.1. In §8.2 the regularity in the interior of the generalized solution is studied; §8.3 is devoted to the existence of strong solutions and §8.4 to uniqueness.

In this chapter we follow [92] and A. Talarczyk [207]. We note that in the special case when $\mathcal{O}$ is a half-space, some regularity results up to the boundary were proved by E. Priola [190], [193], [194], [195], [191].

8.1 Introduction

Let $\mathcal{O}$ be an open subset of a separable Hilbert space H. The present chapter is devoted to the Kolmogorov equation in the set $\mathcal{O}$, with the Dirichlet boundary condition

$$\begin{cases} D_t u(t,x) = \frac{1}{2}\mathrm{Tr}[Q^{\frac{1}{2}}D^2u(t,x)Q^{\frac{1}{2}}] + \langle x, A^*Du(t,x)\rangle & \text{if } x \in \mathcal{O}, t > 0, \\ u(0,x) = \varphi(x) & \text{if } x \in \mathcal{O}, \\ u(t,x) = 0 & \text{if } x \in \partial\mathcal{O}, t > 0. \end{cases}$$

$$(8.1.1)$$

We shall assume that

$$\begin{cases} (i) & A \text{ is the infinitesimal generator of a } C_0 \text{ semigroup } e^{tA}, \\ (ii) & Q \text{ is self-adjoint and } \int_0^t \mathrm{Tr}[e^{sA}Qe^{sA^*}]ds < \infty, \ t > 0. \end{cases}$$

$$(8.1.2)$$

The second important assumption is the controllability condition (6.2.3),

$$e^{tA}(H) \subset Q_t^{1/2}(H), \quad t > 0. \tag{8.1.3}$$

As we have noted in Chapter 6, by (8.1.3) and the closed graph theorem it follows that the operator Λ_t defined by

$$\Lambda_t = Q_t^{-\frac{1}{2}} e^{tA}, \quad t > 0,$$

is bounded. Moreover, again by the closed graph theorem, it follows that for each $t > 0$, e^{tA} is Hilbert-Schmidt, and consequently, by the semigroup property, e^{tA} is of trace class.

We will also require that for some $0 < \alpha < 1$, $T > 0$,

$$\int_0^T s^{-\alpha} \mathrm{Tr}[e^{sA} Q e^{sA^*}] ds < +\infty. \tag{8.1.4}$$

If $\mathcal{O} = H$ then the generalized solution of the problem (8.1.1) is given by the formula, see Chapter 7,

$$u(t, x) = P_t \varphi(x) = \mathbb{E}[\varphi(X(t, x))], \ t \geq 0, \ x \in H, \tag{8.1.5}$$

where $X(t, x)$ is the solution to the differential stochastic equation

$$dX = AX dt + dW(t), \quad X(0) = x, \tag{8.1.6}$$

given by

$$X(t, x) = e^{tA} x + \int_0^t e^{(t-s)A} dW_s. \tag{8.1.7}$$

The generalized solution to (8.1.1) is defined by a modified formula:

$$u(t, x) = \mathbb{E}[\varphi(X(t, x)) \chi_{\{\tau_{\mathcal{O}}^x > t\}}], \quad x \in \mathcal{O},$$

where

$$\tau_{\mathcal{O}}^x = \inf\{t > 0 : X(t, x) \in \mathcal{O}^c\}$$

is the *exit time* of the process $X(t, x)$ from $\mathcal{O}$. As a counterpart of the semigroup (P_t), it is convenient to introduce a family of linear operators $(P_t^{\mathcal{O}})$, corresponding to the Dirichlet problem in $\mathcal{O}$, and acting on Borel functions defined on $\mathcal{O}$ in the following way:

$$P_t^{\mathcal{O}} \varphi(x) = \mathbb{E}[\varphi(X(t, x)) \chi_{\{\tau_{\mathcal{O}}^x > t\}}], \quad x \in \mathcal{O}. \tag{8.1.8}$$

One can show that the family $(P_t^{\mathcal{O}})$ forms a semigroup $P_{t+s}^{\mathcal{O}} = P^{\mathcal{O}} R_t P_s^{\mathcal{O}}$ which will be called the *restricted semigroup*.

We will prove that, under weak requirements, the function $P_t^{\mathcal{O}} \varphi$ is regular in $\mathcal{O}$, for each $t > 0$. We will follow the papers by G. Da Prato, B. Goldys and J. Zabczyk [92] and by A. Talarczyk [207]. Following Talarczyk [207] we prove also that the generalized solution is the classical one and unique.

8.2   Regularity of the generalized solution

We know by Chapter 6 that, for $t > 0$, the function $P_t \varphi$ is differentiable on H an arbitrary number of times. A similar result holds also for the restricted semigroup.

The main result of the present section is the following, taken from [92] and A. Talarczyk [207].

Theorem 8.2.1 *Assume that* (8.1.2) *and* (8.1.3) *hold and there exist* $t_0, C > 0$ *and* $\delta > 0$ *such that*

$$\|\Lambda_t\| \leq Ct^{-\delta}, \quad t \in (0, t_0). \tag{8.2.1}$$

Then, for arbitrary $\varphi \in B_b(\mathcal{O})$ *and* $t > 0$, *the function* $P_t^{\mathcal{O}} \varphi$ *is continuously differentiable in* $\mathcal{O}$ *an arbitrary number of times.*

The proof will be based on several results of independent interest. The following proposition, due to E. B. Dynkin, see [108], is valid for general transition semigroups.

Lemma 8.2.2 *Let* $\varphi \in B_b(\mathcal{O})$. *Define*

$$\hat{P}_t^{\mathcal{O}} \varphi(x) = \begin{cases} P_t^{\mathcal{O}} \varphi(x) & \text{if } x \in \mathcal{O}, \\ 0 & \text{if } x \in \mathcal{O}^c. \end{cases}$$

Let $t \geq s > 0$, $x \in \mathcal{O}$ *and* $\varphi \in B_b(\mathcal{O})$, *then*

$$\left| P_t^{\mathcal{O}} \varphi(x) - P_s \left(\hat{P}_{t-s}^{\mathcal{O}} \varphi \right)(x) \right| \leq \|\varphi\|_{\mathcal{O}} P(\tau_{\mathcal{O}}^x \leq s). \tag{8.2.2}$$

Proof. We have

$$\begin{aligned} P_t^{\mathcal{O}} \varphi(x) &= \mathbb{E}(\varphi(X(t,x)) : X(r,x) \in \mathcal{O} \quad \text{for all } r \in [0,t]) \\ &= \mathbb{E}(\varphi(X(t,x)) : X(r,x) \in \mathcal{O} \quad \text{for all } r \in [s,t]) \\ &= \mathbb{E}(\varphi(X(t,x)) : X(r,x) \in \mathcal{O} \quad \text{for some } r \in [0,s] \\ &\qquad\qquad\qquad\qquad \text{and } X(r,x) \in \mathcal{O} \quad \text{for all } r \in [s,t]) \\ &= I_1 + I_2. \end{aligned}$$

By the Markov property, see [102],

$$
\begin{aligned}
I_1 &= \mathbb{E}(\varphi(X(t,x)) : X(r,x) \in \mathcal{O} \text{ for all } r \in [s,t]) \\
&= \mathbb{E}(P^{\mathcal{O}}_{t-s}\varphi(X(s,x)) : X(s,x) \in \mathcal{O}) = P_s\left(\hat{P}^{\mathcal{O}}_{t-s}\varphi\right)(x).
\end{aligned}
$$

Since

$$
\begin{aligned}
|I_2| &\le |\varphi|_0 \mathbb{P}(X(r,x) \in \mathcal{O}^c \text{ for some } r \in [0,s] \\
&\qquad\quad \text{and } X(r,x) \in \mathcal{O} \text{ for all } r \in [s,t]) \\
&\le |\varphi|_0 \mathbb{P}(X(r,x) \in \mathcal{O}^c \text{ for some } r \in [0,s]) \\
&\le |\varphi|_0 \mathbb{P}(\tau^x_{\mathcal{O}} \le s),
\end{aligned}
$$

the proof is complete. $\square$

We need the following interpolatory result.

Lemma 8.2.3 *Let U be an open subset of H. For $x \in U$ denote $d_U(x) = \mathrm{dist}(x, U^c) \wedge 2$. Then*

$$
\|Dg(x)\| \le \frac{4}{d_U(x)} \|g\|_U^{\frac{1}{2}} \left(\|g\|_U + \|D^2 g\|_U\right)^{\frac{1}{2}} \qquad \text{if } g \in C^2(U),
$$

$$
\|D^2 g(x)\| \le \left(\frac{8}{d_U(x)}\right)^{\frac{3}{2}} \left(\|g\|_U\right)^{\frac{1}{4}} \left(\|g\|_U + \|D^2 g\|_U\right)^{\frac{1}{4}}
$$

$$
\times \left(\|Dg\|_U + \|D^3 g\|_U\right)^{\frac{1}{2}} \quad \text{if } g \in C^3(U),
$$

$$
\|D^3 g(x)\| \le \left(\frac{16}{d_U(x)}\right)^{\frac{7}{4}} \left(\|g\|_U\right)^{\frac{1}{8}} \left(\|g\|_U + \|D^2 g\|_U\right)^{\frac{1}{8}}
$$

$$
\times \left(\|Dg\|_U + \|D^3 g\|_U\right)^{\frac{1}{4}} \left(\|D^2 g\|_U + \|D^4 g\|_U\right)^{\frac{1}{2}}
$$

$$
\text{if } g \in C^4(U).
$$

Proof. Assume that the closure of the open ball,

$$
B(x,r) = \{y \in H : \|y - x\| < r\},
$$

is contained in U and $|h| = r$. By the mean value theorem, for each $t \in [0,1]$ there exists $z \in B(x,r)$ such that

$$
g(x + th) = g(x) + t\langle Dg(x), h\rangle + \frac{1}{2} t^2 \langle D^2 g(z)h, h\rangle. \tag{8.2.3}
$$

From (8.2.3)

$$\left| \left\langle Dg(x), \frac{h}{|h|} \right\rangle \right| \leq \frac{2\|g\|_U}{t|h|} + \frac{1}{2}t|h|\|D^2g\|_U$$

$$\leq \left(\frac{2\|g\|_U}{r} \right)\frac{1}{t} + t\left(\frac{1}{2}r\|D^2g\|_U \right),$$

and therefore, for all $t \in (0,1)$,

$$|Dg(x)| \leq \left(\frac{1\|g\|_U}{r} \right)\frac{1}{t} + t\left(\frac{1}{2}r\|D^2g\|_U \right)$$

$$\leq \frac{\alpha}{t} + t\beta.$$

Setting $t = \sqrt{\frac{\alpha}{\alpha+\beta}}$ one gets that $\frac{\alpha}{t} + t\beta \leq 2\sqrt{\alpha(\alpha+\beta)}$. Consequently

$$|Dg(x)| \leq 2\sqrt{\frac{2\|g\|_U}{r}}\sqrt{\frac{2\|g\|_U}{r} + \frac{1}{2}r\|D^2g\|_U}$$

$$\leq \frac{4}{r}\sqrt{\|g\|_U}\sqrt{\|g\|_U + \frac{1}{4}r^2\|D^2g\|_U}.$$

Since r was an arbitrary positive number smaller than $d_U(x)$, the estimate follows.

Now let V be an open ball with center at x and radius $d_U(x)/2$. We apply the first inequality to V and to functions of the form $f_v(x) = \langle Dg(x), v \rangle$, $v \in H$, and we see that

$$\|D^2g(x)\| \leq \frac{4}{d_V(x)}\sqrt{\|Dg\|_V}\sqrt{\|Dg\|_V + \|D^3g\|_V}$$

$$\leq \frac{8}{d_U(x)}\sqrt{\|Dg\|_V}\sqrt{\|Dg\|_U + \|D^3g\|_U} . \qquad (8.2.4)$$

For $y \in V$ we have

$$\|Dg(y)\| \leq \frac{4}{d_U(y)}\sqrt{\|g\|_U}\sqrt{\|g\|_U + \|D^2g\|_U} .$$

Now, since $d_U(y) \geq d_U(x)/2$, we obtain

$$\|Dg\|_V \leq \frac{8}{d_U(x)}\sqrt{\|g\|_U}\sqrt{\|g\|_U + \|D^2g\|_U} .$$

Combining this with (8.2.4) we get the second inequality of the lemma. The proof of the last one is similar. $\square$

The following result plays an essential rôle in the estimates of the exit probabilities and therefore in the proof of the theorem. It is an improved version of a result from [92] with a simpler proof based on the arguments of [182] and on the properties of Gaussian measures.

Lemma 8.2.4 *Assume that condition (8.1.4) holds. Then*

(i) *the process $Z(t) = \int_0^t e^{(t-s)A} dW_s$ has a continuous version and for each $T > 0$ there exist positive constants C_1, C_2 such that for all $t \in (0, T]$ and $r > 0$,*

$$\mathbb{P}\left(\sup_{s \le t} \|Z(s)\| \ge r\right) \le C_1 e^{-C_2 \frac{r^2}{t^\alpha}}, \qquad (8.2.5)$$

(ii) *for all $x \in \mathcal{O}$ and $T > 0$ there exist positive constants r_0, t_0, C_3, C_4 (dependent on x, T and $\mathcal{O}$) such that for all $t \in (0, t_0]$,*

$$\sup_{y \in B(x, r_0)} \mathbb{P}\left(\tau_{\mathcal{O}}^y \le t\right) \le C_3 e^{\frac{-C_4}{t^\alpha}}. \qquad (8.2.6)$$

Proof. The proof of the continuity of Z, as well as the formula (8.2.5), is a consequence of the following formula:

$$Z(t) = \frac{\sin \beta \pi}{\pi} \int_0^t (t - s)^{\beta - 1} e^{(t-s)A} Y(s) ds, \qquad t \ge 0, \qquad (8.2.7)$$

where

$$Y(s) = \int_0^t (s - \sigma)^{-\beta} e^{(s-\sigma)A} dW(\sigma), \qquad s \ge 0, \qquad (8.2.8)$$

see [101] and [102], valid for $\beta \in (0, \frac{1}{2})$. We prove only formula (8.2.5) leaving the proof of continuity of Z to the reader, see e.g. [101]. Define

$$\beta = \frac{\alpha}{2}, \quad n_0 = \left[\frac{1}{\alpha}\right] + 1, \quad q_0 = 2n_0, \quad p_0 = \frac{q_0}{q_0 - 1}.$$

Then $p_0^{-1} + q_0^{-1} = 1$. Moreover if $n \ge n_0$, $n \in \mathbb{N}$ and $q = 2n$, $p = q(q - 1)^{-1}$ then by the Hölder inequality

$$\|Z(t)\| \le \frac{\sin \beta \pi}{\pi} \int_0^t (t - s)^{\beta - 1} \|e^{(t-s)A}\| \, \|Y(s)\| ds$$

$$\le \frac{\sin \beta \pi}{\pi} \left(\int_0^t (t - s)^{(\beta-1)p} \|e^{(t-s)A}\|^p ds\right)^{1/p} \left(\int_0^t \|Y(s)\|^q ds\right)^{1/q}.$$

Therefore, for all $u \in (0, T]$

$$\sup_{t \leq u} \|Z(t)\| \leq cu^{\frac{2(\beta-1)p+1}{p}} \left(\int_0^u \|Y(s)\|^q ds \right)^{2/q}$$

where $c = \left(\frac{\sin \beta \pi}{\pi} \right)^2 M^2$, $M = \sup_{t \leq T} \|e^{tA}\|$. For arbitrary $\lambda > 0$, $n \geq n_0$, $u \in [0, T]$,

$$\frac{\sup_{t \leq u} \|Z(t)\|^2}{c\lambda u^\alpha} \leq \left(\frac{1}{u} \int_0^u \left(\frac{\|Y(s)\|^2}{\lambda} \right)^n ds \right)^{1/n},$$

$$\mathbb{E} \left(\frac{\sup_{t \leq u} \|Z(t)\|^2}{c\lambda u^\alpha} \right)^n \leq \frac{1}{u} \int_0^u \mathbb{E} \left(\frac{\|Y(s)\|^2}{\lambda} \right)^n ds.$$

Consequently

$$\mathbb{E} \left(\sum_{n \geq n_0} \frac{1}{n!} \left(\frac{\sup_{t \leq u} \|Z(t)\|^2}{c\lambda u^\alpha} \right)^n \right) \leq \frac{1}{u} \int_0^u \mathbb{E} \left(e^{\frac{\|Y(s)\|^2}{\lambda}} \right) ds. \qquad (8.2.9)$$

Now if $\mathbb{E}\|Y(T)\|^2 \leq \frac{\lambda}{2}$ and $s \in (0, T]$ we have

$$\mathbb{E}(e^{\frac{\|Y(s)\|^2}{\lambda}}) \leq \frac{1}{\sqrt{1 - \frac{2}{\lambda} \mathbb{E}\|Y(s)\|^2}} \qquad (8.2.10)$$

$$\leq \frac{1}{\sqrt{1 - \frac{2}{\lambda} \mathbb{E}\|Y(T)\|^2}} = C(\lambda).$$

Note that $\mathbb{E}\|Y(T)\|^2 = \int_0^T t^{-\alpha} \|e^{tA} Q^{1/2}\|_{HS}^2 dt$. Moreover, for $n = 0, 1, \dots,$ $n_0 - 1$

$$\mathbb{E} \left(\frac{\sup\{\|Z(t)\|^2 : t \leq u\}}{c\lambda u^\alpha} \right)^n \leq \left(\mathbb{E} \left(\frac{\sup\{\|Z(t)\|^2 : t \leq u\}}{c\lambda u^\alpha} \right)^{n_0} \right)^{\frac{n}{n_0}}$$

and

$$\sum_{n=0}^{n_0-1} \frac{1}{n!} \mathbb{E} \left(\frac{\sup\{\|Z(t)\|^2 : t \leq u\}}{c\lambda u^\alpha} \right)^n$$

$$\leq \sum_{n=0}^{n_0-1} \frac{1}{n!} \left(\left[\mathbb{E} \left(\frac{\sup\{\|Z(t)\|^2 : t \leq u\}}{c\lambda u^\alpha} \right)^{n_0} \right]^{\frac{n}{n_0}} \right)^n$$

$$\leq e^{\left(\mathbb{E} \left(\frac{\sup\{\|Z(t)\|^2 : t \leq u\}}{c\lambda u^\alpha} \right)^{n_0} \right)^{\frac{1}{n_0}}}. \qquad (8.2.11)$$

Since

$$\mathbb{E}\left(\frac{\sup\{\|Z(t)\|^2 : t \leq u\}}{c\lambda u^\alpha}\right)^{n_0} \leq n_0! \frac{1}{u} \int_0^u \mathbb{E}\left(\frac{1}{n_0!}\left(\frac{\|Y(s)\|^2}{\lambda}\right)^{n_0}\right) ds$$

$$\leq n_0! \frac{1}{u} \int_0^u \mathbb{E}\left(e^{\frac{\|Y(s)\|^2}{\lambda}}\right) ds$$

$$\leq n_0! C(\lambda)$$

we have, taking into account (8.2.9), (8.2.10) and (8.2.11),

$$\mathbb{E}\left(\frac{\sup\{\|Z(t)\|^2 : t \leq u\}}{c\lambda u^\alpha}\right) \leq C(\lambda) + e^{(n_0! C(\lambda))^{1/n_0}}.$$

Finally, by Chebyshev's inequality

$$\mathbb{P}\left(\sup_{t \leq u} \|Z(t)\| \geq r\right) = \mathbb{P}\left(e^{\frac{\sup\{\|Z(t)\|^2 : t \leq u\}}{c\lambda u^\alpha}} \geq e^{\frac{r^2}{c\lambda u^\alpha}}\right)$$

$$\leq e^{-\frac{r^2}{c\lambda u^\alpha}} \mathbb{E}\left(e^{\frac{\sup\{\|Z(t)\|^2 : t \leq u\}}{c\lambda u^\alpha}}\right)$$

$$\leq e^{-\frac{r^2}{c\lambda u^\alpha}}\left(C(\lambda) + e^{(n_0! C(\lambda))^{1/n_0}}\right)$$

and the proof of the estimate (8.2.5) is complete.

To prove (8.2.6) it is enough to show that if (8.1.4) holds and $a \in \mathcal{O}, r > 0$, and $T > 0$ are such that

$$B(a,r) \subset \mathcal{O}, \text{ and } \sup_{t \leq T} |e^{tA}a - a| \leq \frac{r}{4},$$

then for $M = \sup_{t \leq T} \|e^{tA}\|$ and $r_0 = r/4M$, there exist positive constants C_1, C_2 such that for $s \in [0, T]$,

$$\sup_{x \in B(a,r_0)} \mathbb{P}(\tau^x_{B(a,r)} \leq s) \leq C_1 e^{-C_2 \frac{r^2}{s^\alpha}}. \tag{8.2.12}$$

Notice that for $x \in B(u, r_0)$

$$\|X(t,x) - a\| = \|e^{tA}x + Z(t) - a\| \leq \|e^{tA}(x - a)\| + \|e^{tA}a - a\| + \|Z(t)\|$$

$$\leq 1/2r + \|Z(t)\|.$$

Therefore, for $x \in B(a, r_0)$ and $s \leq T$,

$$\mathbb{P}\left(\tau^x_{B(a,r)} \leq s\right) \leq \mathbb{P}\left(\sup_{t \leq s} \|X(t,x) - a\| \geq r\right) \leq \mathbb{P}\left(\sup_{t \leq s} \|Z(t)\| \geq \frac{r}{2}\right),$$

and it is enough to apply (8.2.6). □

The case of the first derivative was treated in [92]. The proof for higher derivatives is analogous. We show, following [207], the existence of the second derivative of $P_t^{\mathcal{O}}\varphi$, its continuity and boundedness on small balls contained in $\mathcal{O}$.

Fix $x \in \mathcal{O}$ and $0 < t < T$. Let r_0, t_0 be as in Lemma 8.2.4 (ii). Taking t_0 smaller if necessary we can assume that for all $s \leq t_0$ assumptions (8.3.2) and (8.2.1) are satisfied. In what follows C will always denote a positive constant dependent only on x, T and the set $\mathcal{O}$. This constant can be different in different expressions below.

For $k \in \mathbb{N}$, $k \geq 1$ we define functions ψ_k as follows:

$$\psi_k(y) = P_{\frac{t}{k}}\hat{P}^{\mathcal{O}}_{\frac{k-1}{k}t}\varphi(y) \quad \text{for} \quad y \in H. \qquad (8.2.13)$$

By Lemmas 8.2.2 and 8.2.4 the sum $\psi_1(y) + \sum_{k=2}^{\infty}(\psi_k(y) - \psi_{k-1}(y))$ converges uniformly to $P_t^{\mathcal{O}}\varphi(y)$ for $y \in B(x, r_0)$. Each ψ_k is in $C_b^{\infty}(H)$, hence to prove that $P_t^{\mathcal{O}}\varphi$ is twice differentiable at x it suffices to show that

$$\left\|D^2\psi_1\right\|_{B(x,\frac{r_0}{2})} + \sum_{k=2}^{\infty}\left\|D^2(\psi_k - \psi_{k-1})\right\|_{B(x,\frac{r_0}{2})} < \infty. \qquad (8.2.14)$$

For each $k \geq 1$ the norm $\left\|D^2\psi_k\right\|_{B(x,r_0/2)}$ is finite. For k such that $\frac{t}{k} < t_0$ and $k > 2$ we will use Lemma 8.2.3 with $g_k = \psi_k - \psi_{k-1}$, $U = B(x, r_0)$. By formula (6.2.12) and assumption (8.2.1) we have for $y \in B(x, \frac{r_0}{2})$,

$$\left\|Dg_k(y)\right\| \leq \|\varphi\|_{\mathcal{O}}\left(\left\|\Lambda_{\frac{t}{k}}\right\| + \left\|\Lambda_{\frac{t}{k-1}}\right\|\right) \leq C\|\varphi\|_{\mathcal{O}}\left(\tfrac{k}{t}\right)^{\delta},$$

$$\left\|D^2g_k(y)\right\| \leq C\|\varphi\|_{\mathcal{O}}\left(\tfrac{k}{t}\right)^{2\delta},$$

$$\left\|D^3g_k(y)\right\| \leq C\|\varphi\|_{\mathcal{O}}\left(\tfrac{k}{t}\right)^{3\delta}.$$

As a consequence of Lemma 8.2.3 we get

$$\left\|D^2(\psi_k - \psi_{k-1})(y)\right\| \leq C\left(\|\psi_k - \psi_{k-1}\|_{B(x,r_0)}\right)^{\frac{1}{4}}\|\varphi\|_{\mathcal{O}}^{\frac{3}{4}}\left(\tfrac{k}{t}\right)^{2\delta},$$

which, by Lemmas 8.2.2 and 8.2.4(ii), can be estimated by

$$C\|\varphi\|_{\mathcal{O}}\, e^{\frac{-1}{C}\left(\frac{k-1}{t}\right)^{\alpha}}\left(\tfrac{k}{t}\right)^{2\delta}.$$

Thus the series in (8.2.14) is convergent and for $y \in B(x, \frac{r_0}{2})$,

$$D^2 P_t^{\mathcal{O}} \varphi(y) = D^2 \psi_1(y) + \sum_{k=2}^{\infty} D^2 \left(\psi_k - \psi_{k-1} \right)(y). \qquad (8.2.15)$$

The required regularity of $P_t^{\mathcal{O}} \varphi$ is proved. $\square$

8.3   Existence theorems

For $x \in H$ denote by $\tau_{\mathcal{O}}^x$ the first exit time of the process $X(t, x)$ from $\mathcal{O}$,

$$\tau_{\mathcal{O}}^x = \inf\{t > 0 : X(t, x) \in \mathcal{O}^c\}.$$

By the Blumenthal 0-1 law the probability $\mathbb{P}(\tau_{\mathcal{O}}^x = 0)$ is either 0 or 1. We say that a point x is regular if $\mathbb{P}(\tau_{\mathcal{O}}^x = 0) = 1$, otherwise we call it irregular.

Definition 8.3.1 *We say that a continuous and bounded function $u(t, x)$ on $[0, \infty) \times \mathcal{O}$ is a strong solution of (8.1.1) if it satisfies*

(i) *for each $t > 0$ the Fréchet derivatives $Du(t, x)$ and $D^2 u(t, x)$ exist, are continuous on $x \in \mathcal{O}$, and locally bounded as functions of (t, x), i.e. for all $x \in \mathcal{O}$ and $t > 0$ there exist $t_1 \in (0, t)$ and $r > 0$ such that*

$$\sup_{(s,y) \in [t_1, T] \times B(x, r)} \{|Du(s, y)| + \|D^2 u(s, y)\|\} < \infty, \qquad (8.3.1)$$

(ii) *for each $t > 0$, $Du(t, x) \in D(A^*)$, $|A^* Du(t, x)|$ and $\left\| Q^{\frac{1}{2}} D^2 u(t, x) Q^{\frac{1}{2}} \right\|_1$ are locally bounded, and for fixed $t > 0$, $Q^{\frac{1}{2}} D^2 u(t, x) Q^{\frac{1}{2}}$ is continuous as a function of $x \in \mathcal{O}$ into the space of linear trace class operators on H,*

(iii) *for each $x \in \mathcal{O}$, $D_t u(t, x)$ exists for $t > 0$ and is a continuous function of t,*

(iv) *equation (8.1.1) is satisfied in the classical sense,*

(v) *$u(0, x) = \varphi(x)$ for $x \in \mathcal{O}$,*

(vi) *for any sequences (x_n) of points in $\mathcal{O}$ converging to a regular point x of the boundary, and (t_n), $t_n > 0$, converging to $t > 0$, $u(t_n, x_n)$ converges to zero.*

Remark 8.3.2 *By the compactness of* $\{x\} \times [\gamma, T]$ *for* $0 < \gamma < T < \infty$, *it follows that* (8.3.1) *is equivalent to*

$$\forall x \in \mathcal{O}, \ \forall \, 0 < \gamma < T < \infty, \ \exists r > 0,$$

$$\sup_{(s,y)\in[\gamma,T]\times B(x,r)} \{|Du(s,y)| + \|D^2 u(s,y)\|\} < \infty.$$

Our aim is to prove the following two existence theorems due to A. Talarczyk [207]. The first theorem deals with bounded initial functions φ, while the second deals with continuous ones.

Theorem 8.3.3 *Assume that conditions* (8.1.3), (8.1.4) *and* (8.2.1) *are satisfied. Assume in addition that* $e^{tA} \subset D(A)$ *and there exist* $t_0 > 0, C > 0$ *and* $\delta \geq 0$ *such that for all* $t \in (0, t_0]$

$$\|Ae^{tA}\| \leq Ct^{-\delta}. \tag{8.3.2}$$

Then for each $\varphi \in B_b(\mathcal{O})$ *the function* $u(t,x) = P_t^{\mathcal{O}}\varphi(x)$ *satisfies (i)-(iv) of Definition 8.3.1. Moreover,* $D^2 P_t^{\mathcal{O}}\varphi(x) \in D(A^*)$ *and* $A^* D^2 P_t^{\mathcal{O}}\varphi(x)$ *is a bounded operator.*

Theorem 8.3.4 *Assume that in addition to the assumptions of Theorem 8.3.3 the initial function* φ *is continuous. Then* $u(t,x) = P_t^{\mathcal{O}}\varphi(x)$, $t \geq 0$, $x \in H$, *is a strong solution of* (8.1.1).

Remark 8.3.5 If $e^{tA}(H) \subset D(A)$ then, since A is closed, Ae^{tA} is a bounded operator. Consequently, $e^{tA}A$ can be extended to a bounded operator, this extension is equal Ae^{tA}, $e^{tA^*}(h) \in D(A^*)$ and $\|A^* e^{tA^*}\| = \|Ae^{tA}\|$.

Assumption (8.3.2) is satisfied for analytic semigroups with $\delta = 1$ (see [104]).

Example 8.3.6 Let

$$H = L^2(0,1), \quad A = \frac{d^2}{d\xi^2}, \quad D(A) = H^2(0,1) \cap H_0^1(0,1), \quad Q = I$$

and let $\mathcal{O}$ be an open set in H. The operator A has eigenvalues $-n^2\pi^2$, $n \in \mathbb{N}$. The semigroup e^{tA} generated by A, and the operator Q_t, have the same eigenvectors as A. The corresponding eigenvalues are $e^{-n^2\pi^2 t}$ and $\frac{1-e^{-2n^2\pi^2 t}}{2n^2\pi^2}$, respectively. All the assumptions of Theorem 8.3.4 are satisfied. (8.1.4) holds for all $0 < \alpha < \frac{1}{2}$, (8.3.2) is satisfied for $\delta = 1$ and (8.2.1) with $\delta = \frac{1}{2}$. Thus for each $\varphi \in C_b(\mathcal{O})$ the function $P_t^{\mathcal{O}}\varphi(x)$ is a strong solution to (8.1.1). If $\mathcal{O} = B(0, r)$ is a ball in H then, as shown in [92], every $z \in \partial\mathcal{O}$ such that $\langle Qz, z \rangle > 0$, $z \in D(A)$ or $z \in D((-A)^{\frac{1}{2}})$ is regular, and consequently, for $t > 0$,

$$\lim_{\substack{x \in \mathcal{O} \\ x \to z}} P_t^{\mathcal{O}}\varphi(x) = 0.$$

But in [92] it was also shown that there exists a dense set of irregular points.

The proofs of both theorems follow the work of A. Talarczyk [207]. We refer to [207] for more details

We need one more lemma to prove that for arbitrary $\varphi \in B_b(\mathcal{O})$, $x \in \mathcal{O}$ and $t > 0$ the operator $Q^{\frac{1}{2}} D^2 P_t^{\mathcal{O}}\varphi(x)Q^{\frac{1}{2}}$ is of trace class and that $DP_t^{\mathcal{O}}\varphi(x)$ is in the domain of A^*. This result was proved in Chapter 6.

Lemma 8.3.7 *Assume that (8.1.4) and (8.1.3) hold.*
  (i) We have

$$DP_t\phi(x) = \int_H e^{tA^*} D\phi \left(e^{tA}x + y \right) N_{Q_t}(dy) \quad if \, \phi \in C_b^1(H),$$

$$D^2 P_t\phi(x) = \int_H e^{tA^*} D^2\phi \left(e^{tA}x + y \right) e^{tA} N_{Q_t}(dy) \quad if \, \phi \in C_b^2(H). \quad (8.3.3)$$

  (ii) $DP_t\phi(x) \in D(A^*)$ for all $t > 0$, $x \in H$ and $\phi \in B_b(H)$ if and only if $e^{tA}(H) \in D(A)$.
  (iii) The operator $D^2 P_t\phi$ is of trace class for all $t > 0$, $x \in H$ and $\phi \in B_b(H)$.

Proof of Theorem 8.3.3.
  Step 1. Proof of (i) in the definition of the strong solution .
  This is a direct corollary of the regularity theorem from the previous section which states in particular that for all $t > 0$, $P_t^{\mathcal{O}}\varphi(x)$ is of class C^2 with respect to the space variable $x \in \mathcal{O}$, with locally bounded first and second derivatives.
  Step 2. We will prove that if (8.1.4) is satisfied then

$$\sum_{k=1}^{\infty} \left(\frac{t}{k} \right)^2 \left\| e^{\frac{t}{k}A} Q^{\frac{1}{2}} \right\|_{HS}^2 < \infty. \quad (8.3.4)$$

Let $N > 0$ and $a \in \mathbb{R}$ be such that $\left\| e^{\sigma A} \right\| \le N e^{a\sigma}$, $\sigma \ge 0$. Denote $M = \sup_{\sigma \le T} \left\| e^{\sigma A} \right\| \vee 1$. From (8.1.4) we have

$$\infty > \int_0^t \sigma^{-\alpha} \left\| e^{\sigma A} Q^{\frac{1}{2}} \right\|_{HS}^2 \, d\sigma \ge t^{-\alpha} \sum_{k=1}^{\infty} \int_{\frac{t}{k+1}}^{\frac{t}{k}} \left\| e^{\sigma A} Q^{\frac{1}{2}} \right\|_{HS}^2 \, d\sigma$$

$$\ge t^{-\alpha} \sum_{k=1}^{\infty} \int_0^{\frac{t}{k} - \frac{t}{k+1}} \left\| e^{(\frac{t}{k+1} + \sigma) A} Q^{\frac{1}{2}} \right\|_{HS}^2 \, d\sigma.$$

By the semigroup property

$$\left\| e^{\frac{t}{k} A} Q^{\frac{1}{2}} \right\|_{HS} \le \left\| e^{(\frac{t}{k} - \frac{t}{k+1} - \sigma) A} \right\| \left\| e^{(\frac{t}{k+1} + \sigma) A} Q^{\frac{1}{2}} \right\|_{HS}$$

$$\le M \left\| e^{(\frac{t}{k+1} + \sigma) A} Q^{\frac{1}{2}} \right\|_{HS}$$

for $0 \le \sigma \le \frac{t}{k(k+1)}$. Thus we get

$$\frac{1}{M^2} \sum_{k=1}^{\infty} \frac{t}{k(k+1)} \left\| e^{\frac{t}{k} A} Q^{\frac{1}{2}} \right\|_{HS}^2 < \infty$$

and (8.3.4) follows.

Step 3. Now we will show that for arbitrary $t > 0$ and $x \in \mathcal{O}$ the operator $Q^{\frac{1}{2}} D^2 P_t^{\mathcal{O}} \varphi(x) Q^{\frac{1}{2}}$ is of trace class and $DP_t^{\mathcal{O}} \varphi(x) \in D(A^*)$, which means that the right hand side of (8.1.1) is well defined for $u(t,x) = P_t^{\mathcal{O}} \varphi(x)$. We also prove here local boundedness of $A^* D P_t^{\mathcal{O}} \varphi(x)$ and $\left\| Q^{\frac{1}{2}} D^2 P_t^{\mathcal{O}} \varphi(x) Q^{\frac{1}{2}} \right\|_1$. Fix $x \in \mathcal{O}$, $0 < t < T$ and $k_0 \ge 2$ such that for all $k > k_0$, $\left| e^{\frac{t}{k} A} x - x \right| < \frac{r_0}{4}$ and $\frac{t}{k} < t_0$. To prove that $Q^{\frac{1}{2}} D^2 P_t^{\mathcal{O}} \varphi(x) Q^{\frac{1}{2}}$ is of trace class it suffices to show that

$$\left\| Q^{\frac{1}{2}} D^2 \psi_{k_0}(x) Q^{\frac{1}{2}} \right\|_1 + \sum_{k=k_0+1}^{\infty} \left\| Q^{\frac{1}{2}} D^2 \left(\psi_k - \psi_{k-1} \right)(x) Q^{\frac{1}{2}} \right\|_1 < \infty, \qquad (8.3.5)$$

where ψ_k are defined in (8.2.13). Then we will also have

$$\mathrm{Tr}[Q^{\frac{1}{2}} D^2 P_t^{\mathcal{O}} \varphi(x) Q^{\frac{1}{2}}] = \mathrm{Tr}[Q^{\frac{1}{2}} D^2 \psi_{k_0}(x) Q^{\frac{1}{2}}]$$

$$+ \sum_{k=k_0+1}^{\infty} \mathrm{Tr}[Q^{\frac{1}{2}} D^2 \left(\psi_k - \psi_{k-1} \right)(x) Q^{\frac{1}{2}}].$$

Set

$$\Psi_k(x) = P_{\frac{t}{2k}} \hat{P}^{\mathcal{O}}_{\frac{k-1}{k}t} \varphi(x) - P_{\frac{t}{k-1} - \frac{t}{2k}} \hat{P}^{\mathcal{O}}_{\frac{k-2}{k-1}t} \varphi(x). \tag{8.3.6}$$

By the smoothing property of the global semigroup P_t we see that $\Psi_k \in C^2_b(H)$. Applying Lemma 8.3.7 to

$$D^2 \left(\psi_k - \psi_{k-1} \right)(x) = D^2 \left(P_{\frac{t}{2k}} \Psi_k \right)(x)$$

we get

$$D^2 \left(\psi_k - \psi_{k-1} \right)(x) = \int_H e^{\frac{t}{2k} A^*} D^2 \Psi_k (e^{\frac{t}{2k} A} x + y) e^{\frac{t}{2k} A} N_{Q_{\frac{t}{2k}}}(dy), \tag{8.3.7}$$

and consequently we have

$$\left\| Q^{\frac{1}{2}} D^2 \left(\psi_k - \psi_{k-1} \right)(x) Q^{\frac{1}{2}} \right\|_1$$

$$\leq \left\| e^{\frac{t}{2k} A} Q^{\frac{1}{2}} \right\|_{HS}^2 \int_H \left\| D^2 \Psi_k \left(e^{\frac{t}{2k} A} x + y \right) \right\| N_{Q_{\frac{t}{2k}}}(dy). \tag{8.3.8}$$

First we estimate the integral in (8.3.8). From the definition of Ψ_k we see that

$$\left\| D^2 \Psi_k \right\|_H \leq C_2 \left\| \varphi \right\|_{\mathcal{O}} \left(\left\| \Lambda_{\frac{t}{2k}} \right\|^2 + \left\| \Lambda_{\frac{t}{k-1} - \frac{t}{2k}} \right\|^2 \right)$$

$$\leq C \left\| \varphi \right\|_{\mathcal{O}} \left(\frac{k}{t} \right)^{2\delta}. \tag{8.3.9}$$

Thus by (8.3.9) and Lemma 8.2.4(i),

$$\int_{|y| \geq \frac{r_0}{4}} \left\| D^2 \Psi_k \left(e^{\frac{t}{2k} A} x + y \right) \right\| N_{Q_{\frac{t}{2k}}}(dy)$$

$$\leq C \left\| \varphi \right\|_{\mathcal{O}} \left(\frac{k}{t} \right)^{2\delta} P \left(\left| \int_0^{\frac{t}{2k}} e^{(\frac{t}{2k} - u) A} dW_u \right| \geq \frac{r_0}{4} \right)$$

$$\leq C \left\| \varphi \right\|_{\mathcal{O}} \left(\frac{k}{t} \right)^{2\delta} e^{\frac{-1}{C} (\frac{k}{t})^\alpha}. \tag{8.3.10}$$

If $|y| < \frac{r_0}{4}$ then $e^{\frac{t}{2k} A} x + y \in B(x, \frac{r_0}{2})$, since $\left| e^{\frac{t}{2k} A} x - x \right| < \frac{r_0}{4}$ and by Lemmas 8.2.2 and 8.2.4

$$\sup_{z \in B(x, r_0)} |\Psi_k(z)| \leq 2 \left\| \varphi \right\|_{\mathcal{O}} \sup_{z \in B(x, r_0)} P \left(\tau^z_{\mathcal{O}} \leq \frac{t}{k-1} - \frac{t}{2k} \right)$$

$$\leq C \left\| \varphi \right\|_{\mathcal{O}} e^{\frac{-1}{C} (\frac{k}{t})^\alpha}. \tag{8.3.11}$$

Using Lemma 8.2.3, estimates (8.2.1) for Λ_t and (8.3.11) we obtain

$$\int_{|y|<\frac{r_0}{4}} \left\| D^2 \Psi_k \left(e^{\frac{t}{2k}A} x + y \right) \right\| N_{Q_{\frac{t}{2k}}} (dy)$$

$$\leq C \int_{|y|<\frac{r_0}{4}} \|\Psi_k\|_{B(x,r_0)}^{\frac{1}{4}} \left(\|\Psi_k\|_H + \|D^2\Psi_k\|_H \right)^{\frac{1}{4}}$$

$$\times \left(\|D\Psi_k\|_H + \|D^3\Psi_k\|_H \right)^{\frac{1}{2}} N_{Q_{\frac{t}{2k}}} (dy)$$

$$\leq C \|\varphi\|_0 \left(\frac{k}{t} \right)^{2\delta} e^{\frac{-1}{C}(\frac{k}{t})^\alpha}. \tag{8.3.12}$$

Applying (8.3.10) and (8.3.12) to (8.3.8), we get

$$\left\| Q^{\frac{1}{2}} D^2 \left(\psi_k - \psi_{k-1} \right)(x) Q^{\frac{1}{2}} \right\|_1 \leq C \|\varphi\|_0 \left\| e^{\frac{t}{2k}A} Q^{\frac{1}{2}} \right\|_{HS}^2 \left(\frac{k}{t} \right)^{2\delta} e^{\frac{-1}{C}(\frac{k}{t})^\alpha}. \tag{8.3.13}$$

In view of (8.3.4) the sum over k of these terms is finite, since $k^\gamma e^{\frac{-1}{C}k^\alpha} \to 0$ as $k \to \infty$ for each $\gamma \in \mathbb{R}$. This completes the proof of (8.3.5).

To prove that $DP_t^{\mathcal{O}} \varphi(x)$ belongs to the domain of A^* it suffices to show that the series

$$A^* D\psi_{k_0}(x) + \sum_{k=k_0+1}^{\infty} A^* D \left(\psi_k - \psi_{k-1} \right)(x) \tag{8.3.14}$$

is convergent in H. Estimates are similar to the previous ones and they are omitted.

By (8.3.3) we also have that for each $v \in H$, $D^2 P_t^{\mathcal{O}} \varphi(x) v \in D(A^*)$. Since A^* is closed it follows that $A^* D^2 P_t^{\mathcal{O}} \varphi(x)$ is a bounded operator.

From the above discussion we easily obtain, see [207], that for all $x \in \mathcal{O}$ and $0 < t_1 < T < \infty$ there exists $r > 0$ such that both $\left\| Q^{\frac{1}{2}} D^2 P_t^{\mathcal{O}} \varphi(y) Q^{\frac{1}{2}} \right\|_1$ and $|A^* DP_t^{\mathcal{O}} \varphi(y)|$ are uniformly bounded for $(t, y) \in [t_1, T] \times B(x, r)$. Moreover, $Q^{\frac{1}{2}} D^2 P_t^{\mathcal{O}} \varphi(x) Q^{\frac{1}{2}}$ is continuous as a function of $x \in \mathcal{O}$ into the space of trace class operators on H.

Thus we get (ii) in Definition 8.3.1.

Step 4. In this step we show that for a fixed $x \in \mathcal{O}$ the right derivative of $P_t^{\mathcal{O}} \varphi(x)$ with respect to t exists for each $t > 0$ and satisfies

$$D_t^+ P_t^{\mathcal{O}} \varphi(x) = \frac{1}{2} \text{Tr}[Q^{\frac{1}{2}} D^2 P_t^{\mathcal{O}} \varphi(x) Q^{\frac{1}{2}}] + \langle x, A^* D P_t^{\mathcal{O}} \varphi(x) \rangle. \tag{8.3.15}$$

Let $h \in (0, t)$, then

$$\frac{P_{t+h}^{\mathcal{O}} \varphi(x) - P_t^{\mathcal{O}} \varphi(x)}{h} = \frac{P_h \hat{P}_t^{\mathcal{O}} \varphi(x) - P_t^{\mathcal{O}} \varphi(x)}{h} + \frac{P_{t+h}^{\mathcal{O}} \varphi(x) - P_h \hat{P}_t^{\mathcal{O}} \varphi(x)}{h} \tag{8.3.16}$$

and by Lemmas 8.2.2 and 8.2.4(ii) the second term in the right hand side of (8.3.16) converges to zero as $h \to 0$. Let $M = \sup_{s \leq 2t} \left\| e^{sA} \right\|$. By Lemma 8.2.4(i),

$$\frac{1}{h} \int_{|y| \geq \frac{r_0}{8M}} \left(\hat{P}_t^{\mathcal{O}} \varphi \left(e^{hA} x + y \right) - P_t^{\mathcal{O}} \varphi(x) \right) N_{Q_h}(dy)$$

$$\leq 2 \left\| \varphi \right\|_{\mathcal{O}} \frac{1}{h} C e^{\frac{-1}{Ch^\alpha}} \xrightarrow[h \to 0]{} 0$$

and (8.3.16) takes the form

$$\frac{P_{t+h}^{\mathcal{O}} \varphi(x) - P_t^{\mathcal{O}} \varphi(x)}{h}$$

$$= \frac{1}{h} o(h) + \frac{1}{h} \int_{|y| < \frac{r_0}{8M}} \left(\hat{P}_t^{\mathcal{O}} \varphi \left(e^{hA} x + y \right) - P_t^{\mathcal{O}} \varphi(x) \right) N_{Q_h}(dy). \tag{8.3.17}$$

Suppose that h is so small that for each $u \leq h$, $\left| e^{uA} x - x \right| < \frac{r_0}{8M}$. If $|y| < \frac{r_0}{8M}$ then $e^{hA} x + y$ is in the ball $B \left(x, \frac{r_0}{4M} \right)$ and here $P_t^{\mathcal{O}} \varphi$ is three times continuously differentiable, with bounded derivatives. We use Taylor's formula to see that the right hand side of (8.3.17) is equal to

$$\frac{1}{h} o(h) + \frac{1}{h} \int_{|y| < \frac{r_0}{8M}} \left\langle D P_t^{\mathcal{O}} \varphi(x), e^{hA} x - x + y \right\rangle N_{Q_h}(dy)$$

$$+ \frac{1}{2h} \int_{|y| < \frac{r_0}{8M}} \left\langle \left(D^2 P_t^{\mathcal{O}} \varphi(x) \right) \left(e^{hA} x - x + y \right), e^{hA} x \quad x + y \right\rangle N_{Q_h}(dy)$$

$$+ \frac{1}{6h} \int_{|y| < \frac{r_0}{8M}} \left\langle D^3 P_t^{\mathcal{O}} \varphi(x + \theta_{hy}) \left(e^{hA} x - x + y \right) \left(e^{hA} x - x + y \right), \right.$$

$$\left. \left(e^{hA} x - x + y \right) \right\rangle N_{Q_h}(dy)$$

$$= \frac{1}{h} o(h) + I_1(h) + I_2(h) + I_3(h), \tag{8.3.18}$$

where $\theta_{hy} = \eta_{hy}\left(e^{hA}x - x + y\right)$ for some $0 \leq \eta_{hy} \leq 1$. Notice that if $|y| < \frac{r_0}{8M}$ then $|\theta_{hy}| < \frac{r_0}{4M}$.

Since the measure N_{Q_h} is symmetric we have

$$I_1(h) = \frac{1}{h}\left\langle DP_t^{\mathcal{O}}\varphi(x), e^{hA}x - x \right\rangle \mathbb{P}\left(|Z(h)| \leq \frac{r_0}{8M}\right), \qquad (8.3.19)$$

where $Z(h) = \int_0^h e^{(h-u)A}dW_u$. Moreover,

$$\frac{1}{h}\left\langle DP_t^{\mathcal{O}}\varphi(x), e^{hA}x - x \right\rangle = \left\langle DP_t^{\mathcal{O}}\varphi(x), \frac{1}{h}A\int_0^h e^{uA}x\,du \right\rangle$$

$$= \left\langle A^*DP_t^{\mathcal{O}}\varphi(x), \frac{1}{h}\int_0^h e^{uA}x\,du \right\rangle \xrightarrow[h\to 0]{} \left\langle A^*DP_t^{\mathcal{O}}\varphi(x), x \right\rangle \qquad (8.3.20)$$

and

$$\lim_{h\to 0}\mathbb{P}\left(|Z(h)| \leq \frac{r_0}{8M}\right) = 1.$$

Thus from (8.3.19) we get

$$\lim_{h\to 0} I_1(h) = \left\langle A^*DP_t^{\mathcal{O}}\varphi(x), x \right\rangle. \qquad (8.3.21)$$

Next, $I_2(h)$ in (8.3.18) can be written as

$$I_2(h) = \frac{1}{2h}\left\langle D^2P_t^{\mathcal{O}}\varphi(x)\left(e^{hA}x - x\right), e^{hA}x - x\right\rangle \mathbb{P}\left(|Z(h)| \leq \frac{r_0}{8M}\right)$$

$$+ \frac{1}{2h}\int_H \left\langle D^2P_t^{\mathcal{O}}\varphi(x)y, y\right\rangle N_{Q_h}(dy)$$

$$- \frac{1}{2h}\int_{|y|\geq\frac{r_0}{M}} \left\langle D^2P_t^{\mathcal{O}}\varphi(x)y, y\right\rangle N_{Q_h}(dy)$$

$$= I_{21}(h) + I_{22}(h) + I_{23}(h). \qquad (8.3.22)$$

As in (8.3.20) we get

$$I_{21}(h) = \frac{1}{2}\left\langle A^*D^2P_t^{\mathcal{O}}\varphi(x)\left(e^{hA}x - x\right), \frac{1}{h}\int_0^h e^{uA}x\,du \right\rangle \mathbb{P}\left(|Z(h)| \leq \frac{r_0}{8M}\right).$$

Recall that by Step 3, $A^*D^2P_t^{\mathcal{O}}\varphi(x)$ is a bounded operator. Since $e^{hA}x - x \to 0$ and $\frac{1}{h}\int_0^h e^{uA}x\,du \to x$ we have

$$\lim_{h\to 0} I_{21}(h) = 0. \qquad (8.3.23)$$

Finally, by Hölder's inequality we get

$$I_{23}(h) \leq \left\| D^2 P_t^{\mathcal{O}} \varphi(x) \right\| \frac{1}{2h} \left(\int_{y \geq \frac{r_0}{8M}} N_{Q_h}(dy) \right)^{\frac{1}{2}} \left(\int_H |y|^4 N_{Q_h}(dy) \right)^{\frac{1}{2}}.$$

By Corollary 2.17 in [102], for each $m \in \mathbb{N}$ there exists a constant $C_m > 0$ such that

$$\int_H |y|^{2m} N_{Q_h}(dy) \leq C_m \left(\mathrm{Tr}\, Q_h \right)^m. \tag{8.3.24}$$

Thus by Lemma 8.2.4 and (8.3.24),

$$I_{23}(h) \leq C \left\| D^2 P_t^{\mathcal{O}} \varphi(x) \right\| \frac{1}{h} e^{\frac{-1}{Ch^\alpha}} \mathrm{Tr}\, Q_h \xrightarrow[h \to 0]{} 0. \tag{8.3.25}$$

The term $I_{22}(h)$ in (8.3.22) can be written as follows:

$$\begin{aligned} I_{22} = {} & \frac{1}{2} \mathrm{Tr}[Q^{\frac{1}{2}} D^2 P_t^{\mathcal{O}} \varphi(x) Q^{\frac{1}{2}}] \\ & + \frac{1}{2} \left(\frac{1}{h} \mathrm{Tr}[Q_h D^2 P_t^{\mathcal{O}} \varphi(x)] - \mathrm{Tr}[Q^{\frac{1}{2}} D^2 P_t^{\mathcal{O}} \varphi(x) Q^{\frac{1}{2}}] \right). \end{aligned} \tag{8.3.26}$$

We will show that the last term in (8.3.26) converges to zero as $h \to 0$.

By an elementary transformation, see [207], we obtain the estimate

$$\begin{aligned} & \left| \frac{1}{h} \mathrm{Tr}[Q_h D^2 P_t^{\mathcal{O}} \varphi(x)] - \mathrm{Tr}[Q^{\frac{1}{2}} D^2 P_t^{\mathcal{O}} \varphi(x) Q^{\frac{1}{2}}] \right| \\ & \leq \frac{1}{h} \int_0^h \left| \mathrm{Tr}[Q^{\frac{1}{2}} e^{uA^*} D^2 P_t^{\mathcal{O}} \varphi(x) e^{uA} Q^{\frac{1}{2}}] - \mathrm{Tr}[Q^{\frac{1}{2}} D^2 P_t^{\mathcal{O}} \varphi(x) Q^{\frac{1}{2}}] \right| du. \end{aligned} \tag{8.3.27}$$

It suffices to prove that the integrand in (8.3.27) tends to zero as $h \to 0$. Let $M = \sup_{t \in [0,T]} \| e^{tA} \|$ and $\kappa_0 > 2 \min\{2, T/t_0\}$ be such that for any $h \leq T/t_0$, we have $|e^{hA}x - x| < \kappa_0/8$ and for all $k > k_0$ we have $\left| e^{\frac{t}{k}A}x - x \right| < \frac{r_0}{8M}$. We can use the expansion (8.2.15) of $D^2 P_t^{\mathcal{O}} \varphi(x)$ but starting from k_0 instead of 1. Both series are absolutely convergent. Thus we can estimate the integrand in(8.3.27) by

$$\sum_{i=1}^{\infty} \left| \left\langle \left(Q^{\frac{1}{2}} e^{uA^*} D^2 \psi_{k_0}(x) e^{uA} Q^{\frac{1}{2}} - Q^{\frac{1}{2}} D^2 \psi_{k_0} Q^{\frac{1}{2}} \right) e_i, e_i \right\rangle \right|$$

$$+ \sum_{k=k_0}^{\infty} \sum_{i=1}^{\infty} \left| \left\langle \left[Q^{\frac{1}{2}} e^{uA^*} D^2 \left(\psi_k - \psi_{k-1} \right) (x) e^{uA} Q^{\frac{1}{2}} \right. \right. \right.$$

$$\left. \left. \left. - Q^{\frac{1}{2}} D^2 \left(\psi_k - \psi_{k-1} \right) (x) Q^{\frac{1}{2}} \right] e_i, e_i \right\rangle \right| \qquad (8.3.28)$$

where (e_i) is an arbitrary orthonormal basis of H.

But, for all $v \in H$, $\left| e^{uA} v - v \right| \to 0$ if $h \to 0$, hence each term of (8.3.28) converges to zero as well. By the dominated convergence theorem (8.3.28) converges to zero. Putting it together with (8.3.27) and (8.3.26) we see that

$$\lim_{h \to 0} I_{22}(h) = \frac{1}{2} \operatorname{Tr}[Q^{\frac{1}{2}} D^2 P_t^{\mathcal{O}} \varphi(x) Q^{\frac{1}{2}}]. \qquad (8.3.29)$$

Applying (8.3.23), (8.3.25) and (8.3.29) to (8.3.22), we obtain

$$\lim_{h \to 0} I_2(h) = \frac{1}{2} \operatorname{Tr}[Q^{\frac{1}{2}} D^2 P_t^{\mathcal{O}} \varphi(x) Q^{\frac{1}{2}}]. \qquad (8.3.30)$$

We will show finally, that $I_3(h)$ in (8.3.18) tends to 0 as h goes to 0. We will estimate a typical term $I_{31}(h)$ only, for the other estimates we refer to [207]. We have

$$I_{31}(h) = \left| \int_{|y| < \frac{r_0}{8M}} \left\langle D^3 P_t^{\mathcal{O}} \varphi \left(x + \theta_{hy} \right) \left(e^{hA} x - x \right) \left(e^{hA} x - x \right), \right. \right.$$

$$\left. \frac{1}{h} A \int_0^h e^{uA} x \, du \right\rangle N_{Q_h}(dy) \Bigg|. \qquad (8.3.31)$$

From (8.1.5) it is easy to see that for $\phi \in C_b^3(H)$ and $f, g, h \in H$,

$$\left\langle (D^3 P_t \phi(x) f) g, h \right\rangle = \int_H \left\langle \left(e^{tA^*} D^3 \phi(e^{tA} x + y) e^{tA} f \right) e^{tA} g, h \right\rangle. \qquad (8.3.32)$$

The third derivative of $P_t^{\mathcal{O}}\varphi$ in the ball $B\left(x, \frac{r_0}{4M}\right)$ can be written as $D^3\psi_{k_0} + \sum_{k=k_0+1}^{\infty} D^3\left(\psi_{k+1} - \psi_k\right)$. Applying (8.3.32) we get

$$\left| \int_{|y| < \frac{r_0}{8M}} \left\langle \left[D^3 P_{\frac{t}{k_0}} \hat{P}^{\mathcal{O}}_{\frac{k_0-1}{k_0}t} \varphi\left(x + \theta_{hy}\right) \left(e^{hA}x - x\right) \right] \left(e^{hA}x - x\right), \right. \right.$$

$$\left. \left. \frac{1}{h} A \int_0^h e^{uA}x\, du \right\rangle N_{Q_h}(dy) \right|$$

$$\leq \int_{|y| < \frac{r_0}{8M}} \int_H \left| \left\langle e^{\frac{t}{2k_0}A^*} \left[D^3 P_{\frac{t}{2k_0}} \hat{P}^{\mathcal{O}}_{\frac{k_0-1}{k_0}t} \varphi\left(e^{\frac{t}{2k_0}A}(x + \theta_{hy}) + z\right) e^{\frac{t}{2k_0}A} \right. \right. \right.$$

$$\times \left. \left. \left(e^{hA}x - x\right) \right] e^{\frac{t}{2k_0}A} \left(e^{hA}x - x\right), \frac{1}{h} A \int_0^h e^{uA}x\, du \right\rangle \right| N_{Q_{\frac{t}{2k_0}}}(dz) N_{Q_h}(dy)$$

$$\leq C \left(\frac{2k_0}{t}\right)^{4\delta} \|\varphi\|_{\mathcal{O}} \left| e^{hA}x - x \right|^2 \left| \frac{1}{h} \int_0^h e^{uA}x\, du \right| \xrightarrow[h \to 0]{} 0.$$

Applying (8.3.32) to terms containing $(\psi_k - \psi_{k-1})$ we obtain

$$\left| \int_{|y| < \frac{r_0}{8M}} \left\langle D^3\left(\psi_k - \psi_{k-1}\right)(x + \theta_{hy})\left(e^{hA}x - x\right)\left(e^{hA}x - x\right), \right. \right.$$

$$\left. \left. \frac{1}{h} A \int_0^h e^{uA}x\, du \right\rangle N_{Q_h}(dy) \right| \leq C \left(\frac{k}{t}\right)^{\delta} \left| e^{uA}x - x \right|^2 \left| \frac{1}{h} \int_0^h e^{hA}x\, du \right|$$

$$\times \int_{|y| < \frac{r_0}{8M}} \int_H \left\| D^3 \Psi_k \left(e^{\frac{t}{2k}A}(x + \theta_{hy}) + z\right) \right\| N_{Q_{\frac{t}{2k}}}(dz) N_{Q_h}(dy),$$

$$(8.3.33)$$

were Ψ_k is defined by (8.3.6). Recall that $|\theta_{hy}| \leq \frac{r_0}{4M}$, thus

$$\left| e^{\frac{t}{2k}A}(x + \theta_{hy}) - x \right| \leq \frac{r_0}{4} + \frac{r_0}{8}.$$

Processing the inner integral in (8.3.33) as in (8.3.10) and (8.3.12), but taking $|z| < r_0/8$ and $|z| \geq r_0/8$, we see that the expression in (8.3.33) can be estimated by

$$C \left| e^{hA}x - x \right|^2 \left| \frac{1}{h} \int_0^h e^{uA}x\, du \right| \left(\frac{k}{t}\right)^{4\delta} e^{\frac{-1}{C}\left(\frac{k}{t}\right)^{\alpha}}.$$

Hence the sum over k of these terms tends to zero. Since we also have (8.3),
we get that for $I_{31}(h)$ defined in (8.3.31),

$$\lim_{h \to 0} I_{31}(h) = 0 \qquad (8.3.34)$$

Combining this with (8.3.21), (8.3.30) and (8.3.18) we get (8.3.15).

Step 5 To finish the proof of the theorem we need to show that $P_t^{\mathcal{O}} \varphi(x)$
is differentiable with respect to t for $t > 0$.

Fix $x \in \mathcal{O}$. First we notice that by Step 3 for arbitrary $0 < t_1 < T < \infty$
the right hand side of (8.1.1) is uniformly bounded for $t \in [t_1, T]$. Thus
$\frac{\partial^+}{\partial t} P_t^{\mathcal{O}} \varphi(x)$ is bounded for $t \in [t_1, T]$

Now the argument from Step 4 gives the uniform convergence of
$\frac{P_{t+h}^{\mathcal{O}} \varphi(x) - P_t^{\mathcal{O}} \varphi(x)}{h}$ with respect to $t \in [t_1, T]$ as $h \to 0$. Therefore there exists
h_0 such that for each $0 < h < h_0$,

$$\sup_{t \in [t_1, T]} \left| \frac{P_{t+h}^{\mathcal{O}} \varphi(x) - P_t^{\mathcal{O}} \varphi(x)}{h} \right| \leq C + \sup_{t \in [t_1, T]} \left| D_t^+ P_t^{\mathcal{O}} \varphi(x) \right|,$$

hence $P_t^{\mathcal{O}} \varphi(x)$ is continuous in (t_1, T). Since on compact subsets of (t_1, T),
$\frac{\partial^+}{\partial t} P_t^{\mathcal{O}} \varphi(x)$ is the limit of a uniformly convergent sequence of uniformly continuous functions, it follows that $\frac{\partial^+}{\partial t} P_t^{\mathcal{O}} \varphi(x)$ is continuous in (t_1, T).

Fix $t \in (t_1, T)$. For $h > 0$ and $u > 0$, such that $u \leq \frac{t - t_1}{2}$ we have

$$\left| \frac{P_t^{\mathcal{O}} \varphi(x) - P_{t-h}^{\mathcal{O}} \varphi(x)}{h} - D_t^+ P_t^{\mathcal{O}} \varphi(x) \right|$$

$$\leq \left| \frac{P_{t-u+h}^{\mathcal{O}} \varphi(x) - P_{t-u}^{\mathcal{O}} \varphi(x)}{h} - D_t^+ P_{t-u}^{\mathcal{O}} \varphi(x) \right|$$

$$+ \left| D_t^+ P_{t-u}^{\mathcal{O}} \varphi(x) - \frac{\partial^+}{\partial t} P_t^{\mathcal{O}} \varphi(x) \right|$$

$$+ \frac{1}{h} |P_{t-u}^{\mathcal{O}} \varphi - P_{t-h}^{\mathcal{O}} \varphi| + \frac{1}{h} |P_{t-u+h}^{\mathcal{O}} \varphi - P_t^{\mathcal{O}} \varphi|.$$

The first term converges to zero as $h \to 0$ uniformly with respect to u. It
follows from the continuity of $P_t^{\mathcal{O}} \varphi(x)$ and $D_t^+ P_t^{\mathcal{O}} \varphi(x)$ that $P_t^{\mathcal{O}} \varphi(x)$ has a
continuous derivative with respect t. This finishes the proof of (iii) and (iv)
in Definition 8.3.1. $\square$

For the proof of Theorem 8.3.4 we need a result on regular points of the
boundary of $\mathcal{O}$.

Lemma 8.3.8 *Assume that (8.1.4) and (8.1.3) hold. If x is a regular point of $\partial\mathcal{O}$ then for $t > 0$,*

$$\lim_{\substack{y \in \mathcal{O} \\ y \to x}} \mathbb{P}(\tau_\mathcal{O}^y > t) = 0. \tag{8.3.35}$$

Proof. A nonnegative measurable function f on H is called α-excessive if

$$e^{-\alpha t} P_t f \leq f, \quad \forall t \geq 0, \tag{8.3.36}$$
$$\lim_{t \to 0} e^{-\alpha t} P_t f(x) = f(x), \quad \forall x \in H. \tag{8.3.37}$$

Such α-excessive functions were studied in [17] on locally compact spaces, which is not our case; however, some arguments are similar.

Define

$$f(x) = \mathbb{E}[e^{-\tau_\mathcal{O}^x}]. \tag{8.3.38}$$

This is a Borel measurable function, since $\tau_\mathcal{O}^x$ is measurable with respect to x and ω. The function f is 1-excessive. The latter fact is an analogue of a special case of Prop. II.2.8 in [17].

By the Markov property

$$e^{-t} P_t f(x) = \mathbb{E}[e^{-\tau_\mathcal{O}^{x,t}}],$$

where $\tau_\mathcal{O}^{x,t} = \inf\{s > t : X_s^x \in \mathcal{O}^c\}$. It is obvious that $\tau_\mathcal{O}^{x,t} \geq \tau_\mathcal{O}^x$ and for $t \to 0$, $\tau_\mathcal{O}^{x,t} \to \tau_\mathcal{O}^x$. Thus (8.3.36) and (8.3.37) follow with $\alpha = 1$.

A bounded α-excessive function f is lower semicontinuous. By (8.3.36),

$$f(y) - f(x) \geq \left(e^{-\alpha t} P_t f(y) - e^{-\alpha t} P_t f(x)\right) + \left(e^{-\alpha t} P_t f(x) - f(x)\right). \tag{8.3.39}$$

By (8.1.4) and (8.1.3) the semigroup P_t has the smoothing property and $P_t f$ is continuous. We also have (8.3.37), thus f is lower semicontinuous.

The function f defined in (8.3.38) is bounded by 1, is lower semicontinuous and for a regular point $x \in \partial\mathcal{O}$, $f(x) = 1$. Therefore if y tends to x then for each $t > 0$, $\mathbb{P}(\tau_\mathcal{O}^y > t)$ tends to 0. $\square$

Proof of Theorem 8.3.4. In view of Theorem 8.3.3 it remains to show that $u(t, x) = P_t^\mathcal{O} \varphi(x)$ is jointly continuous on $[0, \infty) \times \mathcal{O}$ and (v) and (vi) hold.

$P_t^\mathcal{O} \varphi(x)$ is jointly continuous on $(0, \infty) \times \mathcal{O}$, since $P_t^\mathcal{O} \varphi(x)$ has locally bounded space and time derivatives.

Let $x_n \in \mathcal{O}$, $x_n \to x \in \mathcal{O}$ as $n \to \infty$, and $t_n > 0$, $t_n \to 0$. Fix $\varepsilon > 0$. By the continuity of the function φ and by Lemma 8.2.4 there exist positive constants r, r_0, t_0 such that $0 < r_0 < r < \frac{\text{dist}\,(x, \mathcal{O}^c)}{2}$, and

$$|\varphi(y) - \varphi(x)| \; < \; \frac{\varepsilon}{2} \quad \forall y \in B(x, r),$$

$$\sup_{y \in B(x, r_0)} \mathbb{P}\left(\tau^y_{B(x,r)} \le t\right) \; \le \; Ce^{-\frac{1}{Ct^\alpha}} \quad \forall t \le t_0.$$

Let n_0 be such that for all $n \ge n_0$ we have $|x - x_n| < r_0$, $t_n \le t_0$ and $2\,\|\varphi\|_{\mathcal{O}}\, Ce^{-\frac{1}{Ct_n^\alpha}} < \frac{\varepsilon}{2}$. Then for all $n \ge n_0$,

$$\left|\mathbb{E}\varphi(X(t_n, x_n))1_{\{\tau^{x_n}_{\mathcal{O}} > t_n\}} - \varphi(x)\right|$$

$$\le \mathbb{E}\left(|\varphi\left(X(t_n, x_n)\right) - \varphi(x)|\,1_{\{\tau^{x_n}_{B(x,r)} > t_n\}}\right) + \|\varphi\|_{\mathcal{O}}\,\mathbb{P}\left(\tau^{x_n}_{\mathcal{O}} > t_n, \tau^{x_n}_{B(x,r)} \le t_n\right)$$

$$+ \|\varphi\|_{\mathcal{O}}\,\mathbb{P}\left(\tau^{x_n}_{B(x,r)} \le t_n\right) \le \frac{\varepsilon}{2} + 2\,\|\varphi\|_{\mathcal{O}}\,\mathbb{P}\left(\tau^{x_n}_{B(x,r)} \le t_n\right) \le \varepsilon.$$

Thus we obtain the continuity of u on $[0, \infty) \times \mathcal{O}$ and (v) follows.

Let (t_n) be a sequence of positive numbers converging to $t > 0$ and $(x_n) \subset \mathcal{O}$ a sequence of points converging to some regular $x \in \partial\mathcal{O}$. Then for n large enough, $t_n > \frac{t}{2}$ and

$$\left|\mathbb{E}\left[\varphi\left(X(t_n, x_n)\right) 1_{\tau^{x_n}_{\mathcal{O}} > t_n}\right]\right| \le \|\varphi\|_{\mathcal{O}}\, P\left(\tau^{x_n}_{\mathcal{O}} > \frac{t}{2}\right).$$

By Lemma 8.3.8 the right hand side converges to zero as n tends to infinity, and we get (vi). This concludes the proof that u is a strong solution of (8.1.1). $\square$

8.4 Uniqueness of the solutions

The next theorem provides the following result about the uniqueness of a solution.

Theorem 8.4.1 *Assume that (8.1.4) is satisfied, the solution $X(\cdot, x)$ of (8.1.6) exits from the open set $\mathcal{O}$ only through regular points of the boundary and for all $x \in \mathcal{O}$ and $t > 0$, $\mathbb{P}(\tau^x_{\mathcal{O}} = t) = 0$. If $u(t, x)$ is a strong solution of (8.1.1), then $u(t, x) = \mathbb{E}[\varphi(X(t, x)); \tau^x_{\mathcal{O}} > t]$.*

Remark 8.4.2 There exist sets $\mathcal{O}$ for which all points of $\partial\mathcal{O}$ are regular. The method used in the proof of Theorem 3.1 in [92] shows that for each $\gamma \in \mathbb{R}$ and $a \in D(A^*)$ such that $\langle Qa, a \rangle > 0$, the sets $\{x \in H : \langle x, a \rangle > \gamma\}$ and $\{x \in H : \langle x, a \rangle < \gamma\}$ have regular boundaries. A finite intersection of sets of this form also has a boundary consisting of regular points only.

Proof of Theorem 8.4.1. Fix $x \in \mathcal{O}$ and $t > 0$, $t < T$. In the proof we will always consider processes starting from x, therefore we will write $X(t)$ instead of $X(t, x)$.

Step 1. Let $\varepsilon > 0$, $0 < \gamma < t$, $i \in \mathbb{N}$. First we construct sets $U_{\varepsilon,\gamma,i}$ that increase as i increases and such that

$$|Du(s, y)|, \ \left\|D^2 u(s, y)\right\|, \ \left\|Q^{\frac{1}{2}} D^2 u(s, y) Q^{\frac{1}{2}}\right\|_1, \ |A^* Du(s, y)|,$$

are bounded on $[\gamma, T] \times \overline{U_{\varepsilon,\gamma,i}}$ and that the exit times of $X(t)$ from $U_{\varepsilon,\gamma,i}$ converge to $\tau_{\mathcal{O}}^x$ when i goes to infinity, on a subset of Ω with probability bigger than $1 - \varepsilon$.

For each $\varepsilon > 0$ there exists a compact set L_ε in H such that

$$\mathbb{P}\left(X(s) \in L_\varepsilon, \forall s \in [0, T]\right) \geq 1 - \varepsilon \tag{8.4.1}$$

(see Proposition 2 in [218]). Let $i_0 \in \mathbb{N}$ be such that $\frac{\operatorname{dist}(x, \mathcal{O}^c)}{4} \geq \frac{1}{i_0}$. For $i \geq i_0$ we define an open set $(\mathcal{O}^c)_i = \bigcup_{y \in \mathcal{O}^c} B(y, \frac{2}{i})$. Then $x \notin (\mathcal{O}^c)_i$ and $\operatorname{dist}(\mathcal{O}^c, (\mathcal{O}^c)_i^c) \geq \frac{2}{i}$. Let $K_{\varepsilon,i} = L_\varepsilon \cap (\mathcal{O}^c)_i^c$. $K_{\varepsilon,i}$ is a compact subset of $\mathcal{O}$. By properties (i) and (ii) of a strong solution , for each $y \in K_{\varepsilon,i}$ there exists $r_y > 0$ such that $|Du(s, y)|$, $\left\|D^2 u(s, y)\right\|$, $\left\|Q^{\frac{1}{2}} D^2 u(s, y) Q^{\frac{1}{2}}\right\|_1$ and $|A^* Du(s, y)|$ are bounded on the set $[\gamma, T] \times B(y, r_y)$. By compactness of $K_{\varepsilon,i}$ we can choose a finite covering $\{B(y_j, r_{y_j} \wedge \frac{1}{i})\}_{j=1}^{n_{\varepsilon,\gamma,i}}$ of $K_{\varepsilon,i}$. Let $V_{\varepsilon,\gamma,i} = \bigcup_{j=1}^{n_{\varepsilon,\gamma,i}} B(y_j, r_{y_j} \wedge \frac{1}{i})$. We define sets $U_{\varepsilon,\gamma,i}$ for $i \geq i_0$ inductively, taking $U_{\varepsilon,\gamma,i_0} = V_{\varepsilon,\gamma,i_0}$ and $U_{\varepsilon,\gamma,i} = U_{\varepsilon,\gamma,i-1} \cup V_{\varepsilon,\gamma,i}$.

For each $i \geq i_0$ we have $U_{\varepsilon,\gamma,i} \subset U_{\varepsilon,\gamma,i+1}$. Moreover, $\overline{U_{\varepsilon,\gamma,i}} \subset \mathcal{O}$, $x \in U_{\varepsilon,\gamma,i}$ and $|Du(s, y)|$, $\left\|D^2 u(s, y)\right\|$, $|A^* Du(s, y)|$ and $\left\|Q^{\frac{1}{2}} D^2 u(s, y) Q^{\frac{1}{2}}\right\|_1$ are bounded on $[\gamma, T] \times \overline{U_{\varepsilon,\gamma,i}}$.

Denote by $\tau_{\varepsilon,\gamma,i}$ the first exit time of X from $U_{\varepsilon,\gamma,i}$, that is $\tau_{\varepsilon,\gamma,i} = \inf\{s > 0 : X(s) \in U_{\varepsilon,\gamma,i}^c\}$. We will show that

$$\lim_{i \to \infty} \tau_{\varepsilon,\gamma,i} \wedge T = \tau_{\mathcal{O}}^x \wedge T \quad \text{on the set} \quad \{\omega : X(s, \omega) \in L_\varepsilon, \forall s \in [0, T]\}. \tag{8.4.2}$$

Let ω be such that $X(s,\omega) \in L_\varepsilon, \forall s \in [0,T]$. It is obvious that $\tau_{\varepsilon,\gamma,i}(\omega) \le \tau_{\mathcal{O}}^x(\omega)$ with a strict inequality if $\tau_{\mathcal{O}}^x(\omega) \ne \infty$. Consequently, if $\lim_{i\to\infty} \tau_{\varepsilon,\gamma,i}(\omega) \wedge T = T$ then (8.4.2) holds. Suppose that $\lim_{i\to\infty} \tau_{\varepsilon,\gamma,i}(\omega) = \sigma(\omega) < T$. We observe that $X(\tau_{\varepsilon,\gamma,i}(\omega)) \in (\mathcal{O}^c)_i$. This holds because $X(\tau_{\varepsilon,\gamma,i}(\omega)) \in U_{\varepsilon,\gamma,i}^c$ and $K_{\varepsilon,i} = L_\varepsilon \cap (\mathcal{O}^c)_i^c \subset U_{\varepsilon,\gamma,i}$. Thus $X(\tau_{\varepsilon,\gamma,i}(\omega)) \in (\mathcal{O}^c)_i$ or $X(\tau_{\varepsilon,\gamma,i}(\omega)) \in L_\varepsilon^c$ but our assumption on ω excludes the second case. Hence, by the construction of the set $(\mathcal{O}^c)_i$ we see that $\mathrm{dist}(X(\tau_{\varepsilon,\gamma,i}(\omega)), \mathcal{O}^c) < \frac{2}{i}$. By the continuity of X and the distance function we obtain

$$\mathrm{dist}(X(\sigma(\omega)), \mathcal{O}^c) = \lim_{i\to\infty} \mathrm{dist}(X(\tau_{\varepsilon,\gamma,i}(\omega)), \mathcal{O}^c) = 0.$$

Since $\mathcal{O}^c$ is closed, $X(\sigma(\omega)) \in \mathcal{O}^c$, which means that $\sigma(\omega) = \tau_{\mathcal{O}}^x(\omega)$.

   Step 2. Proof of the identity

$$\mathbb{E}u\left(t - (t - \gamma) \wedge \tau_{\varepsilon,\gamma,i}, X((t - \gamma) \wedge \tau_{\varepsilon,\gamma,i})\right) = u(t,x). \qquad (8.4.3)$$

To prove (8.4.3) one usually applies Itô's formula to $u(t-s, X(s))$. Since in general $X(s)$ is not a strong solution to (8.1.6) we will apply Itô's formula to approximating processes. We will only define approximating sequences, for the complete proof of (8.4.3) we refer to [207].

   Let (e_l) be an orthonormal basis of H and let (β_l) be a sequence of independent one dimensional standard Wiener processes. Write $W_n(t) = \sum_{l=1}^n e_l\beta_l(t)$ and $W(t) = \sum_{l=1}^\infty e_l\beta_l(t)$. Let $A_k = AkR(k, A)$ be the Yosida approximations of A. Let $X_{n,k}$, X_n and X be the mild solutions to the equations

$$dX_{n,k}(t) = A_k X_{n,k}(t)dt + Q^{\frac{1}{2}}dW_n(t), \quad X_{n,k}(0) = x, \qquad (8.4.4)$$

$$dX_n(t) = AX_n(t)dt + Q^{\frac{1}{2}}dW_n(t), \quad X_n(0) = x, \qquad (8.4.5)$$

$$dX(t) = AX(t)dt + Q^{\frac{1}{2}}dW(t), \quad X(0) = x, \qquad (8.4.6)$$

respectively. Of course the process X has the same law as the process satisfying (8.1.6) and $X_{n,k}$ is a strong solution to (8.4.4).

   By Theorem 2.2.6 in [76] we have

$$\lim_{k\to\infty} \sup_{s\le T} \mathbb{E}\,|X_{n,k}(s) - X_n(s)|^2 = 0,$$

$$\lim_{n\to\infty} \sup_{s\le T} \mathbb{E}\,|X_n(s) - X(s)|^2 = 0. \qquad (8.4.7)$$

Let us fix temporary ε, γ and i and define exit times

$$\tau_{n,k} = \inf\{s > 0 : X_{n,k}(s) \in U_{\varepsilon,\gamma,i}^c\},$$

$$\tau_n = \inf\{s > 0 : X_n(s) \in U_{\varepsilon,\gamma,i}^c\},$$

$$\tau = \inf\{s > 0 : X(s) \in U_{\varepsilon,\gamma,i}^c\}.$$

Denote $\sigma = \tau_{n,k} \wedge \tau_n \wedge \tau$. Since $X_{n,k}$ is a strong solution to (8.4.4), by Itô's formula we get

$$u(t - (t-\gamma) \wedge \sigma, X_{n,k}((t-\gamma) \wedge \sigma)) = u(t,x)$$
$$- \int_0^{(t-\gamma)\wedge\sigma} D_t u(t-r, X_{n,k}(r))\, dr$$
$$+ \int_0^{(t-\gamma)\wedge\sigma} \langle Du(t-r, X_{n,k}(r)), AkR(k,A)X_{n,k}(r)\rangle\, dr$$
$$+ \int_0^{(t-\gamma)\wedge\sigma} \left\langle Du(t-r, X_{n,k}(r)), Q^{\frac{1}{2}} dW_n(r)\right\rangle$$
$$+ \frac{1}{2} \int_0^{(t-\gamma)\wedge\sigma} \mathrm{Tr}[D^2 u(t-r, X_{n,k}(r))(Q^{\frac{1}{2}}I_n)(Q^{\frac{1}{2}}I_n)^*]\, dr,$$

$$(8.4.8)$$

where I_n denotes the orthogonal projection on $\mathrm{span}\{e_1, \dots, e_n\}$. The fact that u satisfies (8.1.1) allows us to write (8.4.8) in the form

$$u(t - (t-\gamma) \wedge \sigma, X_{n,k}((t-\gamma) \wedge \sigma)) = u(t,x)$$
$$+ \int_0^{(t-\gamma)\wedge\sigma} \langle A^* Du(t-r, X_{n,k}(r)), kR(k,A)X_{n,k}(r) - X_{n,k}(r)\rangle\, dr$$
$$+ \int_0^{(t-\gamma)\wedge\sigma} \left\langle Du(t-r, X_{n,k}(r)), Q^{\frac{1}{2}} dW_n(r)\right\rangle$$
$$+ \frac{1}{2} \int_0^{(t-\gamma)\wedge\sigma} \mathrm{Tr}[Q^{\frac{1}{2}} D^2 u(t-r, X_{n,k}(r))Q^{\frac{1}{2}}[I_n - I]]\, dr. \qquad (8.4.9)$$

After taking the expectations of both sides of (8.4.9) the term with the stochastic integral disappears since $|Du(s,y)|$ is bounded on $[\gamma, T] \times \overline{U}_{\varepsilon,\gamma,i}$. Hence we get

$$\mathbb{E}u(t - (t-\gamma) \wedge \sigma, X_{n,k}((t-\gamma) \wedge \sigma)) = u(t,x)$$
$$+ \mathbb{E}\int_0^{(t-\gamma)\wedge\sigma} \langle A^* Du(t-r, X_{n,k}(r)), kR(k,A)X_{n,k}(r) - X_{n,k}(r)\rangle\, dr$$
$$+ \mathbb{E}\frac{1}{2} \int_0^{(t-\gamma)\wedge\sigma} \mathrm{Tr}[Q^{\frac{1}{2}} D^2 u(t-r, X_{n,k}(r))Q^{\frac{1}{2}}[I_n - I]]\, dr. \qquad (8.4.10)$$

Passing in (8.4.10) to the limit as $k \to \infty$ and $n \to \infty$, see [207], one arrives at (8.4.3).

Step 3. We will show that if i goes to infinity, and γ goes to zero and finally ε goes to zero, the left hand side of (8.4.3) converges to $\mathbb{E}\varphi(X(t))1_{\tau_{\mathcal{O}}^x > t}$, which is the desired conclusion. We have

$$
\begin{aligned}
&\big|\mathbb{E}u\left(t-(t-\gamma)\wedge\tau_{\varepsilon,\gamma,i},X((t-\gamma)\wedge\tau_{\varepsilon,\gamma,i})\right)-\mathbb{E}\varphi(X(t))1_{\tau_{\mathcal{O}}^x>t}\big|\\
&\leq \big|\mathbb{E}u(\gamma,X(t-\gamma))1_{\tau_{\varepsilon,\gamma,i}\geq t-\gamma}1_{X\in L_\varepsilon}-\mathbb{E}\varphi(X(t))1_{\tau_{\mathcal{O}}^x>t}1_{X\in L_\varepsilon}\big|\\
&\quad+\big|\mathbb{E}u(t-\tau_{\varepsilon,\gamma,i},X(\tau_{\varepsilon,\gamma,i}))1_{\tau_{\varepsilon,\gamma,i}<t-\gamma}1_{X\in L_\varepsilon}\big|+2\,\|u\|_{[0,T]\times\mathcal{O}}\,\mathbb{P}(X\notin L_\varepsilon)\\
&=I_1(\varepsilon,\gamma,i)+I_2(\varepsilon,\gamma,i)+I_3(\varepsilon).
\end{aligned}\tag{8.4.11}
$$

Observe that

$$
\begin{aligned}
I_1(\varepsilon,\gamma,i)\leq\;&\mathbb{E}\,|u(\gamma,X(t-\gamma))-\varphi(X(t))|\,1_{\tau_{\mathcal{O}}^x>t}\\
&+\|u\|_{[0,T]\times\mathcal{O}}\,\mathbb{E}\,\big|1_{\tau_{\varepsilon,\gamma,i}\geq t-\gamma}-1_{\tau_{\mathcal{O}}^x>t}\big|\,1_{X\in L_\varepsilon}.
\end{aligned}\tag{8.4.12}
$$

By (8.4.2) and the fact that $\tau_{\varepsilon,\gamma,i}<\tau_{\mathcal{O}}^x$ if $\tau_{\mathcal{O}}^x<\infty$ we see that

$$
\begin{aligned}
&\limsup_{i\to\infty}\mathbb{E}\,\big|1_{\tau_{\varepsilon,\gamma,i}\geq t-\gamma}-1_{\tau_{\mathcal{O}}^x>t}\big|\,1_{X\in L_\varepsilon}\\
&\leq\limsup_{i\to\infty}\mathbb{E}\,\big|1_{\tau_{\varepsilon,\gamma,i}\geq t-\gamma}-1_{\tau_{\mathcal{O}}^x>t-\gamma}\big|\,1_{X\in L_\varepsilon}+\mathbb{P}(t\geq\tau_{\mathcal{O}}^x>t-\gamma)\\
&=\mathbb{P}(t\geq\tau_{\mathcal{O}}^x>t-\gamma).
\end{aligned}\tag{8.4.13}
$$

Moreover, by the assumption on $\tau_{\mathcal{O}}^x$,

$$
\lim_{\gamma\to0}\mathbb{P}(t\geq\tau_{\mathcal{O}}^x>t-\gamma)=\mathbb{P}(\tau_{\mathcal{O}}^x=t)=0.\tag{8.4.14}
$$

By (8.4.13) and (8.4.14) the second term on the right hand side of (8.4.12) converges to zero when we let $i\to\infty$ and then $\gamma\to0$. The first term on the right hand side of (8.4.12) does not depend on i and converges to zero if γ goes to zero by the joint continuity and property (v) of the strong solution. Thus from (8.4.12) we obtain

$$
\limsup_{\gamma\to0}\limsup_{i\to\infty}I_1(\varepsilon,\gamma,i)=0.\tag{8.4.15}
$$

By (8.4.2) and property (vi)

$$
\lim_{i\to\infty}I_2(\varepsilon,\gamma,i)=0.\tag{8.4.16}
$$

We also have that

$$
\lim_{\varepsilon\to0}I_3(\varepsilon)\leq\lim_{\varepsilon\to0}C\varepsilon=0.\tag{8.4.17}
$$

By (8.4.15), (8.4.16) and (8.4.17) applied to (8.4.11) we obtain

$$\limsup_{\varepsilon \to 0} \limsup_{\gamma \to 0} \limsup_{i \to \infty} |\mathbb{E}u\left(t - (t - \gamma) \wedge \tau_{\varepsilon,\gamma,i}, X\left((t - \gamma) \wedge \tau_{\varepsilon,\gamma,i}\right)\right)$$

$$-\mathbb{E}\varphi(X_t)1_{\tau_{\mathcal{O}}^x > t}| = 0.$$

Consequently from (8.4.3) it follows that

$$u(t, x) = \mathbb{E}[\varphi(X(t))1_{\tau_{\mathcal{O}}^x > t}].$$

$\square$

Part II

THEORY IN SOBOLEV SPACES

Chapter 9

L^2 and Sobolev spaces

We are given a Gaussian measure $\mu = N_Q$, where $Q \in L_1^+(H)$. We shall assume for simplicity that Ker $Q = \{0\}$. We shall denote by (e_k) a complete orthonomal system in H and by (λ_k) a sequence of positive numbers such that $Qe_k = \lambda_k e_k$, $k \in \mathbb{N}$. For any $x \in H$ we set $x_k = \langle x, e_k \rangle$, $k \in \mathbb{N}$.

For any $p \geq 1$ we shall denote by $L^p(H, \mu)$ the Banach space of all p-integrable equivalence classes of functions from H into $\mathbb{R}$, with norm

$$\|\varphi\|_{L^p(H,\mu)} = \left(\int_H |\varphi(x)|^p \mu(dx) \right)^{1/p}, \ \varphi \in L^p(H, \mu).$$

In particular $L^2(H, \mu)$ is a real Hilbert space with the inner product

$$\langle \varphi, \psi \rangle_{L^2(H,\mu)} = \int_H \varphi(x)\psi(x)\mu(dx), \ \varphi, \psi \in L^2(H, \mu).$$

We shall denote by $E(H)$ the linear space spanned by all exponential functions, that is all functions φ of the form

$$\varphi(x) = e^{\langle h,x \rangle}, \ h \in H.$$

We know by Proposition 1.2.5 that $E(H)$ is dense in $L^p(H, \mu)$, $p \geq 1$.

In §9.1 we shall study several properties of the space $L^2(H, \mu)$, including the Itô-Wiener decomposition. Then in §9.2 we shall define the Sobolev spaces $W^{1,2}(H, \mu)$ and $W^{2,2}(H, \mu)$ and prove compactness of the embedding of $L^2(H, \mu)$ into $W^{1,2}(H, \mu)$. Finally, in §9.3 we shall define the *Malliavin derivative* and the Sobolev spaces $W_Q^{1,2}(H, \mu)$ and $W_Q^{2,2}(H, \mu)$.

9.1 Itô-Wiener decomposition

9.1.1 Real Hermite polynomials

Let us consider the analytic function

$$F(t,\xi) = e^{-\frac{t^2}{2}+t\xi}, \quad t,\xi \in \mathbb{R}.$$

Then the Hermite polynomials H_n, $n \in \{0\} \cup \mathbb{N}$, are defined through the formula

$$F(t,\xi) = \sum_{n=0}^{\infty} \frac{t^n}{\sqrt{n!}} H_n(\xi), \quad t,\xi \in \mathbb{R}. \tag{9.1.1}$$

Proposition 9.1.1 *For any $n \in \{0\} \cup \mathbb{N}$ we have*

$$H_n(\xi) = \frac{(-1)^n}{\sqrt{n!}} e^{\frac{\xi^2}{2}} D_\xi^n \left(e^{-\frac{\xi^2}{2}} \right), \quad \xi \in \mathbb{R}. \tag{9.1.2}$$

Proof. In fact

$$e^{-\frac{(t-\xi)^2}{2}} = \sum_{n=0}^{\infty} \frac{t^n}{n!} D_t^n \left[e^{-\frac{(t-\xi)^2}{2}} \right]_{t=0}$$

$$= \sum_{n=0}^{\infty} \frac{t^n}{n!} (-1)^n D_\xi^n \left(e^{-\frac{\xi^2}{2}} \right), \quad t,\xi \in \mathbb{R}.$$

Multiplying both sides by $e^{\frac{\xi^2}{2}}$, we find

$$F(t,\xi) = \sum_{n=0}^{\infty} \frac{t^n}{n!} (-1)^n e^{\frac{\xi^2}{2}} D_\xi^n \left(e^{-\frac{\xi^2}{2}} \right), \quad t,\xi \in \mathbb{R},$$

which yields (9.1.2). $\square$

It follows by (9.1.2) that, for all $n \in \mathbb{N}$, H_n is a polynomial of degree n with positive leading coefficient. We have in particular

$$H_0(\xi) = 1, \; H_1(\xi) = \xi, \; H_2(\xi) = \frac{1}{\sqrt{2}} (\xi^2 - 1), \; H_3(\xi) = \frac{1}{\sqrt{6}} (\xi^3 - 3\xi).$$

Let us prove some useful identities.

Proposition 9.1.2 *Let $n \in \mathbb{N}$, then fror all $\xi \in \mathbb{R}$ we have*

$$\xi H_n(\xi) = \sqrt{n+1}\, H_{n+1}(\xi) + \sqrt{n}\, H_{n-1}(\xi), \quad (9.1.3)$$
$$D_\xi H_n(\xi) = \sqrt{n}\, H_{n-1}(\xi), \quad (9.1.4)$$
$$D_\xi^2 H_n(\xi) - \xi D_\xi H_n(\xi) = -n H_n(\xi),. \quad (9.1.5)$$

Proof. Equations (9.1.3) and (9.1.4) follow from the identities

$$D_t F(t,\xi) = (\xi - t)F(t,\xi) = \sum_{n=1}^{\infty} \sqrt{n}\, \frac{t^{n-1}}{\sqrt{(n-1)!}}\, H_n(\xi), \quad t,\xi \in \mathbb{R},$$

and

$$D_\xi F(t,\xi) = t F(t,\xi) = \sum_{n=1}^{\infty} \frac{t^n}{\sqrt{n!}}\, H_n'(\xi), \quad t,\xi \in \mathbb{R}.$$

Equation (9.1.5) is a consequence of (9.1.3) and (9.1.4). $\square$

Finally, we prove orthonormality and completeness of the system of Hermite polynomials on $L^2(\mathbb{R}, N_1)$.

Proposition 9.1.3 *The system $(H_n)_{n \in \{0\} \cup \mathbb{N}}$ is orthonormal and complete on $L^2(\mathbb{R}, N_1)$.*

Proof. Let us first prove orthonormality. We have, for $t, s, \xi \in \mathbb{R}$,

$$F(t,\xi)F(s,\xi) = e^{-\frac{1}{2}(t^2+s^2)+\xi(t+s)} = \sum_{m,n=0}^{\infty} \frac{t^m}{\sqrt{m!}}\, \frac{s^n}{\sqrt{n!}}\, H_n(\xi)\, H_m(\xi).$$

Integrating both sides of this equality with respect to N_1, and taking into account the identity

$$\frac{1}{\sqrt{2\pi}} \int_{\mathbb{R}} e^{-\frac{\xi^2}{2}} e^{-\frac{1}{2}(t^2+s^2)+\xi(t+s)}\, d\xi = e^{ts},$$

we find

$$e^{ts} = \sum_{m,n=0}^{\infty} \frac{t^m}{\sqrt{m!}}\, \frac{s^n}{\sqrt{n!}} \int_{\mathbb{R}} H_n(\xi)\, H_m(\xi)\, N_1(d\xi),$$

which yields

$$\int_{\mathbb{R}} H_n(\xi)\, H_m(\xi)\, N_1(d\xi) = \delta_{n,m}.$$

It remains to prove completeness. Let $f \in L^2(\mathbb{R}, N_1)$ be such that

$$\int_{\mathbb{R}} f(\xi) H_n(\xi) N_1(d\xi) = 0, \quad \forall n \in \{0\} \cup \mathbb{N}.$$

It follows that

$$\int_{\mathbb{R}} f(\xi)F(t,\xi)e^{-\frac{\xi^2}{2}}\,d\xi = \int_{\mathbb{R}} f(\xi)e^{-\frac{(t-\xi)^2}{2}}\,d\xi = 0.$$

Taking Fourier transforms we find $f = 0$. $\square$

9.1.2   Chaos expansions

Let us introduce Hermite polynomials on $L^2(H,\mu)$. Consider the set Γ of all mappings $\gamma : \mathbb{N} \to \{0\} \cup \mathbb{N}$, $n \to \gamma_n$, such that $|\gamma| := \sum_{k=1}^{\infty} \gamma_k < +\infty$. Note that if $\gamma \in \Gamma$ then $\gamma_n = 0$ for all n, except possibly a finite number. For any $\gamma \in \Gamma$ we define the *Hermite polynomial*

$$H_\gamma(x) = \prod_{k=1}^{\infty} H_{\gamma_k}(W_{e_k}(x)), \quad x \in H,$$

where the mapping W was defined in §1.2.4. This definition is meaningful since all factors, with the exception of a finite number, are equal to $H_0(W_{e_k}(x)) = 1$, $x \in H$.

We are going to prove that $(H_\gamma)_{\gamma \in \Gamma}$ is a complete orthonormal system in $L^2(H,\mu)$. For this we need a lemma.

Lemma 9.1.4 *Let $h, g \in H$ with $|h| = |g| = 1$ and let $n, m \in \mathbb{N} \cup \{0\}$. Then we have*

$$\int_H H_n(W_h)H_m(W_g)\,d\mu = \delta_{n,m}[\langle h,g\rangle]^n. \tag{9.1.6}$$

Proof. For any $t, s \in \mathbb{R}$ we have

$$\int_H F(t, W_h(x))F(s, W_g(x))\mu(dx) = e^{-\frac{t^2+s^2}{2}} \int_H e^{tW_h(x)+sW_g(x)}\,\mu(dx)$$

$$= e^{-\frac{t^2+s^2}{2}} \int_H e^{W_{th+sg}(x)}\,\mu(dx) = e^{-\frac{t^2+s^2}{2}} e^{\frac{1}{2}|th+sg|^2} = e^{ts\langle h,g\rangle},$$

because $|h| = |g| = 1$. It follows that

$$e^{ts\langle h,g\rangle} = \sum_{m,n=0}^{\infty} \frac{t^n s^m}{\sqrt{n!m!}} \int_H H_n(W_h(x))H_m(W_g(x))\,\mu(dx),$$

which implies (9.1.6). $\square$

We can now prove the result.

Theorem 9.1.5 *The system $(H_\gamma)_{\gamma\in\Gamma}$ is orthonormal and complete on $L^2(H,\mu)$.*

Proof. Let $\gamma,\eta\in\Gamma$. We have

$$\int_H H_\gamma(x)H_\eta(x)\mu(dx) = \prod_{n=1}^\infty \int_H H_{\gamma_n}(W_{e_n}(x))H_{\eta_n}(W_{e_n}(x))\mu(dx) = \delta_{\gamma,\eta},$$

in virtue of Lemma 9.1.4, where $\delta_{\eta,\gamma}=\prod_{n=1}^\infty \delta_{\eta_n,\gamma_n}$. So the system $(H_\gamma)_{\gamma\in\Gamma}$ is orthonormal. To prove that it is complete, let $\varphi\in L^2(H,\mu)$ be such that

$$\int_H \varphi(x)H_\gamma(x)\mu(dx) = 0, \quad \forall\gamma\in\Gamma.$$

We have to show that $\varphi=0$. Taking into account Proposition 1.2.5 it is enough to show that

$$\int_H e^{\langle f,x\rangle}\varphi(x)\mu(dx) = 0, \quad \forall f\in H. \tag{9.1.7}$$

Also it is enough to prove (9.1.7) for f of the form

$$f = \sum_{k=1}^n \lambda_k^{-1/2}t_k e_k, \ (t_1,\ldots,t_n)\in\mathbb{R}^n, \ n\in\mathbb{N}. \tag{9.1.8}$$

If f is of the form (9.1.8) we have

$$\int_H e^{\langle f,x\rangle}\varphi(x)\mu(dx) = \int_H e^{\sum_{k=1}^n t_k W_{e_k}(x)}\varphi(x)\mu(dx)$$

$$= e^{\sum_{k=1}^n \frac{t_k^2}{2}}\int_H F(t_1,W_{e_1(x)})\ldots F(t_n,W_{e_n(x)})\varphi(x)\mu(dx).$$

On the other hand

$$\int_H F(t_1,W_{e_1})\ldots F(t_n,W_{e_n})\varphi d\mu$$

$$= \sum_{j_1,\ldots,j_n=1}^\infty \frac{t_1^{j_1}}{\sqrt{j_1!}}\ldots\frac{t_n^{j_n}}{\sqrt{j_n!}}\int_H H_{j_1}(W_{e_1})\ldots H_{j_n}(W_{e_n})\varphi\,d\mu = 0,$$

and (9.1.7) follows. □

We finally introduce the *Itô-Wiener decomposition* of $L^2(H,\mu)$. For all $n\in\{0\}\cup\mathbb{N}$, we set $\Gamma_n=\{\gamma\in\gamma: |\gamma|=n\}$ and denote by $L_n^2(H,\mu)$ the closed subspace of $L^2(H,\mu)$ spanned by $\{H_\gamma: \gamma\in\Gamma_n\}$. In particular $L_0^2(H,\mu)$ is the set of all constant functions in $L^2(H,\mu)$. Moreover, we denote by Π_n the orthogonal projector on $L_n^2(H,\mu)$, $n\in\{0\}\cup\mathbb{N}$.

Exercise 9.1.6 Prove that $L_1^2(H,\mu)$ is the closed subspace of $L^2(H,\mu)$ generated by all functions $\{W_f : f \in H\}$.

Theorem 9.1.7 *We have*

$$L^2(H,\mu) = \bigoplus_{n=0}^{\infty} L_n^2(H,\mu).$$

Proof. Note first that, in view of Lemma 9.1.4, the subspaces $L_n^2(H,\mu)$, $n \in \mathbb{N} \cup \{0\}$, are pairwise orthogonal. Assume that $\varphi \in L^2(H,\mu)$ is such that, for any $x \in H$ with $|h| = 1$ and for any $n \in \mathbb{N}$,

$$\int_H \varphi(x) H_n(W_h(x))\, \mu(dx) = 0. \tag{9.1.9}$$

Arguing as in the proof of Theorem 9.1.5 we get that $\varphi = 0$. $\square$

We end this subsection by computing some projections, needed later. For this we prove two lemmas.

Lemma 9.1.8 *Let $f,g \in H$ be such that $|f| = |g| = 1$, $s \in \mathbb{R}$ and $n \in \mathbb{N} \cup \{0\}$. Then we have*

$$I := \int_H e^{sW_f(x)} H_n(W_g(x))\mu(dx) = \frac{1}{\sqrt{n!}}\, s^n e^{\frac{s^2}{2}}\, \langle f,g\rangle^n. \tag{9.1.10}$$

Proof. We have, taking into account (9.1.6)

$$
\begin{aligned}
I &= e^{\frac{s^2}{2}} \int_H F(s, W_f(x)) H_n(W_g(x))\mu(dx) \\
&= e^{\frac{s^2}{2}} \sum_{k=0}^{\infty} \frac{s^k}{\sqrt{k!}} \int_H H_k(W_f(x)) H_n(W_g(x))\mu(dx) \\
&= \frac{1}{\sqrt{n!}}\, s^n e^{\frac{s^2}{2}}\, \langle f,g\rangle^n. \quad \square
\end{aligned}
$$

Lemma 9.1.9 *Let $f,g \in H$ be such that $|f| = |g| = 1$, and $n \in \mathbb{N} \cup \{0\}$. Then we have*

$$J := \int_H W_f^n(x) H_n(W_g(x))\mu(dx) = \sqrt{n!}\, [\langle f,g\rangle]^n. \tag{9.1.11}$$

Proof. By Lemma 9.1.8 it follows that

$$\frac{1}{\sqrt{n!}} \, s^n e^{\frac{s^2}{2}} \, [\langle f, g \rangle]^n = \sum_{h=0}^{\infty} \int_H \frac{s^h W_f^h(x)}{h!} \, H_n(W_g(x)) \mu(dx),$$

and the conclusion follows differentiating n times with respect to s and setting $s = 0$. $\square$

Now we compute the projections $\Pi_n(W_f^n)$, $n \in \mathbb{N}$.

Proposition 9.1.10 *Let $f \in H$ such that $|f| = 1$ and let $n \in \mathbb{N}$. Then we have*

$$\Pi_n(W_f^n) = \sqrt{n!} \, H_n(W_f). \qquad (9.1.12)$$

Proof. First notice that $\sqrt{n!} \, H_n(W_f) \in L_n^2(H, \mu)$. Thus it is enough to show that for all $g \in H$ such that $|g| = 1$, we have

$$\int_H [W_f^n(x) - \sqrt{n!} \, H_n(W_f(x))] H_n(W_g(x)) \mu(dx) = 0,$$

and this follows from (9.1.6) and (9.1.11). $\square$

Finally we compute the projection $\Pi_n(e^{sW_f})$.

Proposition 9.1.11 *Let $f \in H$ with $|f| = 1$. Then we have*

$$\Pi_n\left(e^{sW_f}\right) = \frac{1}{\sqrt{n!}} \, s^n e^{\frac{s^2}{2}} \, H_n(W_f). \qquad (9.1.13)$$

Proof. It is enough to show that for all $g \in H$, such that $|g| = 1$, we have

$$\int_H \left(e^{sW_f(x)} - \frac{s^n}{\sqrt{n!}} \, e^{\frac{s^2}{2}} \, H_n(W_f) \right) H_n(W_g(x)) \mu(dx) = 0.$$

This follows using once again (9.1.6) and (9.1.10). $\square$

9.1.3 The space $L^2(H, \mu; H)$

Let us consider the space $L^2(H, \mu; H)$ of all equivalence classes of measurable functions $F : H \to H$ such that

$$\int_H |F(x)|^2 \mu(dx) < +\infty.$$

$L^2(H, \mu; H)$, endowed with the scalar product

$$\langle F, G \rangle_{L^2(H,\mu;H)} = \int_H \langle F(x), G(x) \rangle \mu(dx),$$

is a Hilbert space.

For any $F \in L^2(H, \mu; H)$ we set $F_k(x) = \langle F(x), e_k \rangle$, $k \in \mathbb{N}$. Then we have

$$F(x) = \sum_{k=1}^{\infty} F_k(x) e_k, \quad \mu\text{-a.e.}$$

Proposition 9.1.12 $(H_\gamma \otimes e_h)_{\gamma \in \Gamma, h \in \mathbb{N}}$ *is a complete orthonormal system on* $L^2(H, \mu; H)$.

Proof. In fact let $F \in L^2(H, \mu; H)$ be such that for any $\gamma \in \Gamma$ and $h \in \mathbb{N}$

$$\int_H \langle F(x), H_\gamma \otimes e_h \rangle d\mu = 0.$$

Then we have for any $\gamma \in \Gamma$

$$\int_H F_h(x) H_\gamma(x) \mu(dx) = 0,$$

which yields $F_h(x) = 0$, μ-a.e. for all $h \in \mathbb{N}$, so that $f = 0$. $\square$

9.2   Sobolev spaces

For any $k \in \mathbb{N}$ we consider the partial derivative in the direction e_k, defined as

$$D_k \varphi(x) = \lim_{\varepsilon \to 0} \frac{1}{\varepsilon} \left(\varphi(x + \varepsilon e_k) - \varphi(x) \right), \quad x \in H, \ \varphi \in E(H).$$

If $\varphi(x) = e^{\langle f, x \rangle}$ with $f \in H$, we have obviously

$$D_k \varphi(x) = f_k e^{\langle f, x \rangle}, \quad x \in H.$$

Moreover we denote by Λ_0 the linear span of $\{H_\gamma \otimes e_h : \gamma \in \Gamma, \ h \in \mathbb{N}\}$ and by D the linear operator

$$D : E(H) \subset L^2(H, \mu) \to L^2(H, \mu; H), \ \varphi \to D\varphi.$$

By Proposition 9.1.12, Λ_0 is dense in $L^2(H, \mu; H)$.

We want to prove that D and D_h, $h \in \mathbb{N}$, are closable operators in $L^2(H, \mu)$. To this purpose we need some integration by parts formulae.

Lemma 9.2.1 *Let* $\varphi, \psi \in E(H)$. *Then the following identity holds:*

$$\int_H D_k \varphi(x) \psi(x) \mu(dx) + \int_H \varphi(x) D_k \psi(x) \mu(dx) = \frac{1}{\lambda_k} \int_H x_k \varphi(x) \psi(x) \mu(dx).$$

$$(9.2.1)$$

Proof. Since $E(H)$ is dense in $L^2(H, \mu)$ it is enough to prove (9.2.1) for

$$\varphi(x) = e^{\langle f, x \rangle}, \quad \psi(x) = e^{\langle g, x \rangle}, \quad x \in H,$$

where $f, g \in H$. In this case we have

$$\int_H D_k \varphi(x) \psi(x) \mu(dx) = \int_H f_k e^{\langle f+g, x \rangle} \mu(dx) = f_k e^{\frac{1}{2}\langle Q(f+g), f+g \rangle}, \quad (9.2.2)$$

$$\int_H \varphi(x) D_k \psi(x) \mu(dx) = \int_H g_k e^{\langle f+g, x \rangle} \mu(dx) = g_k e^{\frac{1}{2}\langle Q(f+g), f+g \rangle}, \quad (9.2.3)$$

$$\int_H \varphi(x) x_k \psi(x) \mu(dx) = \int_H x_k e^{\langle f+g, x \rangle} \mu(dx)$$

$$= \frac{d}{dt} \int_H e^{\langle f+g+te_k, x \rangle} \mu(dx) \Big|_{t=0} = \frac{d}{dt} e^{\frac{1}{2}\langle Q(f+g+te_k), f+g+te_k \rangle} \Big|_{t=0} \quad (9.2.4)$$

$$= \lambda_k (f_k + g_k) e^{\frac{1}{2}\langle Q(f+g), f+g \rangle}.$$

Now summing up (9.2.2) and (9.2.3), we find (9.2.4). $\square$

Proposition 9.2.2 D_k *is closable for all* $k \in \mathbb{N}$.

Proof. Let $(\varphi_n) \subset E(H)$ be such that

$$\varphi_n \to 0, \quad \text{in } L^2(H, \mu), \quad D_k \varphi_n \to f \text{ in } L^2(H, \mu).$$

We have to show that $f = 0$. If $g \in E(H)$, then by (9.2.1) we have

$$\int_H D_k \varphi_n \, g \, d\mu + \int_H \varphi_n \, D_k g \, d\mu = \frac{1}{\lambda_k} \int_H x_k \varphi_n \, g \, d\mu.$$

By taking the limit as $n \to \infty$ we obtain $\int_H f g d\mu = 0$, which yields $f = 0$ since $E(H)$ is dense in $L^2(H, \mu)$. $\square$

If φ belongs to the domain of the closure of D_k, which we shall still denote by D_k, we shall say that $D_k \varphi$ belongs to $L^2(H, \mu)$.

We come now to the operator D. We first state an easy consequence of Lemma 9.2.1.

Lemma 9.2.3 *Let* $\varphi \in E(H)$ *and* $G \in \Lambda_0$. *Then we have*

$$\int_H \langle D\varphi, G \rangle d\mu + \int_H \text{div } G \, \varphi \, d\mu = \int_H \langle x, Q^{-1} G(x) \rangle d\mu. \quad (9.2.5)$$

Now we are ready to prove the following result.

Proposition 9.2.4 D *is closable.*

Proof. Let $(\varphi_n) \subset E(H)$ be such that

$$\varphi_n \to 0 \quad \text{in } L^2(H, \mu), \quad D\varphi_n \to F \quad \text{in } L^2(H, \mu; H).$$

We have to show that $F = 0$. If $G \in \Lambda_0$, then by (9.2.5) we have

$$\int_H \langle D\varphi_n, G \rangle d\mu + \int_H \varphi_n \operatorname{div} G(x)\, d\mu = \int_H \langle x, Q^{-1}G(x) \rangle \varphi d\mu.$$

As $n \to \infty$ we obtain $\int_H \langle F, G \rangle d\mu = 0$, which yields $F = 0$ since Λ_0 is dense in $L^2(H, \mu; H)$. $\square$

If φ belongs to the domain of the closure of D, which we shall still denote by D, we shall say that $D\varphi$ belongs to $L^2(H, \mu; H)$.

9.2.1   The space $W^{1,2}(H, \mu)$

$W^{1,2}(H, \mu)$ is the linear space of all functions $\varphi \in L^2(H, \mu)$ such that $D\varphi \in L^2(H, \mu; H)$. $W^{1,2}(H, \mu)$, endowed with the inner product

$$\langle \varphi, \psi \rangle_{W^{1,2}(H,\mu)} = \langle \varphi, \psi \rangle_{L^2(H,\mu)} + \int_H \langle D\varphi(x), D\psi(x) \rangle \mu(dx),$$

is a Hilbert space.

If $\varphi \in W^{1,2}(H, \mu)$ it is clear that $D_k\varphi \in L^2(H, \mu)$ for all $k \in \mathbb{N}$. We shall set

$$D\varphi(x) = \sum_{k=1}^{\infty} D_k\varphi(x)e_k, \ x \in H.$$

Since

$$|D\varphi(x)|^2 = \sum_{k=1}^{\infty} |D_k\varphi(x)|^2, \quad \text{a.e. in } H,$$

the series above is convergent for almost all $x \in H$. We call $D\varphi(x)$ the *gradient* of φ at x. Notice that if $\varphi \in W^{1,2}(H, \mu)$, one has

$$\langle D\varphi(x), e_k \rangle = D_k\varphi(x), \ \forall k \in \mathbb{N}, \quad \text{a.e.}$$

Let us state an immediate consequence of Lemma 9.2.1 and Proposition 9.2.4.

Lemma 9.2.5 *Let $\varphi, \psi \in W^{1,2}(H, \mu)$, and let $k \in \mathbb{N}$. Then the following identity holds.*

$$\int_H D_k\varphi(x)\psi(x)\mu(dx) + \int_H \varphi(x)D_k\psi(x)\mu(dx) = \frac{1}{\lambda_k}\int_H x_k\varphi(x)\psi(x)\mu(dx).$$
(9.2.6)

Remark 9.2.6 Let $n \in \mathbb{N}$, $g : \mathbb{R}^n \to \mathbb{R}$ be such that $g \in W^{1,2}(\mathbb{R}^n, N_{Q_n})$, where $Q_n = \operatorname{diag}\{\lambda_1, \dots, \lambda_n\}$. Setting $\varphi(x) = g(x_1, \dots, x_n)$, it is easy to check that $\varphi \in W^{1,2}(H, \mu)$.

9.2.2 Some additional summability results

We are going to prove the useful result that if $D_k\varphi \in L^2(H, \mu)$, then $x_k D_k\varphi \in L^2(H, \mu)$. For this we need a lemma.

Lemma 9.2.7 *For all $\zeta \in E(H)$ we have*

$$\int_H x_k^2 \zeta^2(x)\mu(dx) \le 2\lambda_k \int_H \zeta^2(x)\mu(dx) + 4\lambda_k^2 \int_H (D_k\zeta(x))^2\mu(dx). \quad (9.2.7)$$

Proof. Let $\varphi(x) = x_k$, and $\psi(x) = \zeta^2(x)$, $x \in H$. Then by (9.2.1) it follows that

$$\int_H x_k^2 \zeta^2(x)\mu(dx) = 2\lambda_k \int_H x_k\zeta(x)D_k\zeta(x)\mu(dx) + \lambda_k \int_H \zeta^2(x)\mu(dx).$$

Consequently

$$\int_H x_k^2 \zeta^2(x)\mu(dx) \le \frac{1}{2}\int_H x_k^2 \zeta^2(x)\mu(dx) + 2\lambda_k^2 \int_H (D_k\zeta)^2(x)\mu(dx)$$

$$+ \lambda_k \int_H \zeta^2(x)\mu(dx),$$

which yields the conclusion. $\square$

We are now ready to prove the following result.

Proposition 9.2.8 *Let $k \in \mathbb{N}$ and $D_k\varphi \in L^2(H, \mu)$. Then $x_k\varphi \in L^2(H, \mu)$ and the following estimate holds:*

$$\int_H x_k^2 \varphi^2(x)\mu(dx) \le 2\lambda_k \int_H \varphi^2(x)\mu(dx) + 4\lambda_k^2 \int_H (D_k\varphi(x))^2\mu(dx). \quad (9.2.8)$$

Proof. Let $(\varphi_n) \subset E(H)$ be such that

$$\varphi_n \to \varphi, \quad D_k \varphi_n \to D_k \varphi, \quad \text{in } L^2(H, \mu).$$

Then by Lemma 9.2.7 we have

$$\int_H x_k^2 \varphi_n^2(x) \mu(dx) \leq 2\lambda_k \int_H \varphi_n^2(x) \mu(dx) + 4\lambda_k^2 \int_H (D_k \varphi_n(x))^2 \mu(dx).$$

Letting n tend to infinity, we find (9.2.8). $\square$

The next corollary follows easily from Lemma 9.2.1.

Corollary 9.2.9 *Let $k \in \mathbb{N}$ and $D_k \varphi, D_k \psi \in L^2(H, \mu)$. Then the following identity holds:*

$$\int_H D_k \varphi(x) \psi(x) \mu(dx) + \int_H \varphi(x) D_k \psi(x) \mu(dx) = \frac{1}{\lambda_k} \int_H x_k \varphi(x) \psi(x) \mu(dx).$$
$$(9.2.9)$$

Let us prove finally the following result.

Proposition 9.2.10 *Let $\zeta \in W^{1,2}(H, \mu)$. Then the function*

$$H \to \mathbb{R}, \quad x \to |x|\zeta(x),$$

belongs to $L^2(H, \mu)$ and the following estimate holds:

$$\int_H |x|^2 \zeta^2(x) \mu(dx) \leq 2 \operatorname{Tr} Q \int_H \zeta^2(x) \mu(dx) + 4 \|Q\|^2 \int_H |D\zeta(x)|^2 \mu(dx).$$
$$(9.2.10)$$

Proof. By (9.2.7) it follows that

$$\int_H x_k^2 \zeta^2(x) \mu(dx) \leq 2\lambda_k \int_H \zeta^2(x) \mu(dx) + 4\|Q\|^2 \int_H (D_k \zeta(x))^2 \mu(dx).$$

Summing up over k yields (9.2.10). $\square$

9.2.3 Compactness of the embedding $W^{1,2}(H, \mu) \subset L^2(H, \mu)$

We start by giving a characterization of functions belonging to $W^{1,2}(H, \mu)$ in terms of the complete orthonormal basis $(H_\gamma)_{\gamma \in \Gamma}$.

For any $\varphi \in L^2(H,\mu)$ we have

$$\varphi = \sum_{\gamma \in \Gamma} \varphi_\gamma H_\gamma,$$

where $\varphi_\gamma = \langle \varphi, H_\gamma \rangle_{L^2(H,\mu)}$. We recall that for any $\gamma = (\gamma_n)$, the Hermite polynomial H_γ is defined by

$$H_\gamma = \prod_{n=0}^{\infty} H_{\gamma_n}(W_{e_n}).$$

This is equivalent to

$$H_\gamma(x) = \prod_{n=0}^{\infty} H_{\gamma_n}\left(\frac{x_n}{\sqrt{\lambda_n}}\right), \quad x \in H.$$

For any $i \in \mathbb{N} \cup \{0\}$ we set

$$H_\gamma^{(i)} = \prod_{\substack{n=0 \\ n \neq i}}^{\infty} H_{\gamma_n}(W_{e_n}).$$

In view of Remark 9.2.6, for any $\gamma \in \Gamma$ and, due to (9.1.4), for any $k \in \mathbb{N}$ we have $H_\gamma \in W^{1,2}(H,\mu)$ and

$$D_k H_\gamma = \sqrt{\frac{\gamma_k}{\lambda_k}} \, H_{\gamma_k - 1}(W_{e_k}) H_\gamma^{(k)}. \tag{9.2.11}$$

In the formula above we have made the convention that $H_{-1}(W_{e_k}) = 0$.

Remark 9.2.11 Notice that the set of elements of $L^2(H,\mu)$,

$$\left\{ H_{\gamma_k - 1}(W_{e_k}) H_\gamma^{(k)} : \gamma \in \Gamma, \ \gamma_k > 0 \right\},$$

is orthonormal.

The following result was proved independently by G. Da Prato, P. Malliavin and D. Nualart [95] and S. Peszat [181].

Theorem 9.2.12 *A function $\varphi \in L^2(H,\mu)$ belongs to $W^{1,2}(H,\mu)$ if and only if*

$$\sum_{\gamma \in \Gamma} \langle \gamma, \lambda^{-1} \rangle \, |\varphi_\gamma|^2 < +\infty. \tag{9.2.12}$$

where

$$\langle \gamma, \lambda^{-1} \rangle = \sum_{k=1}^{\infty} \gamma_k \lambda_k^{-1}. \qquad (9.2.13)$$

Moreover if (9.2.12) holds, we have

$$\|\varphi\|_{W^{1,2}(H,\mu)}^2 = \|\varphi\|_{L^2(H,\mu)}^2 + \sum_{\gamma \in \Gamma} \langle \gamma, \lambda^{-1} \rangle \, |\varphi_\gamma|^2. \qquad (9.2.14)$$

Finally, the embedding of $W^{1,2}(H,\mu)$ in $L^2(H,\mu)$ is compact.

Proof. Assume first that $\varphi \in W^{1,2}(H,\mu)$ and let $k \in \mathbb{N}$. By a direct verification, when $\varphi \in E(H)$

$$D_k \varphi = \sum_{\gamma \in \Gamma} \varphi_\gamma \sqrt{\frac{\gamma_k}{\lambda_k}} \, H_{\gamma_k - 1}(W_{e_k}) H_\gamma^{(k)}. \qquad (9.2.15)$$

Now let $(\varphi^{(n)}) \subset E(H)$ be such that

$$\varphi^{(n)} \to \varphi, \quad D_k \varphi^{(n)} \to D_k \varphi \quad \text{in } L^2(H,\mu).$$

Then, recalling Remark 9.2.11, we have

$$\int_H |D_k \varphi(x)|^2 \mu(dx) = \lim_{n \to \infty} \int_H |D_k \varphi_n(x)|^2 \mu(dx)$$

$$= \lim_{n \to \infty} \sum_{\gamma \in \Gamma} |\varphi_{n,\gamma}|^2 \frac{\gamma_k}{\lambda_k} = \sum_{\gamma \in \Gamma} |\varphi_\gamma|^2 \frac{\gamma_k}{\lambda_k},$$

which yields (9.2.14).

Conversely, assume that $\varphi \in L^2(H,\mu)$ is such that (9.2.12) holds. Then for any $k \in \mathbb{N}$ we have obviously

$$\sum_{\gamma \in \Gamma} |\varphi_\gamma|^2 \frac{\gamma_k}{\lambda_k} < +\infty.$$

For any $n \in \mathbb{N}$ let φ^n be defined by

$$\varphi^n = \sum_{\substack{\gamma \in \Gamma_n \\ \gamma_l = 0 \text{ if } l \geq n}} \varphi_\gamma H_\gamma.$$

It follows that $D_k\varphi^n \in L^2(H, \mu)$ and

$$D_k\varphi^n = \sum_{\substack{\gamma \in \Gamma_n \\ \gamma_l = 0 \text{ if } l \geq n}} \varphi_\gamma \sqrt{\frac{\gamma_k}{\lambda_k}} H_{\gamma_k - 1}(W_{e_k}) H_\gamma^{(k)}.$$

Again by Remark 9.2.11, we find

$$\int_H |D_k\varphi^n(x)|^2 \mu(dx) = \sum_{\substack{\gamma \in \Gamma_n \\ \gamma_l = 0 \text{ if } l \geq n}} |\varphi_\gamma^n|^2 \frac{\gamma_k}{\lambda_k}.$$

Consequently

$$\int_H |D_k\varphi^n(x)|^2 \mu(dx) \leq \sum_{\gamma \in \Gamma} |\varphi_\gamma|^2 \frac{\gamma_k}{\lambda_k}.$$

This implies that the sequence $(D_k\varphi^n)$ is bounded in $L^2(H, \mu)$. Then there exists a subsequence $(D_k\varphi^{n_m})$ weakly convergent to some element ψ_k. Since any linear closed operator is also closed with respect to the weak topology, we can conclude that $\psi_k = D_k\varphi$. Therefore $D_k\varphi \in L^2(H, \mu)$, and

$$\int_H |D_k\varphi(x)|^2 \mu(dx) \leq \sum_{\gamma \in \Gamma} |\varphi_\gamma|^2 \frac{\gamma_k}{\lambda_k}.$$

Now, summing up over k, we see that $\varphi \in W^{1,2}(H, \mu)$, as required.

It remains to prove compactness of the embedding above. Let us introduce the following linear closed operator B on $L^2(H, \mu)$:

$$B\varphi = \sum_{\gamma \in \Gamma} \langle \gamma, \lambda^{-1} \rangle \varphi_\gamma H_\gamma, \quad \varphi \in D(B) = W^{1,2}(H, \mu). \tag{9.2.16}$$

It is enough to notice that the operator B^{-1} is compact. In fact for any $C > 0$ the set $\{\gamma \in \Gamma : \langle \gamma, \lambda^{-1} \rangle < C\}$ is finite, as is easily checked. This implies that any subsequence of the set $\{\langle \gamma, \lambda^{-1} \rangle : \gamma \in \Gamma\}$ tends to 0 as $|\gamma|$ tends to infinity. Consequently the sequence of all eigenvalues of B is infinitesimal and B^{-1} is compact . $\square$

9.2.4 The space $W^{2,2}(H, \mu)$

The proof of the following lemma is straightforward, it is left as an exercise to the reader.

Lemma 9.2.13 *Let* $h, k \in \mathbb{N}$, *then the linear operator* $D_h D_k$, *defined in* $E(H)$, *is closable.*

If φ belongs to the domain of the closure of $D_h D_k$, which we shall still denote by $D_h D_k$, we shall say that $D_h D_k \varphi$ belongs to $L^2(H, \mu)$.

Now we define $W^{2,2}(H, \mu)$ as the space of all functions $\varphi \in W^{1,2}(H, \mu)$ such that $D_h D_k \varphi \in L^2(H, \mu)$ for all $h, k \in \mathbb{N}$ and

$$\sum_{h,k=1}^{\infty} \int_H |D_h D_k \varphi(x)|^2 \mu(dx) < +\infty.$$

$W^{2,2}(H, \mu)$ is a Hilbert space with the inner product

$$\langle \varphi, \psi \rangle_{W^{2,2}(H,\mu)} = \langle \varphi, \psi \rangle_{W^{1,2}(H,\mu)} + \sum_{h,k=1}^{\infty} \int_H \langle D_h D_k \varphi(x), D_h D_k \psi(x) \rangle \mu(dx).$$

If $\varphi \in W^{2,2}(H, \mu)$ we can define a Hilbert-Schmidt operator $D^2 \varphi(x)$ on H for almost any $x \in H$ by setting

$$\langle D^2 \varphi(x) \alpha, \beta \rangle = \sum_{h,k=1}^{\infty} D_h D_k \varphi(x) \alpha_h \beta_k, \quad \alpha, \beta \in H.$$

Remark 9.2.14 If H is infinite dimensional the embedding $W^{2,2}(H, \mu) \subset W^{1,2}(H, \mu)$, is not compact. In fact setting $\varphi^{(n)}(x) = x_n$, $n \in \mathbb{N}$, we have

$$\|\varphi^{(n)}\|_{L^2(H,\mu)} = \lambda_n \to 0 \text{ as } n \to \infty,$$

and

$$\|\varphi^{(n)}\|_{W^{1,2}(H,\mu)} = \|\varphi^{(n)}\|_{W^{2,2}(H,\mu)} = 1 + \lambda_n.$$

Consequently, $\varphi^{(n)} \to 0$ in $L^2(H, \mu)$, and no subsequence of (φ^n) tends to 0 in $W^{1,2}(H, \mu)$.

Let us give some properties of $W^{2,2}(H, \mu)$.

We first note that, for any $\gamma \in \Gamma$ and for any $k \in \mathbb{N}$, we have $D_k^2 H_\gamma \in L^2(H, \mu)$ and

$$D_k^2 H_\gamma = \begin{cases} \dfrac{\sqrt{\gamma_k(\gamma_k - 1)}}{\lambda_k} H_{\gamma_k - 2}(W_{e_k}) H_\gamma^{(k)} & \text{if } \gamma_k \neq 0, \\[2mm] 0 \text{ otherwise,} \end{cases} \tag{9.2.17}$$

and so, since $\{H_{\gamma_k - 2}(W_{e_k})H_\gamma^{(k)} : \gamma \in \Gamma, \gamma > 0\}$ is an orthonormal set,

$$\int_H |D_k^2 \varphi(x)|^2 \mu(dx) = \sum_{\gamma \in \Gamma} \frac{\gamma_k(\gamma_k - 1)}{\lambda_k^2} |\varphi_\gamma|^2. \qquad (9.2.18)$$

Proceeding in the same way, we find for $h \neq k$,

$$\int_H |D_h D_k \varphi(x)|^2 \mu(dx) = \sum_{\gamma \in \Gamma} \frac{\gamma_h \gamma_k}{\lambda_k \lambda_k} |\varphi_\gamma|^2. \qquad (9.2.19)$$

Proceeding as in the proof of Theorem 9.2.12 we get the next result.

Theorem 9.2.15 *A function* $\varphi \in L^2(H, \mu)$ *belongs to* $W^{2,2}(H, \mu)$ *if and only if*

$$\sum_{\gamma \in \Gamma} \langle \gamma, \lambda^{-1} \rangle^2 \, |\varphi_\gamma|^2 - \sum_{\gamma \in \Gamma} \langle \gamma, \lambda^{-2} \rangle^2 \, |\varphi_\gamma|^2 < +\infty. \qquad (9.2.20)$$

9.3   The Malliavin derivative

We shall denote by $\mathcal{D}_Q$ the linear operator

$$\mathcal{D}_Q : E(H) \subset L^2(H, \mu) \to L^2(H, \mu; H), \quad \varphi \to Q^{1/2} D\varphi.$$

Proposition 9.3.1 $\mathcal{D}_Q$ *is closable.*

Proof. Let $(\varphi_n) \subset E(H)$ and $F \in L^2(H, \mu; H)$ be such that

$$\varphi_n \to 0 \text{ in } L^2(H, \mu), \quad \mathcal{D}_Q \varphi_n \to F \text{ in } L^2(H, \mu; H).$$

If $G \in \Lambda_0$, then $Q^{1/2}G \in \Lambda_0$ and by (9.2.5), replacing G with $Q^{1/2}G$, we have

$$\int_H \langle \mathcal{D}_Q \varphi_n(x), G(x) \rangle \mu(dx) \; + \; \int_H \varphi_n(x) \, \mathrm{div} \, [Q^{1/2}G(x)] \mu(dx)$$

$$= \int_H \langle x, Q^{-1/2}G(x) \rangle \mu(dx).$$

As $n \to \infty$ we obtain

$$\int_H \langle F(x), G(x) \rangle \mu(dx) = 0,$$

which yields $F = 0$ since Λ_0 is dense in $L^2(H, \mu; H)$. $\square$

If φ belongs to the domain of the closure of $\mathcal{D}_Q$, which we shall still denote by $\mathcal{D}_Q$, we shall say that $\mathcal{D}_Q\varphi$ belongs to $L^2(H, \mu; H)$.

$\mathcal{D}_Q\varphi$ is called the *Malliavin derivative* of φ.

Now we define $W_Q^{1,2}(H, \mu)$ as the linear space of all functions $\varphi \in L^2(H, \mu)$ such that $\mathcal{D}_Q\varphi \in L^2(H, \mu; H)$.

$W_Q^{1,2}(H, \mu)$, endowed with the inner product

$$\langle \varphi, \psi \rangle_{W_Q^{1,2}(H,\mu)} = \langle \varphi, \psi \rangle_{L^2(H,\mu)} + \int_H \langle \mathcal{D}_Q\varphi(x), \mathcal{D}_Q\psi(x) \rangle \mu(dx),$$

is a Hilbert space.

Finally we define $W_Q^{2,2}(H, \mu)$ as the space of all functions $\varphi \in W_Q^{1,2}(H, \mu)$ such that $D_h D_k \varphi \in L^2(H, \mu)$ for all $h, k \in \mathbb{N}$ and

$$\sum_{h,k=1}^{\infty} \int_H \lambda_h \lambda_k |D_h D_k \varphi(x)|^2 \mu(dx) < +\infty.$$

$W_Q^{2,2}(H, \mu)$ is a Hilbert space with the inner product

$$\langle \varphi, \psi \rangle_{W_Q^{2,2}(H,\mu)} = \langle \varphi, \psi \rangle_{W_Q^{1,2}(H,\mu)}$$

$$+ \sum_{h,k=1}^{\infty} \int_H \lambda_h \lambda_k \langle D_h D_k \varphi(x), D_h D_k \psi(x) \rangle \mu(dx).$$

Chapter 10

Ornstein-Uhlenbeck semigroups on $L^p(H, \mu)$

In this chapter we study the Ornstein-Uhlenbeck semigroup (R_t), introduced in Chapter 6, acting on spaces $L^p(H, \mu)$, $p \geq 1$, were μ is an invariant measure. First we prove in §10.1 that (R_t) extends uniquely to a contraction semigroup on $L^p(H, \mu)$. Next §10.2 is devoted to the study of the infinitesimal generator L_p of (R_t), and to the corresponding Dirichlet form. In §10.3 the important case when (R_t) is strong Feller is studied in detail. It is shown that (R_t) can be expressed as an integral operator with respect to μ, and that it is hypercontractive. In §10.4 we give a representation formula for (R_t) in terms of the second quantization operator . This formula allows us to find an explicit expression for the adjoint of L_2. Then §10.5 is devoted to Poincaré and logarithmic Sobolev inequalities and to their consequences as spectral gap and exponential convergence to equilibrium. Finally, §10.6 includes some supplements.

We shall use the notation $\mathcal{E}_A(H)$ to mean the space of all real parts of the functions $(e^{i\langle x,h\rangle})_{h\in D(A^*)}$. Frequently we shall consider functions for simplicity as $\varphi_h(x) = e^{i\langle x,h\rangle}$ instead of their real parts. In these cases the standard complexification of the space X is assumed. It is easy to see that $\mathcal{E}_A(H)$ is dense in $L^2(H, \mu)$ for any probability measure μ on $(H, \mathcal{B}(H))$.

10.1   Extension of (R_t) to $L^p(H, \mu)$

The following conditions are assumed throughout the chapter.

$$
\begin{cases}
(i) \;\; A : D(A) \subset H \to H \text{ is the generator of a } C_0 \text{ semigroup } (e^{tA}) \text{ in } H \\
\quad\; \text{and there exist } M, \omega > 0 \text{ such that } \|e^{tA}\| \le M e^{-\omega t}, \;\; t \ge 0. \\
(ii) \;\; Q \in L^+(H). \\
(iii) \;\; \text{For all } s > 0, \text{ we have } e^{sA} Q e^{sA^*} \in L_1(H) \text{ and} \\
\qquad\quad \displaystyle\int_0^t \mathrm{Tr}[e^{sA} Q e^{sA^*}] ds < +\infty, \quad t \ge 0.
\end{cases}
$$

$$(10.1.1)$$

Let us recall the definition of the Ornstein-Uhlenbeck semigroup (R_t) introduced in Chapter 6:

$$
R_t \varphi(x) = \int_H \varphi(y) N_{e^{tA}x, Q_t}(dy) = \int_H \varphi(e^{tA}x + y) N_{Q_t}(dy), \;\; \varphi \in B_b(H),
$$

$$(10.1.2)$$

where

$$
Q_t x = \int_0^t e^{sA} Q e^{sA^*} x \, ds, \; t \ge 0, \; x \in H,
$$

and $B_b(H)$ represents the set of all Borel and bounded mappings from H into $\mathbb{R}$. We set also

$$
Q_\infty x = \int_0^\infty e^{sA} Q e^{sA^*} x \, ds, \; x \in H.
$$

Note that $Q_\infty \in L_1^+(H)$. We have in fact

$$
Q_\infty x = \sum_{k=0}^\infty \int_k^{k+1} e^{sA} Q e^{sA^*} x \, ds = \sum_{k=0}^\infty e^{kA} Q_1 e^{kA^*} x, \;\; x \in H.
$$

This implies that $Q_\infty \in L_1(H)$ and

$$
\mathrm{Tr} Q_\infty \le M^2 \, \mathrm{Tr}[Q_1] \sum_{k=0}^\infty e^{-2\omega k} < +\infty.
$$

Proposition 10.1.1 *Assume that (10.1.1) holds. Then the Gaussian measure $\mu = N_{Q_\infty}$ is invariant for the semigroup (R_t), that is*

$$
\int_H R_t \varphi(x) \mu(dx) = \int_H \varphi(y) \mu(dy), \quad \varphi \in C_b(H).
$$

$$(10.1.3)$$

Moreover for any $\varphi \in C_b(H)$ *and any* $x \in H$ *we have*

$$\lim_{t \to +\infty} R_t \varphi(x) = \int_H \varphi(x) \mu(dx). \tag{10.1.4}$$

Proof. It is enough to show (10.1.3) and (10.1.4) for $\varphi \in \mathcal{E}_A(H)$. Let in fact $\varphi_h(x) = e^{i\langle h,x \rangle}$, $x \in H$. Then we have

$$\int_H R_t e^{i\langle h,x \rangle} \mu(dx) = \exp \left\{ -\frac{1}{2} \langle Q_t h, h \rangle - \frac{1}{2} \langle Q_\infty e^{tA^*} h, e^{tA^*} h \rangle \right\}.$$

Since clearly $Q_t + e^{tA} Q_\infty e^{tA^*} = Q_\infty$, this implies

$$\int_H R_t \varphi_h(x) \mu(dx) = \int_H \varphi_h(x) \mu(dx),$$

and (10.1.3) is proved.

Finally, let us prove (10.1.4) for $\varphi = \varphi_h$, $h \in H$. We have

$$\lim_{t \to +\infty} R_t \varphi_h(x) = \lim_{t \to +\infty} e^{i\langle e^{tA} h, x \rangle - \frac{1}{2} \langle Q_t h, h \rangle} = e^{-\frac{1}{2} \langle Q_\infty h, h \rangle} = \int_H \varphi_h d\mu. \quad \square$$

Throughout the chapter we shall denote by μ the Gaussian measure N_{Q_∞}.

We shall also consider the linear operator

$$D_Q : \mathcal{E}_A(H) \subset L^2(H, \mu) \to L^2(H, \mu; H), \quad \varphi \to Q^{1/2} D\varphi.$$

Proposition 10.1.2 *Assume that* $\operatorname{Ker} Q = \{0\}$ *and that* $Q^{1/2}(H) \supset Q_\infty^{1/2}(H)$. *Then* D_Q *is closable.*

Proof. We first notice that by the assumption and the closed graph theorem, the linear operator $K := Q^{-1/2} Q_\infty^{1/2}$ is bounded. Moreover its adjoint K^* is the closure of $Q_\infty^{1/2} Q^{-1/2}$. Now let $(\varphi_n) \subset \mathcal{E}_A(H)$ and let $F \in L^2(H, \mu; H)$ be such that

$$\lim_{n \to \infty} \varphi_n = 0 \quad \text{in } L^2(H, \mu), \quad \lim_{n \to \infty} Q^{1/2} D\varphi_n = F \quad \text{in } L^2(H, \mu; H).$$

Then it follows that

$$\lim_{n \to \infty} Q_\infty^{1/2} D\varphi_n = \lim_{n \to \infty} K^* Q^{1/2} D\varphi_n = K^* F \quad \text{in } L^2(H, \mu; H).$$

Since the operator D is closable, by Proposition 9.2.4, we find $K^* F = 0$ and so $F = 0$. Therefore D_Q is closable. $\square$

If φ belongs to the domain of the closure of D_Q, which we still denote by D_Q, we shall say that $D_Q\varphi$ belongs to $L^2(H, \mu)$.

Now we define $W_Q^{1,2}(H, \mu)$ as the linear space of all functions $\varphi \in L^2(H, \mu)$ such that $D_Q\varphi \in L^2(H, \mu; H)$. $W_Q^{1,2}(H, \mu)$, endowed with the inner product

$$\langle \varphi, \psi \rangle_{W_Q^{1,2}(H,\mu)} = \langle \varphi, \psi \rangle_{L^2(H,\mu)} + \int_H \langle D_Q\varphi(x), D_Q\psi(x) \rangle \mu(dx),$$

is a Hilbert space. $W_Q^{2,2}(H, \mu)$ can be defined in a similar way.

Remark 10.1.3 The assumption that $\|e^{tA}\| \le Me^{-\omega t}$, $t \ge 0$, for some $\omega > 0$ is not necessary for the existence of an invariant measure for (R_t). The necessary and sufficient condition for this is that

$$Q_\infty = \int_0^{+\infty} e^{sA}Qe^{sA^*} ds \in L_1(H),$$

see G. Da Prato and J. Zabczyk [101].

Proposition 10.1.4 Q_∞ is the unique symmetric and positive solution to the following equation:

$$\langle Q_\infty x, A^*y \rangle + \langle Q_\infty A^*x, y \rangle = -\langle Qx, y \rangle, \quad x, y \in D(A^*). \qquad (10.1.5)$$

This is called the *Lyapunov equation*.
Proof of Proposition 10.1.4. Let $x, y \in D(A^*)$. Then we have

$$\langle Q_\infty x, A^*y \rangle = \int_0^{+\infty} \langle e^{sA}Qe^{sA^*}x, A^*y \rangle ds$$

$$= \int_0^{+\infty} \left\langle Qe^{sA^*}x, \frac{d}{ds}e^{sA^*}y \right\rangle ds.$$

Integrating by parts gives

$$\langle Q_\infty x, A^*y \rangle = -\langle Qx, y \rangle - \langle Q_\infty A^*x, y \rangle,$$

which proves (10.1.5).

Let us prove uniqueness. Let $X \in L^+(H)$ be such that

$$\langle Xx, A^*y \rangle + \langle XA^*x, y \rangle = -\langle Qx, y \rangle, \quad x, y \in D(A^*).$$

If $x \in D(A^*)$ we have

$$\frac{d}{dt} \langle X e^{tA^*} x, e^{tA^*} x \rangle = -\langle Q e^{tA^*} x, e^{tA^*} x \rangle.$$

Integrating between 0 and t gives

$$\langle X e^{tA^*} x, e^{tA^*} x \rangle = \langle X x, x \rangle - \langle Q_t x, x \rangle.$$

Letting t tend to $+\infty$ we find $X = Q_\infty$ as required. $\square$

Let us prove now that (R_t) can be extended to $L^p(H, \mu)$, $p \geq 1$.

Theorem 10.1.5 *Assume that (10.1.1) hold. Then for all $t > 0$, the operator R_t, defined by (10.1.2), is uniquely extendible to a bounded linear operator on $L^p(H, \mu)$, $p \geq 1$, which we still denote by R_t. Moreover (R_t) is a strongly continuous semigroup of contractions on $L^p(H, \mu)$.*

Proof. Let $t > 0$ and $f \in UC_b(H)$. Since for $x \in H$ we have by the Hölder inequality, $|R_t \varphi(x)|^p \leq R_t(|\varphi|^p)(x)$, it follows that

$$\int_H |R_t \varphi(x)|^p \mu(dx) \leq \int_H R_t(|\varphi|^p)(x) \mu(dx) = \int_H |\varphi(x)|^p \mu(dx),$$

in view of the invariance of μ. Since $UC_b(H)$ is dense in $L^p(H, \mu)$, R_t is uniquely extendible to $L^p(H, \mu)$. Therefore

$$\|R_t \varphi\|_{L^p(H, \mu)} \leq \|\varphi\|_{L^p(H, \mu)}, \quad t \geq 0, \ \varphi \in L^p(H, \mu).$$

Finally strong continuity of (R_t) follows from the dominated convergence theorem. $\square$

We end the introductory part of this section by giving a necessary and sufficient condition in order that (R_t) is symmetric in $L^2(H, \mu)$. We follow here A. Chojnowska-Michalik and B. Goldys [51].

Proposition 10.1.6 *The semigroup (R_t) is symmetric if and only if one of the following conditions holds.*

(i) $Q_\infty e^{tA^*} = e^{tA} Q_\infty$, $t \geq 0$.

(ii) $Q e^{tA^*} = e^{tA} Q$, $t \geq 0$.

If (R_t) is symmetric we have

$$Q_\infty = -\frac{1}{2} A^{-1} Q. \tag{10.1.6}$$

Proof. Let $h, k \in H$ and $\varphi(x) = e^{i\langle x, h\rangle}$, $\psi(x) = e^{i\langle x, k\rangle}$, $x \in H$. Then we have

$$R_t\varphi(x) = \exp\left\{i\langle h, e^{tA}x\rangle - \tfrac{1}{2}\langle Q_t h, h\rangle\right\}, \ t \geq 0, \ x \in H,$$

$$R_t\psi(x) = \exp\left\{i\langle k, e^{tA}x\rangle - \tfrac{1}{2}\langle Q_t k, k\rangle\right\}, \ t \geq 0, \ x \in H.$$

By a straightforward computation it follows that

$$\int_H (R_t\varphi)\psi d\mu = \exp\left\{-\frac{1}{2}\langle Q_\infty h, h\rangle - \frac{1}{2}\langle Q_\infty k, k\rangle - \langle Q_\infty e^{tA^*} h, k\rangle\right\}.$$

Consequently

$$\int_H (R_t\varphi)\psi d\mu = \int_H (R_t\psi)\varphi d\mu, \ \text{for all } \varphi, \psi \in \mathcal{E}_A(H),$$

if and only if condition (i) holds. Since $\mathcal{E}_A(H)$ is dense in $L^2(H, \mu)$ the conclusion follows.

Assume now that (ii) holds. Then we have for $x \in H$

$$Q_\infty e^{tA^*} x = \int_0^{+\infty} e^{sA} Q e^{(s+t)A^*} x ds = \int_0^{+\infty} e^{(t+2s)A} Q x ds = e^{tA} Q_\infty x,$$

so that (i) holds and (R_t) is symmetric. Let us prove (10.1.6). Let $x \in D(A^*)$. Then by (i) we see that $Q_\infty x \in D(A)$ and

$$Q_\infty A^* x = A Q_\infty x.$$

Now the conclusion follows from the Lyapunov equation (10.1.5).

Assume, finally, that (R_t) is symmetric. Let us prove that (ii) holds. Let in fact $x, y \in D(A^*)$. Then we have recalling (10.1.5)

$$
\begin{aligned}
\langle Q e^{tA^*} x, y\rangle &= -\langle Q_\infty e^{tA^*} x, A^* y\rangle - \langle Q_\infty A^* e^{tA^*} x, y\rangle \\
&= -\langle Q_\infty x, A^* e^{tA^*} y\rangle - \langle Q_\infty A^* x, e^{tA^*} y\rangle = \langle e^{tA} Q x, y\rangle,
\end{aligned}
$$

which yields (ii). $\square$

Corollary 10.1.7 *Assume that*

$$
\left\{
\begin{array}{l}
(i) \quad A : D(A) \subset H \to H \text{ is self-adjoint and there exists } \omega > 0 \text{ such that} \\
\qquad\qquad \langle Ax, x\rangle \leq -\omega|x|^2, \quad x \in D(A), \\
(ii) \ Q e^{tA} = e^{tA} Q, \quad t \geq 0, \\
(iii) \ QA^{-1} \in L_1(H).
\end{array}
\right.
$$

(10.1.7)

Then R_t is symmetric.

Proof. In fact assumptions (10.1.7) clearly imply (10.1.1), since $Q_t = -\frac{1}{2}QA^{-1}(1 - e^{2tA})$ and condition (ii) of Proposition 10.1.6 holds. □
Setting $Q = I$ we find in particular the following result.

Corollary 10.1.8 *Assume that*

$$
\begin{cases}
(i) \quad A : D(A) {\subset} H \to H \text{ is self-adjoint and there exists } \omega > 0 \text{ such that} \\
\qquad \langle Ax, x \rangle \leq -\omega |x|^2, \quad x \in D(A), \\
(ii) \quad A^{-1} \in L_1(H) \text{ and } Q = 1.
\end{cases}
$$

$$(10.1.8)$$

Then R_t is symmetric.

10.1.1   The adjoint of (R_t) in $L^2(H, \mu)$

In general the adjoint of (R_t) in $L^2(H, \mu)$ is not an Ornstein-Uhlenbeck semigroup of the form (10.1.2). However, this happens under the following assumptions.

$$
\begin{cases}
Q_\infty(H) \subset D(A^*). \text{ Moreover the operator} \\
\quad A_1 x = Q_\infty A^* Q_\infty^{-1} x, \ x \in D(A_1) = \{x \in Q_\infty(H) : \ Q_\infty^{-1} x \in D(A^*)\} \\
generates \ a \ C_0 \ semigroup \ given \ by \ e^{tA_1} = Q_\infty e^{tA^*} Q_\infty^{-1}.
\end{cases}
$$

$$(10.1.9)$$

Under this assumption we can consider the Ornstein-Uhlenbeck semigroup

$$S_t \varphi = \int_H \varphi(y) N_{e^{tA_1} x, Q_{1,t}}(dy), \quad t \geq 0, \tag{10.1.10}$$

where

$$Q_{1,t} x = \int_0^t e^{sA_1} Q e^{sA_1^*} x \, ds, \quad x \in H, t \geq 0. \tag{10.1.11}$$

We notice that the linear operator

$$Q_{1,\infty} x = \int_0^{+\infty} e^{sA_1} Q e^{sA_1^*} x \, ds, \quad x \in H,$$

fulfills the Lyapunov equation (10.1.5). Therefore $Q_{1,\infty} = Q_\infty$ by Proposition 10.1.4.

Proposition 10.1.9 *Assume that (10.1.1) and (10.1.9) hold, and let $t > 0$. Then the adjoint of R_t is the operator S_t defined by (10.1.10).*

Proof. It is enough to show that

$$\int_H R_t\varphi \,\psi d\mu = \int_H S_t\psi \,\varphi d\mu,\tag{10.1.12}$$

for $\varphi(x) = e^{i\langle h,x\rangle}$, $\psi(x) = e^{i\langle k,x\rangle}$, $h, k \in H$. By an elementary computation we have in fact

$$\int_H R_t\varphi \,\psi d\mu = \exp\left\{-\frac{1}{2}\langle Q_t h, h\rangle - \frac{1}{2}\langle Q_\infty(e^{tA^*}h + k), (e^{tA^*}h + k)\rangle\right\},$$

$$\int_H S_t\psi \,\varphi d\mu = \exp\left\{-\frac{1}{2}\langle Q_{1,t}k, k\rangle - \frac{1}{2}\langle Q_\infty(e^{tA_1^*}k + h), (e^{tA_1^*}k + h)\rangle\right\}.$$

Therefore (10.1.12) is equivalent to the equality

$$\langle(Q_t + e^{tA}Q_\infty e^{tA^*} - Q_\infty)h, h\rangle + \langle(2Q_\infty e^{tA^*} - e^{tA_1}Q_\infty)h, k\rangle$$

$$-\langle(Q_{1,t} + e^{tA_1}Q_\infty e^{tA_1^*} - Q_\infty)k, k\rangle = 0.$$

Now (10.1.12) follows from the identities

$$Q_t + e^{tA}Q_\infty e^{tA^*} = Q_\infty, \quad Q_\infty e^{tA^*} = e^{tA_1}Q_\infty, \quad Q_{1,t} + e^{tA_1}Q_\infty e^{tA_1^*} = Q_\infty. \quad \square$$

If assumptions (10.1.9) do not hold we can find an expression for the adjoint of (R_t) by using the second quantization operator , see §10.4.

10.2   The infinitesimal generator of (R_t)

We shall denote by L_p the infinitesimal generator of (R_t) on $L^p(H, \mu)$. First we want to show that $\mathcal{E}_A(H)$ is a core for L_p.

Let us define on $\mathcal{E}_A(H)$ the following differential operator:

$$L_0\varphi(x) = \frac{1}{2}\text{Tr}[QD^2\varphi(x)] + \langle x, A^*D\varphi(x)\rangle, \quad \varphi \in \mathcal{E}_A(H), \ x \in H.\tag{10.2.1}$$

Proposition 10.2.1 *Assume that (10.1.1) hold. Then for any $p \geq 1$, L_p is an extension of L_0 and $\mathcal{E}_A(H)$ is a core for L_p.*

Proof. Let $\varphi(x) = e^{i\langle h,x\rangle}$, $h \in H$. Then by (10.1.2) we have

$$R_t\varphi(x) = \int_H e^{i\langle h, e^{tA}x+y\rangle} N_{Q_t}(dy)$$

$$= e^{i\langle h, e^{tA}x\rangle - \frac{1}{2}\langle Q_t h, h\rangle} = e^{i\langle e^{tA^*}h, x\rangle - \frac{1}{2}\langle Q_t h, h\rangle} \in \mathcal{E}_A(H).$$

It follows that $\mathcal{E}_A(H)$ is invariant for (R_t) and moreover

$$\lim_{h \to 0} \frac{1}{h}(R_h \varphi - \varphi) = L_0 \varphi \quad \text{in } L^p(H, \mu).$$

Therefore $D(L_0) \subset D(L_p)$ and L_p is an extension of L_0. Consequently $\mathcal{E}_A(H)$ is a core , because it is invariant for (R_t), is included in $D(L_p)$ and is dense in $L^p(H, \mu)$, see e.g. E. B. Davies [104]. The proof is complete. $\square$

We now are going to study in particular the operator L_2. We start with two basic identities, proved independently by V. I. Bogachev, M. Röckner and B. Schmuland [22] and M. Fuhrman [120].

Proposition 10.2.2 *For all* $\varphi, \psi \in \mathcal{E}_A(H)$ *the following identities hold:*

$$\int_H L_0 \varphi \, \psi d\mu = \int_H \langle Q_\infty D\psi, A^* D\varphi \rangle d\mu, \tag{10.2.2}$$

and

$$\int_H L_0 \varphi \, \varphi d\mu = -\frac{1}{2} \int_H \langle Q^{1/2} D\varphi, Q^{1/2} D\varphi \rangle d\mu. \tag{10.2.3}$$

If in particular (R_t) *is symmetric then* L_0 *is symmetric too and*

$$\int_H L_0 \varphi \, \psi d\mu = -\frac{1}{2} \int_H \langle QD\varphi, D\psi \rangle d\mu, \quad \varphi, \psi \in \mathcal{E}_A(H). \tag{10.2.4}$$

Proof. We start with the proof of (10.2.2). It is enough to take

$$\varphi(x) = e^{i\langle x, h \rangle}, \quad \psi(x) = e^{i\langle x, k \rangle}, \quad h, k \in D(A^*), \quad x \in H.$$

In this case we have by a simple computation,

$$\int_H L_0 \varphi \, \psi d\mu = \left(\langle A^* h, Q_\infty(h - k) \rangle + \frac{1}{2}|Q^{1/2}h|^2 \right) e^{-\frac{1}{2}\langle Q_\infty(h+k), h+k \rangle},$$

and

$$\int_H \langle Q_\infty D\psi, A^* D\varphi \rangle d\mu = -\langle A^* h, Q_\infty k \rangle e^{-\frac{1}{2}\langle Q_\infty(h+k), h+k \rangle}.$$

Therefore (10.2.2) holds since

$$2\langle A^* h, Q_\infty h \rangle + |Q^{1/2}h|^2 = 0,$$

in view of the Lyapunov equation (10.1.5).

Finally, (10.2.3) follows again by (10.1.5) and (10.2.4) by (10.2.3) and again by the Lyapunov equation. $\square$

Identity (10.2.3) will give important information about the domain of L_2, as the following proposition shows.

Proposition 10.2.3 *Assume that* $\operatorname{Ker} Q = \{0\}$ *and that* $Q^{1/2}(H) \supset Q_\infty^{1/2}(H)$. *Then we have* $D(L_2) \subset W_Q^{1,2}(H, \mu)$. *Moreover for any* $\varphi \in D(L_2)$ *we have*

$$\int_H L_2\varphi \, \varphi d\mu = -\frac{1}{2}\int_H \langle QD\varphi, D\varphi\rangle d\mu. \tag{10.2.5}$$

Finally, if (R_t) *is symmetric , we have* $D((-L_2)^{1/2}) = W_Q^{1,2}(H, \mu)$.

Proof. Let $\varphi \in D(L_2)$. Since, by Proposition 10.2.1, $\mathcal{E}_A(H)$ is a core of L_2 there exists a sequence $(\varphi_n) \subset \mathcal{E}_A(H)$ such that

$$\varphi_n \to \varphi, \quad \psi_n := L_0\varphi_n \to L_2\varphi \quad \text{in } L^2(H, \mu).$$

By (10.2.3) it follows that for any $m, n \in \mathbb{N}$,

$$\int_H \langle Q^{1/2}D(\varphi_n - \varphi_m), Q^{1/2}D(\varphi_n - \varphi_m)\rangle d\mu = -2\int_H (L_2\varphi_n - \varphi_m)(\varphi_n - \varphi_m)d\mu.$$

Consequently (φ_n) is a Cauchy sequence in $W_Q^{1,2}(H, \mu)$ and (10.2.5) holds. Finally, if (R_t) is symmetric the final statement follows from the identity

$$\int_H |(-L_2)^{1/2}\varphi|^2 d\mu = \int_H |Q^{1/2}D\varphi|^2 d\mu. \quad \square$$

Remark 10.2.4 Let us consider the bilinear form

$$a(\varphi, \psi) = \int_H \langle Q_\infty D\psi, A^*D\varphi\rangle d\mu, \quad \varphi, \psi \in \mathcal{E}_A(H);$$

a is not in general continuous on $W_Q^{1,2}(H, \mu) \times W_Q^{1,2}(H, \mu)$ and consequently it is not a Dirichlet form.

The form is a Dirichlet form when $Q = I$, because in this case AQ_∞ is bounded, see [72]. For sufficient conditions that ensure continuity see V. I. Bogachev, M. Röckner and B. Schmuland [22] and M. Fuhrman [120].

Thus in general the operator L_2 is neither variational nor the infinitesimal generator of an analytic semigroup, see [120]. We recall that by a result of B. Goldys [127], L_2 generates an analytic semigroup if and only it is variational.

10.2.1 Characterization of the domain of L_2

A precise characterization of $D(L_2)$ is known when (R_t) is symmetric, see
G. Da Prato and B. Goldys [91]. For a similar characterization in $L^p(H, \mu)$
see A. Chojnowska-Michalik and B. Goldys [53].

Let us start with the case when assumptions (10.1.8) hold, that is when
A is self-adjoint, A^{-1} is of trace class and $Q = I$. Notice that in this case
$Q_\infty = -\frac{1}{2} A^{-1}$. We denote by (e_k) a complete orthonormal system in H,
and by (α_k) a sequence of positive numbers such that

$$Ae_k = -\alpha_k e_k, \quad k \in \mathbb{N}.$$

We need the following identity.

Proposition 10.2.5 *Assume that* (10.1.8) *holds, and let* $\varphi \in \mathcal{E}_A(H)$. *Then
we have*

$$\frac{1}{2} \int_H \mathrm{Tr}[(D^2\varphi)^2]d\mu + \int_H |(-A)^{1/2}D\varphi|^2 d\mu = 2 \int_H (L_2\varphi)^2 d\mu. \qquad (10.2.6)$$

Proof. Let $\varphi \in \mathcal{E}_A(H)$, $h \in \mathbb{N}$, and set $f = L_2\varphi$. Then we have

$$D_h f = L_2 D_h \varphi - \alpha_h D_h \varphi.$$

Multiplying both sides of this identity by $D_h\varphi$, integrating in H with respect
to μ, and taking into account (10.2.3) we find

$$\frac{1}{2} \int_H \langle DD_h\varphi, DD_h\varphi \rangle d\mu + \int_H \alpha_h |D_h\varphi|^2 d\mu = - \int_H D_h f \, D_h\varphi \, d\mu.$$

Summing up over h gives

$$\frac{1}{2} \int_H \mathrm{Tr}[(D^2\varphi)^2]d\mu + \int_H |(-A)^{1/2}D\varphi|^2 d\mu = - \int_H \langle Df, D\varphi \rangle d\mu.$$

Now the conclusion follows taking into account (10.2.4). $\square$

Let us introduce the space $W^{1,2}_{(-A)}(H, \mu)$ consisting of all $\varphi \in W^{1,2}(H, \mu)$
such that

$$\int_H |(-A)^{1/2}D\varphi|^2 d\mu = \sum_{k=1}^{\infty} \int_H \alpha_k |D_k\varphi|^2 d\mu < +\infty.$$

It is easy to see that $W^{1,2}_{(-A)}(H, \mu)$ is a Hilbert space.

We can give now, following G. Da Prato [74], a characterization of the
domain of L_2.

Theorem 10.2.6 *Assume that* (10.1.8) *hold. Then we have*

$$D(L_2) = W^{2,2}(H, \mu) \cap W^{1,2}_{(-A)}(H, \mu). \tag{10.2.7}$$

Proof. Let $\varphi \in \mathcal{E}_A(H)$ and $f = L_2\varphi$. Since $\mathcal{E}_A(H)$ is a core for L_2 (Proposition 10.2.1), there exists a sequence $(\varphi_n) \subset \mathcal{E}_A(H)$ such that

$$\varphi_n \to \varphi, \quad L_2\varphi_n \to L_2\varphi \quad \text{in } L^2(H, \mu).$$

Now if $n, m \in \mathbb{N}$ by (10.2.6), it follows that

$$\frac{1}{2} \int_H \text{Tr}[(D^2(\varphi_n - \varphi_m))^2] d\mu + \int_H |(-A)^{1/2} D(\varphi_n - \varphi_m)|^2 d\mu$$

$$= 2 \int_H (L_2(\varphi_n - \varphi_m))^2 d\mu.$$

Consequently the sequence (φ_n) is Cauchy both in $W^{2,2}(H, \mu)$ and in $W^{1,2}_{(-A)}(H, \mu)$, so that

$$D(L_2) \subset W^{2,2}(H, \mu) \cap W^{1,2}_{(-A)}(H, \mu).$$

Let us prove the opposite inclusion. Let $\psi \in W^{2,2}(H, \mu) \cap W^{1,2}_{(-A)}(H, \mu)$. Then it is not difficult to find a sequence $(\psi_n) \subset \mathcal{E}_A(H)$ such that

$$\psi_n \to \psi \quad \text{in } W^{2,2}(H, \mu) \cap W^{1,2}_{(-A)}(H, \mu).$$

Using (10.2.6) again, we see that $\psi \in D(L_2)$. $\square$

Let us now consider the case when assumptions (10.1.7) hold, that is when A and Q commute and $Q_\infty = -\frac{1}{2} QA^{-1}$ is of trace class. Notice that AQ is negative.

Proposition 10.2.7 *Assume that* (10.1.7) *holds, and let* $\varphi \in \mathcal{E}_A(H)$. *Then the following identity holds.*

$$\frac{1}{2} \int_H \text{Tr}[(QD^2\varphi)^2] d\mu + \int_H \langle (-AQ)D\varphi, D\varphi \rangle d\mu = 2 \int_H (L_2\varphi)^2 d\mu. \tag{10.2.8}$$

Proof. Let (e_k) be a complete orthonormal basis in H, let $\varphi \in \mathcal{E}_A(H)$, $h \in \mathbb{N}$, and set $f = L_2\varphi$. Then we have

$$D_h L_2\varphi = L_2 D_h\varphi + \langle A^* D\varphi, e_h \rangle = D_h f.$$

Multiplying both sides of this identity by $D_k\varphi$, integrating in H with respect to μ and taking into account (10.2.4), as R_t is symmetric, we find

$$\frac{1}{2}\int_H \langle QDD_h\varphi, DD_k\varphi\rangle d\mu - \int_H \langle A^*D\varphi, e_h\rangle D_k\varphi d\mu = -\int_H D_h f D_k\varphi d\mu.$$

Multiplying both sides by $Q_{h,k} = \langle Qe_k, e_h\rangle$ and summing up over h, k gives

$$\frac{1}{2}\sum_{h,k=1}^\infty \int_H Q_{h,k}\langle QDD_k\varphi, DD_h\varphi\rangle d\mu - \sum_{h,k=1}^\infty \int_H Q_{h,k}\langle A^*D\varphi, e_h\rangle D_k\varphi d\mu$$

$$= -\sum_{h,k=1}^\infty \int_H Q_{h,k} D_h f D_k\varphi d\mu,$$

which is equivalent to

$$\frac{1}{2}\int_H \mathrm{Tr}[(QD^2\varphi)^2]d\mu + \int_H \langle(-AQ)D\varphi, D\varphi\rangle d\mu = -\int_H \langle QDf, D\varphi\rangle d\mu.$$

Now the conclusion follows taking into account (10.2.3). $\square$

Now we can prove the following result, see G. Da Prato and B. Goldys [91]. The proof is similar to that of Theorem 10.2.6 and it is left to the reader.

Theorem 10.2.8 *Assume* (10.1.7). *Then we have*

$$D(L_2) = \left\{\varphi \in W_Q^{2,2}(H, \mu) : \int_H \langle(-AQ)D\varphi, D\varphi\rangle d\mu < +\infty\right\}. \qquad (10.2.9)$$

Remark 10.2.9 If the semigroup (R_t) is not symmetric we are not able to give a characterization of $D(L_2)$ in general. If H is finite dimensional and Q has a bounded inverse we have, see A. Lunardi [163], G. Da Prato [74],

$$D(L_2) = W^{2,2}(H, \mu).$$

For some results on the infinite dimensional case see [74].

10.3 The case when (R_t) is strong Feller

We assume here that (R_t) is strong Feller. We recall that this is equivalent to the null controllability condition (6.2.3):

$$e^{tA}(H) \subset Q_t^{1/2}(H), \quad \text{for all } t > 0. \qquad (10.3.1)$$

We set as before $\Lambda_t = Q_t^{-1/2} e^{tA}$, $t \geq 0$, and $\mu = N_{Q_\infty}$. We recall that, see Appendix B,

$$Q_t^{1/2}(H) = Q_\infty^{1/2}(H), \ t > 0. \tag{10.3.2}$$

Proposition 10.3.1 *Assume that (10.1.1) and (10.3.1) hold. Let $p \geq 1$, and $\varphi \in L^p(H, \mu)$. Then for any $t > 0$ we have $R_t\varphi \in W^{1,p}(H, \mu)$ and*

$$\|DR_t\varphi\|_{L^p(H,\mu)} \leq \|\Lambda_t\| \, \|\varphi\|_{L^p(H,\mu)}. \tag{10.3.3}$$

Moreover R_t is compact for all $t > 0$.

If, in addition, the mapping $[0, +\infty) \to \mathbb{R}$, $t \to \|\Lambda_t\|$, is Laplace transformable, then $D(L_p) \subset W^{1,p}(H, \mu)$ with continuous embedding.

Proof. We prove (10.3.3) for $p = 2$, the general case follows by interpolation, taking into account that, in view of Theorem 6.2.2, we have

$$\|DR_t\varphi\|_0 \leq \|\Lambda_t\| \|\varphi\|_0, \ \varphi \in B_b(H).$$

Let first $\varphi \in UC_b(H)$. Then by Theorem 6.2.2 we have for any $h \in H$,

$$\langle DR_t\varphi(x), h \rangle = \int_H \varphi(e^{tA}x + y)\langle \Lambda_t h, Q_t^{-1/2} y \rangle \mu(dy).$$

By using Hölder's inequality we find

$$
\begin{aligned}
|\langle DR_t\varphi(x), h \rangle|^2 &\leq \int_H \varphi^2(e^{tA}x + y)\mu(dy) \int_H |\langle \Lambda_t h, Q_t^{-1/2} y \rangle|^2 \mu(dy) \\
&= |\Lambda_t h|^2 R_t(\varphi^2).
\end{aligned}
$$

Due to the arbitrariness of h it follows that

$$|DR_t\varphi(x)|^2 \leq \|\Lambda_t\|^2 R_t(\varphi^2)(x), \quad t > 0, \ x \in H.$$

Now, integrating on H with respect to μ, and recalling the invariance of μ, we find

$$\int_H |DR_t\varphi(x)|^2 \mu(dx) \leq \|\Lambda_t\|^2 \int_H |\varphi(x)|^2 \mu(dx), \quad t > 0.$$

Therefore (10.3.3) follows from the density of $UC_b(H)$ into $L^2(H, \mu)$.

Since the embedding $W^{1,p}(H, \mu) \subset L^p(H, \mu)$ is compact (this result is proved in Theorem 9.2.12 if $p = 2$ and in A. Chojnowska-Michalik and B. Goldys [51] for p arbitrary), R_t is compact as well for any $t > 0$.

Finally, the final statement follows from (10.3.3), taking the Laplace transform. $\square$

We are now going to prove that, for any $t > 0$ and any $x \in H$, the measures $N_{e^{tA}x,Q_t}$ and $\mu = N_{Q_\infty}$ are equivalent. This will allow us to obtain a representation formula for $R_t\varphi$ for all $\varphi \in L^p(H,\mu)$, $p \geq 1$. More precisely we shall prove, following G. Da Prato, M. Fuhrman and J. Zabczyk [90], that

$$R_t\varphi(x) = \int_H k_t(x,y)\varphi(y)\mu(dy), \quad x \in H, \tag{10.3.4}$$

where

$$k_t(x,y) = \frac{dN_{e^{tA}x,Q_t}}{dN_{Q_\infty}}(y), \quad x,y \in H. \tag{10.3.5}$$

Let us introduce the semigroup of linear operators on H

$$T(t) := Q_\infty^{-1/2}e^{tA}Q_\infty^{1/2}, \quad t \geq 0.$$

This definition is meaningful in view of (10.3.1) and (10.3.2) since

$$Q_\infty^{-1/2}e^{tA}Q_\infty^{1/2} = Q_\infty^{-1/2}Q_t^{1/2}\Lambda_tQ_\infty^{1/2}.$$

It is easy to check that $T(\cdot)$ is a strongly continuous semigroup in H; we shall denote its infinitesimal generator by B, so that

$$e^{tB} = Q_\infty^{-1/2}e^{tA}Q_\infty^{1/2}, \quad t \geq 0.$$

The adjoint e^{tB^*} of e^{tB} is given by

$$e^{tB^*} = \overline{Q_\infty^{1/2}e^{tA^*}Q_\infty^{-1/2}} = Q_\infty^{1/2}(\Lambda_t)^*(Q_\infty^{-1/2}Q_t^{1/2})^*, \quad t \geq 0.$$

Lemma 10.3.2 e^{tB}, $t \geq 0$, *is a semigroup of contractions on H.*

Proof. It is enough to check that

$$|Q_\infty^{1/2}e^{tA^*}x| \leq |Q_\infty^{1/2}x|, \quad x \in D(A^*). \tag{10.3.6}$$

In fact, for all $x \in D(A^*)$ we have

$$\frac{d}{dt}|Q_\infty^{1/2}e^{tA^*}x|^2 = 2\langle Q_\infty A^*e^{tA^*}x, e^{tA^*}x\rangle = -\langle Qe^{tA^*}x, e^{tA^*}x\rangle \leq 0,$$

which yields (10.3.6). $\square$

Lemma 10.3.3 *For any $t > 0$ and any $x \in H$, the measures $N_{e^{tA}x, Q_t}$ and $\mu = N_{Q_\infty}$ are equivalent. Moreover, for $x, y \in H$, the density $\dfrac{dN_{e^{tA}x, Q_t}}{dN_{Q_\infty}} :=$ $k_t(x, \cdot)$ is given by*

$$k_t(x, y) = \det(1 - \Theta_t)^{-1/2} \exp\left\{ -\frac{1}{2}\langle(1 - \Theta_t)^{-1}Q_\infty^{-1/2}e^{tA}x, Q_\infty^{-1/2}e^{tA}x\rangle \right.$$

$$+ \langle(1 - \Theta_t)^{-1}Q_\infty^{-1/2}e^{tA}x, Q_\infty^{-1/2}y\rangle$$

$$\left. -\frac{1}{2}\langle\Theta_t(1 - \Theta_t)^{-1}Q_\infty^{-1/2}y, Q_\infty^{-1/2}y\rangle\right\},$$

$$(10.3.7)$$

where

$$\Theta_t = e^{tB}e^{tB^*} = Q_\infty^{1/2}\Lambda_t^*(Q_\infty^{-1/2}Q_t^{1/2})^*(Q_\infty^{1/2}\Lambda_t^*(Q_\infty^{-1/2}Q_t^{1/2})^*)^*, \quad t \geq 0.$$

$$(10.3.8)$$

Proof. We will first prove the special case corresponding to $x = 0$, namely that

$$k_t(0, \cdot) = \det(1 - \Theta_t)^{-1/2}\exp\left\{ -\frac{1}{2}\langle\Theta_t(1 - \Theta_t)^{-1}Q_\infty^{-1/2}y, Q_\infty^{-1/2}y\rangle\right\}.$$

$$(10.3.9)$$

Since Q_∞ is a trace class operator, we see by (10.3.8) that the operator Θ_t is trace class. Moreover, since

$$Q_t = Q_\infty - e^{tA}Q_\infty e^{tA^*} = Q_\infty^{1/2}\left[1 - (Q_\infty^{-1/2}e^{tA})Q_\infty(Q_\infty^{-1/2}e^{tA})^*\right]Q_\infty^{1/2}$$

$$= Q_\infty^{1/2}(1 - \Theta_t)Q_\infty^{1/2}$$

we have

$$(1 - \Theta_t)x = Q_\infty^{-1/2}Q_t Q_\infty^{-1/2}x, \quad x \in Q_\infty^{1/2}(H). \qquad (10.3.10)$$

Therefore, $\langle(1 - \Theta_t)x, x\rangle \geq 0$ for $x \in Q_\infty^{1/2}(H)$, a dense subset of H, and then it follows that $(1 - \Theta_t)$ is nonnegative. Equality (10.3.10) also implies, by standard arguments, that $(1 - \Theta_t)$ is invertible and

$$(1 - \Theta_t)^{-1} = (Q_t^{-1/2}Q_\infty^{1/2})^*Q_t^{-1/2}Q_\infty^{1/2}. \qquad (10.3.11)$$

Define $G = (Q_t^{-1/2}Q_\infty^{1/2})^*Q_t^{-1/2}Q_\infty^{1/2} - 1$. Then

$$G = (1 - \Theta_t)^{-1} - 1 = \Theta_t(1 - \Theta_t)^{-1}, \tag{10.3.12}$$

so that G is trace class and formula (10.3.9) follows from Proposition 1.3.11. To prove the general case, we use the equality

$$k_t(x, \cdot) = \frac{dN_{e^{tA}x,Q_t}}{dN_{Q_t}} \frac{dN_{Q_t}}{dN_{Q_\infty}} = \frac{dN_{e^{tA}x,Q_t}}{dN_{Q_t}} k_t(0, \cdot) \tag{10.3.13}$$

and we notice that, by the Cameron-Martin theorem (see Theorem 1.3.6),

$$\frac{dN_{e^{tA}x,Q_t}}{dN_{Q_t}}(y) = \exp\left(\langle Q_t^{-1/2}e^{tA}x, Q_t^{-1/2}y\rangle - \frac{1}{2}|Q_t^{-1/2}e^{tA}y|^2\right),$$

for N_{Q_t}-a.e. $y \in H$. If $m \in Q_t(H)$, then (10.3.11) implies

$$(1 - \Theta_t)^{-1}Q_\infty^{-1/2}m = Q_\infty^{1/2}Q_t^{-1}m$$

and we have, for $y \in H$, a.e. with respect to N_{Q_∞} and N_{Q_t},

$$\langle Q_t^{-1/2}m, Q_t^{-1/2}y\rangle = \langle Q_t^{-1}m, y\rangle = \langle Q_\infty^{1/2}Q_t^{-1}m, Q_\infty^{-1/2}y\rangle$$

$$= \langle(1 - \Theta_t)^{-1}Q_\infty^{-1/2}m, Q_\infty^{-1/2}y\rangle. \tag{10.3.14}$$

Equation (10.3.11) implies also

$$|Q_t^{-1/2}m|^2 = |(1 - \Theta_t)^{-1/2}Q_\infty^{-1/2}m|^2. \tag{10.3.15}$$

The equalities (10.3.14) and (10.3.15) extend by continuity to every $m \in Q_t^{1/2}(H)$. So we can set $m = e^{tA}x$, and substituting into (10.3.12) and using (10.3.9) proves (10.3.7). $\square$

Remark 10.3.4 Notice that if A is self-adjoint and e^{tA} and Q commute, then $\Theta_t = e^{2tA}$.

10.3.1 Additional regularity properties of (R_t)

The following result was proved by A. Chojnowska-Michalik and B. Goldys [48] using the Cameron-Martin formula. Here we follow the proof given in G. Da Prato, M. Fuhrman and J. Zabczyk [90].

Theorem 10.3.5 *For any $\varphi \in L^p(H, \mu)$, $p > 1$, and $t > 0$, $R_t\varphi$, defined by formula (10.3.4), is a C^∞ function.*

Note that we cannot expect that $R_t\varphi$ belongs to $C_b^\infty(H)$ because even for $\varphi(x) = \langle x, h \rangle$, $h \in H$, we have $R_t\varphi(x) = \langle e^{tA}x, h \rangle$, $x \in H$, and so $R_t\varphi$ is unbounded.

Proof of Theorem 10.3.5. Let us assume that $\varphi \in L^p(H, \mu)$ and $t > 0$. Choose $r \in (1, p)$. Then by the Hölder inequality we have

$$\int_H |k_t(x, y)\varphi(y)|^r \mu(dy) \le \left(\int_H |k_t(x, y)|^{\frac{pr}{p-r}} \mu(dy) \right)^{\frac{p-r}{p}} \left(\int_H |\varphi(y)|^p \mu(dy) \right)^{\frac{r}{p}}.$$

From Proposition 1.3.11 it follows that the first integral in the right hand side is locally bounded, and therefore the family of functions $(k_t(x, \cdot)\varphi(\cdot))_{x \in K}$ is uniformly integrable for any bounded subset K of H. It follows from (10.3.7) that $k_t(x_n, \cdot) \to k_t(x, \cdot)$ in μ-measure whenever $x_n \to x$ in H. Consequently, $R_t\varphi$ is continuous. Let us prove continuity of the first derivative of $R_t\varphi$. We have

$$\langle DR_t\varphi(x), h \rangle$$

$$= \int_H \langle (1 - \Theta_t)^{-1} Q_\infty^{-1/2} e^{tA} h, Q_\infty^{-1/2} y - Q_\infty^{-1/2} e^{tA} x \rangle \varphi(y) k_t(x, y) \mu(dy)$$

$$= \int_H \langle (1 - \Theta_t)^{-1} Q_\infty^{-1/2} e^{tA} h, Q_\infty^{-1/2} y \rangle \varphi(y) k_t(x, y) \mu(dy)$$

$$- \langle (1 - \Theta_t)^{-1} Q_\infty^{-1/2} e^{tA} x, Q_\infty^{-1/2} e^{tA} h \rangle R_t\varphi(x). \tag{10.3.16}$$

By proceeding as before, we have only to show that the function

$$\int_H |\langle (1 - \Theta_t)^{-1} Q_\infty^{-1/2} e^{tA} h, Q_\infty^{-1/2} y \rangle k_t(\cdot, y)|^{\frac{rp}{p-r}} \mu(dy)$$

is locally bounded. This easily follows using the Hölder inequality again. We proceed similarly for the other derivatives. $\square$

The following result is also proved in G. Da Prato, M. Fuhrman and J. Zabczyk [90].

Theorem 10.3.6 Let $\varphi \in L^1(H, \mu)$ and let $R_t\varphi$ be defined by formula (10.3.4).

(i) If H is infinite dimensional, A is self-adjoint and e^{tA} and Q commute, there exist $\varphi \in L^1(H, \mu)$ nonnegative and $x \in H$ such that $R_t\varphi(x) = +\infty$.

(ii) If H is finite dimensional then $R_t\varphi$ is of class C^∞, for any $\varphi \in L^1(H, \mu)$.

Proof. (i) It is enough to show that for any $t > 0$ there exists $x \in H$ such that ess.sup $k_t(x, \cdot) = +\infty$. To this purpose, taking into account the definition of k, we are going to prove that there exists $x \in H$ such that the function

$$F_x(z) = \langle (1 - \Theta_t)^{-1} Q_\infty^{-1/2} e^{tA} x, z \rangle - \frac{1}{2} \langle \Theta_t (1 - \Theta_t)^{-1} z, z \rangle, \quad z \in H,$$

is unbounded. By Lemma 10.3.7 below F_x is bounded for every $x \in H$ if and only if

$$(1 - \Theta_t)^{-1} Q_\infty^{-1/2} e^{tA}(H) \subset \left(\Theta_t (1 - \Theta_t)^{-1} \right)^{1/2}(H). \tag{10.3.17}$$

This holds if and only if there exists a constant $C_t > 0$ such that

$$|(Q_\infty^{-1/2} e^{tA})^* (1 - \Theta_t)^{-1} z|^2 \le C_t \, \langle \Theta_t (1 - \Theta_t)^{-1} z, z \rangle, \quad z \in H. \tag{10.3.18}$$

By the commutativity assumption, due to Remark 10.3.4, this is equivalent to

$$(1 - e^{2tA})^{-2} Q_\infty^{-1} e^{2tA} \le C_t \, e^{2tA} (1 - e^{2tA})^{-1}. \tag{10.3.19}$$

This cannot hold when dim $H = +\infty$ since in this case Q_∞^{-1} is not bounded. Therefore (i) is proved.

Let us prove (ii). As we remarked before the kernel $k_t(x, y)$ is continuous in x and bounded in y, therefore $R_t \varphi$ is continuous. Consider now the first derivative of $R_t \varphi$. Taking into account formula (10.3.16) we see, by elementary properties of exponentials, that it is also continuous. Similarly we get continuity of all derivatives. $\square$

The following lemma is well known. However, we give a proof for the reader's convenience.

Lemma 10.3.7 *Assume that S is a nonnegative symmetric operator in $L(H)$ and that $b \in H$. If*

$$\psi(x) = -\langle Sx, x \rangle + \langle b, x \rangle, \; x \in H. \tag{10.3.20}$$

Then

$$\sup_{x \in H} \psi(x) = \begin{cases} \frac{1}{4} |S^{-1/2} b|^2 & \text{if } b \in S^{1/2}(H), \\ +\infty & \text{if } b \notin S^{1/2}(H). \end{cases} \tag{10.3.21}$$

Proof. If $b \in S^{1/2}(H)$ one can easily check that

$$\psi(x) = \frac{1}{4} |S^{-1/2}b|^2 - \left| S^{1/2}x - \frac{1}{2} S^{-1/2}b \right|^2.$$

Therefore (10.3.21) follows.

If $b \notin S^{1/2}(H)$ we consider for any $\varepsilon > 0$ the function

$$\psi_\varepsilon(x) = -\langle (S + \varepsilon)x, x \rangle + \langle b, x \rangle, \ x \in H.$$

Then

$$\sup_{x \in H} \psi(x) \geq \sup_{x \in H} \psi_\varepsilon(x) = \frac{1}{4} |(S + \varepsilon)^{-1/2}b|^2.$$

Since $b \notin S^{1/2}(H)$ therefore $|(S+\varepsilon)^{-1/2}b|^2 \to +\infty$ as $\varepsilon \to 0$. This completes the proof. $\square$

10.3.2   Hypercontractivity of (R_t)

We give here, following G. Da Prato, M. Fuhrman and J. Zabczyk [90], a direct proof of hypercontractivity of (R_t) (see Theorem 10.3.10). We recall that hypercontractivity was first proved for symmetric Ornstein-Uhlenbeck semigroups by Nelson [175] with a direct combinatorial argument and by L. Gross [139] with log-Sobolev inequalities. In the nonsymmetric case we quote M. Fuhrman, [118] and [121], which adapted an argument due to Neveu [177], and A. Chojnowska-Michalik and B. Goldys [50], which used the second quantization operator. The result we present here can be found, with small variations, in A. Chojnowska-Michalik and B. Goldys [49, Theorem 4], in [117, Theorem 6.5] and in M. Fuhrman [119, Theorem 5]. Although the hypercontractivity estimates are not sharp, the method is simple and can be generalized to more general semigroups, see I. Simão [199], [200].

We start with a lemma.

Lemma 10.3.8 *For all $s \geq 1$, we have*

$$\left(\int_H k_t^s(x, y) N_{Q_\infty}(dy) \right)^{1/s} = \det(1 - \Theta_t)^{-\frac{1}{2} + \frac{1}{2s}} \det(1 + (s - 1)\Theta_t)^{-\frac{1}{2s}}$$

$$\times \exp \left\{ \frac{s - 1}{2} \langle (1 + (s - 1)\Theta_t)^{-1}Q_\infty^{-1/2}e^{tA}x, Q_\infty^{-1/2}e^{tA}x \rangle \right\}. \quad (10.3.22)$$

Proof. By Proposition 1.3.11,

$$\int_H \exp\Big\{ -\frac{s}{2}\langle \Theta_t(1-\Theta_t)^{-1}Q_\infty^{-1/2}y, Q_\infty^{-1/2}y\rangle$$

$$+s\langle(1-\Theta_t)^{-1}Q_\infty^{-1/2}e^{tA}x, Q_\infty^{-1/2}y\rangle\Big\} N_{Q_\infty}(dy)$$

$$= \det(1+s\Theta_t(1-\Theta_t)^{-1})^{-1/2}\exp\Big\{\frac{s^2}{2}\langle(1+s\Theta_t(1-\Theta_t)^{-1})^{-1}$$

$$\times(1-\Theta_t)^{-1}Q_\infty^{-1/2}e^{tA}x, (1-\Theta_t)^{-1}Q_\infty^{-1/2}e^{tA}x\rangle\Big\}.$$

So we obtain

$$\left(\int_H k_t^s(x,y)N_{Q_\infty}(dy)\right)^{1/s}$$

$$= \det(1-\Theta_t)^{-1/2}\det(1+s\Theta_t(1-\Theta_t)^{-1})^{-1/(2s)}$$

$$\times \exp\Big\{\langle V\, Q_\infty^{-1/2}e^{tA}x, Q_\infty^{-1/2}e^{tA}x\rangle\Big\},$$

where the operator V is

$$-\frac{1}{2}(1-\Theta_t)^{-1} + \frac{s}{2}(1-\Theta_t)^{-1}\Big(1+s\Theta_t(1-\Theta_t)^{-1}\Big)^{-1}(1-\Theta_t)^{-1},$$

and it is easily shown that

$$V = \frac{s-1}{2}(1+(s-1)\Theta_t)^{-1}.$$

Finally, we notice that

$$1+s\Theta_t(1-\Theta_t)^{-1} = (1-\Theta_t)^{-1}(1+(s-1)\Theta_t),$$

so that

$$\det(1+s\Theta_t(1-\Theta_t)^{-1})^{-1/(2s)} = \det(1-\Theta_t)^{1/(2s)}\det(1+(s-1)\Theta_t)^{-1/(2s)}$$

and the formula of the lemma follows. □

Lemma 10.3.9 *Let $s,r \geq 1$. Then the integral*

$$I := \int_H \left[\int_H k_t^s(x,y)\mu(dy)\right]^{\frac{r}{s}}\mu(dx)$$

is finite if and only if $(r-1)(s-1) < \|\Theta_t\|^{-1}$. In this case

$$I^{1/r} = \det(1-\Theta_t)^{-\frac{1}{2}+\frac{1}{2s}}\det(1+(s-1)\Theta_t)^{\frac{1}{2r}-\frac{1}{2s}}\det(1+(s-1)(r-1)\Theta_t)^{-\frac{1}{2r}}.$$

Proof. We apply Lemma 10.3.8 and we compute

$$I = \det(1 - \Theta_t)^{-\frac{r}{2} + \frac{r}{2s}} \det(1 + (s - 1)\Theta_t)^{-\frac{r}{2s}}$$

$$\times \int_H \exp\left\{\frac{r(s-1)}{2}\langle(1 + (s-1)\Theta_t)^{-1}Q_\infty^{-1/2}e^{tA}x, Q_\infty^{-1/2}e^{tA}x\rangle\right\} N_{Q_\infty}(dx).$$

Performing the change of variable $x \rightarrow (1 + (s-1)\Theta_t)^{-1/2}Q_\infty^{-1/2}e^{tA}x$, the integral in the right hand side becomes

$$\int_H \exp\left\{\frac{r(s-1)}{2}|x|^2\right\} N_V(dx),$$

where

$$V = (1 + (s-1)\Theta_t)^{-1/2}Q_\infty^{-1/2}e^{tA}Q_\infty(Q_\infty^{-1/2}e^{tA})^*(1 + (s-1)\Theta_t)^{-1/2}$$

$$= (1 + (s-1)\Theta_t)^{-1/2}\Theta_t(1 + (s-1)\Theta_t)^{-1/2} = \Theta_t(1 + (s-1)\Theta_t)^{-1}.$$

By Proposition 1.3.11, the integral is convergent if and only if $r(s - 1) < \|V\|^{-1}$. Let λ_n denote the eigenvalues of Θ_t, with $\|\Theta_t\| = \lambda_1 \geq \lambda_2 \geq \dots$. Then the eigenvalues of V are $\lambda_n(1 + (s-1)\lambda_n)^{-1}$ and it follows that

$$\|V\| = \lambda_1(1 + (s-1)\lambda_1)^{-1} = \|\Theta_t\|(1 + (s-1)\|\Theta_t\|)^{-1}.$$

Convergence of the integral takes place if and only if

$$r(s-1) < \|\Theta_t\|^{-1}(1 + (s-1)\|\Theta_t\|),$$

i.e. $(r-1)(s-1) < \|\Theta_t\|^{-1}$. Assuming that this inequality holds, we obtain, again by Proposition 1.3.11,

$$I = \det(1 - \Theta_t)^{-\frac{r}{2} + \frac{r}{2s}} \det(1 + (s-1)\Theta_t)^{-\frac{r}{2s}} \det(1 - r(s-1)V)^{-\frac{1}{2}}.$$

Since

$$1 - r(s-1)V = 1 - r(s-1)\Theta_t(1 + (s-1)\Theta_t)^{-1}$$

$$= (1 + (s-1)\Theta_t)^{-1}(1 - (s-1)(r-1)\Theta_t),$$

the conclusion follows. $\square$

We now prove

Theorem 10.3.10 *Let* $t > 0$ *and* $p, q \in [1, \infty)$ *be such that*

$$q - 1 < (p-1)\|Q_\infty^{-1/2}e^{tA}Q_\infty^{1/2}\|^{-2}.$$

Then the operator R_t *has continuous extension to an operator from* $L^p(H, \mu)$ *into* $L^q(H, \mu)$.

Proof. We have to estimate

$$\int_H |R_t\varphi(x)|^q \mu(dx) = \int_H \left| \int_H k_t(x, y)\varphi(y)\mu(dy) \right|^q \mu(dx),$$

for all bounded φ. By the Hölder inequality we have

$$\int_H |R_t\varphi(x)|^q \mu(dx)$$

$$\leq \int_H \left(\int_H k_t(x, y)^{p'} \mu(dy) \right)^{q/p'} \left(\int_H |\varphi(y)|^p \mu(dy) \right)^{q/p} \mu(dx),$$

where $\frac{1}{p} + \frac{1}{p'} = 1$. To show the result it is enough to prove that the integral

$$\int_H \left(\int_H k_t(x, y)^{p'} \mu(dy) \right)^{q/p'} \mu(dx)$$

is finite. By Lemma 10.3.9 this happens if $(p' - 1)(q - 1) < \|\Theta_t\|^{-1}$. This finishes the proof, since $p' - 1 = (p-1)^{-1}$ and $\|\Theta_t\| = \|Q_\infty^{-1/2}e^{tA}Q_\infty^{1/2}\|^2$. $\square$

Remark 10.3.11 Since the kernel k is not symmetric, one may wonder what happens if the order of integration with respect to x and y is reversed in Lemma 10.3.9. We have the following analogue of Lemma 10.3.9 and Lemma 10.3.8 whose proof is very similar and is therefore omitted.

Proposition 10.3.12 *For all* $r \geq 1$, *we have*

$$\left(\int_H k_t^r(x, y)\mu(dx) \right)^{1/r} = \det(1 - \Theta_t)^{-\frac{1}{2} + \frac{1}{2r}} \det(1 + (r-1)\Theta_t)^{-\frac{1}{2r}}$$

$$\times \exp\left\{ \frac{r-1}{2} \langle \Theta_t(1 + (r-1)\Theta_t)^{-1} Q_\infty^{-1/2} y, Q_\infty^{-1/2} y \rangle \right\}.$$

Let $s \geq 1$. *Then*

$$\left[\int_H \left[\int_H k_t^r(x, y)\mu(dx) \right]^{\frac{s}{r}} \mu(dy) \right]^{1/s}$$

$$= \det(1 - \Theta_t)^{-\frac{1}{2} + \frac{1}{2r}} \det(1 + (r-1)\Theta_t)^{\frac{1}{2s} - \frac{1}{2r}} \det(1 + (s-1)(r-1)\Theta_t)^{-\frac{1}{2s}},$$

the integral being finite if and only if $(r-1)(s-1) < \|\Theta_t\|^{-1}$.

10.4 A representation formula for (R_t) in terms of the second quantization operator

We follow here A. Chojnowska-Michalik and B. Goldys, see [50]. We set again $\mu = N_{Q_\infty}$.

10.4.1 The second quantization operator

Let us define an embedding of $L(H)$ into $L^2(H, \mu)$. We set

$$L_\mu(H) = \{T \in L(H) : \exists S \in L(H) \text{ such that } T = SQ_\infty^{1/2}\}.$$

It is easy to see that $L_\mu(H)$ is dense in $L(H)$, with respect to the pointwise convergence. Given (T_n) and T in $L(H)$, we say that $T_n \to T$ pointwise if $\lim_{n \to \infty} T_n x = Tx$, $x \in H$.

Let us define a linear mapping

$$F : L_\mu(H) \to L^2(H, \mu), \ T \to F(T),$$

where

$$F(T)x = Q_\infty^{1/2} Sx, \text{ and } S \text{ is such that } T = SQ_\infty^{1/2}.$$

It is easy to see that this definition does not depend on S. Moreover

$$\int_H |F(T)x|^2 \mu(dx) = \text{Tr}[T^* T Q_\infty].$$

Consequently F is extendible by density to all $L(H)$. We shall still denote the extension by F.

In the following we shall often write

$$F(T)x = Q_\infty^{1/2} T Q_\infty^{-1/2} x, \quad x \in H.$$

Now we define the second quantization operator. For any Banach space E we denote by $\mathcal{K}(E)$ the set of all operators $T \in L(E)$ such that $\|T\|_{L(E)} \leq 1$. For any $p \geq 1$ we define a linear mapping

$$\Gamma : \mathcal{K}(H) \to L(L^p(H, \mu)), \ T \to \Gamma(T),$$

by setting

$$\Gamma(T)\varphi(x) = \int_H \varphi\left(F(T^*)x + F(\sqrt{1 - T^*T})y\right) \mu(dy).$$

In particular we have $\Gamma(1)\varphi = \varphi$, and for any constant κ such that $|\kappa| \leq 1$,

$$\Gamma(\kappa)\varphi(x) = \int_H \varphi\left(\kappa x + \sqrt{1 - |\kappa|^2}y\right)\mu(dy).$$

and $\Gamma(T^*) = (\Gamma(T))^*$.

Proposition 10.4.1 $\Gamma(T) \in \mathcal{K}(L^p(H, \mu))$, $p \geq 1$.

Proof. By Hölder's inequality we have

$$|\Gamma(T)\varphi(x)|^p \leq \int_H \varphi^p\left(F(T^*)x + F(\sqrt{1 - T^*T})y\right)\mu(dy),$$

and so

$$\int_H |\Gamma(T)\varphi(x)|^p\mu(dx) \leq \int_{H \times H} \varphi^p\left(F(T^*)x + F(\sqrt{1 - T^*T})y\right)\mu(dx)\mu(dy).$$

Now the conclusion follows from the identity

$$\int_{H \times H} \psi\left(F(T^*)x + F(\sqrt{1 - T^*T})y\right)\mu(dx)\mu(dy)$$

$$= \int_H \psi(x)\mu(dx), \quad \psi \in C_b(H). \tag{10.4.1}$$

It is enough to check (10.4.1) for $\psi(x) = e^{i\langle h,x\rangle}$, $h \in H$. We have in fact in this case

$$\int_{H \times H} \psi\left(F(T^*)x + F(\sqrt{1 - T^*T})y\right)\mu(dx)\mu(dy)$$

$$= \int_H e^{i\langle h,F(T^*)x\rangle}\mu(dx)\int_H e^{i\langle h,F(\sqrt{1-T^*T})y\rangle}\mu(dy)$$

$$= \int_H e^{iW_{TQ_\infty^{1/2}h}(x)}\mu(dx)\int_H e^{iW_{\sqrt{1-T^*T}Q_\infty^{1/2}(y)}}\mu(dy)$$

$$= e^{-\frac{1}{2}\langle Q_\infty h,h\rangle} = \int_H \psi(x)\mu(dx). \quad \square$$

10.4.2   The adjoint of (R_t)

For any contraction $T \in L(H)$ we define a linear contraction on $L^2(H, \mu)$ by setting

$$\Gamma(T)\varphi(x) = \int_H \varphi(Q_\infty^{1/2} T^* Q_\infty^{-1/2} x + Q_\infty^{1/2} \sqrt{1 - T^*T} \, Q_\infty^{-1/2} y) \mu(dy),$$
$$(10.4.2)$$

for all $\varphi \in L^2(H, \mu)$.

Theorem 10.4.2 *Assume that hypotheses* (10.1.1) *and* (10.3.1) *hold. Then, for all $t > 0$ and $\varphi \in L^2(H, \mu)$ we have*

$$R_t = \Gamma(e^{tB^*}) = \Gamma(Q_\infty^{1/2} e^{tA^*} Q_\infty^{-1/2}). \qquad (10.4.3)$$

Proof. We first remark that (10.4.2) is equivalent to

$$\Gamma(T)\varphi(x) = \int_H \varphi(Q_\infty^{1/2} T^* Q_\infty^{-1/2} x + z) N_{Q_\infty^{1/2}(1 - T^*T)Q_\infty^{1/2}}(dz), \qquad (10.4.4)$$

for all $\varphi \in L^2(H, \mu)$. Then, setting in (10.4.4) $T = Q_\infty^{1/2} e^{tA^*} Q_\infty^{-1/2}$ we obtain

$$\begin{aligned}
\Gamma(Q_\infty^{1/2} e^{tA^*} Q_\infty^{-1/2})\varphi(x) &= \int_H \varphi(e^{tA} x + z) N_{[Q_\infty - e^{tA} Q_\infty e^{tA^*}]}(dz) \\
&= \int_H \varphi(e^{tA} x + z) N_{Q_t}(dz). \quad \square
\end{aligned}$$

Remark 10.4.3 Theorem 10.4.2 allows us to obtain an explicit formula for the adjoint semigroup of (R_t), namely $R_t^* = [\Gamma(e^{tB})]$.

10.5   Poincaré and log-Sobolev inequalities

Let us consider the Dirichlet form

$$a(\varphi, \psi) = -\frac{1}{2} \int_H L_2 \varphi \, \psi d\mu, \quad \varphi, \psi \in C_b^1(H). \qquad (10.5.1)$$

We say that the *Poincaré inequality* holds if there exists $\omega > 0$ such that

$$\int_H |\varphi - \overline{\varphi}|^2 d\mu \le \frac{1}{2\omega} a(\varphi, \varphi), \quad \varphi \in C_b^1(H), \qquad (10.5.2)$$

where

$$\overline{\varphi} = \int_H \varphi(x)\mu(dx).$$

Notice that, due to (10.2.5), we have that (10.5.2) is equivalent to

$$\int_H |\varphi - \overline{\varphi}|^2 d\mu \leq \frac{1}{2\omega} \int_H |Q^{1/2}D\varphi|^2 d\mu, \quad \varphi \in C_b^1(H). \tag{10.5.3}$$

We say that the *logarithmic Sobolev inequality* holds if there exists $\omega > 0$, such that

$$\int_H \varphi^2 \log(\varphi^2) d\mu \leq \frac{1}{\omega} a(\varphi, \varphi) + \|\varphi\|_2^2 \log(\|\varphi\|_2^2), \quad \varphi \in C_b^1(H), \tag{10.5.4}$$

or, equivalently,

$$\int_H \varphi^2 \log(\varphi^2) d\mu \leq \frac{1}{2\omega} \int_H |Q^{1/2}D\varphi|^2 d\mu + \|\varphi\|_2^2 \log(\|\varphi\|_2^2), \quad \varphi \in C_b^1(H). \tag{10.5.5}$$

We want now to show that when the Poincaré inequality (10.5.3) holds the spectrum $\sigma(L_2)$ of L_2 consists of 0 and a set included in the half-space $\{\lambda \in \mathbb{C} : \text{Re } \lambda \leq -\omega\}$ (*spectral gap*). The spectral gap implies an exponential convergence of $R_t\varphi$ to the equilibrium

$$\int_H |R_t\varphi - \overline{\varphi}|^2 d\mu \leq e^{-\omega t} \int_H |\varphi|^2 d\mu, \quad \varphi \in L^2(H, \mu). \tag{10.5.6}$$

We have in fact the result

Proposition 10.5.1 *Assume that (10.1.1) and (10.5.3) hold. Then we have*

$$\sigma(L_2)\backslash\{0\} \subset \{\lambda \in \mathbb{C} : \text{Re } \lambda \leq -\omega\}, \tag{10.5.7}$$

and (10.5.6) holds.

Proof. Let us consider the space

$$L_0^2(H, \mu) = \{\varphi \in L^2(H, \mu) : \overline{\varphi} = 0\}.$$

Clearly $L_0^2(H, \mu)$ is an invariant subspace of (R_t). Moreover, by (10.5.3)

$$\langle L_2\varphi, \varphi \rangle_{L^2(H,\mu)} = -\frac{1}{2} \int_H |C^{1/2}D\varphi|^2 d\mu \leq -\omega\|\varphi\|_{L^2(H,\mu)}^2, \quad \varphi \in L_0^2(H, \mu). \tag{10.5.8}$$

This yields (10.5.7) by the Hille-Yosida theorem. Finally, let us prove (10.5.6). Note that by (10.5.8) it follows that

$$\int_H |R_t\varphi|^2 d\mu \le e^{-\omega t} \int_H |\varphi|^2 d\mu, \quad \varphi \in L_0^2(H,\mu).$$

Therefore for any $\varphi \in L^2(H,\mu)$ we have

$$\int_H |R_t\varphi - \overline{\varphi}|^2 d\mu = \int_H |R_t(\varphi - \overline{\varphi})|^2 d\mu \le e^{-\omega t} \int_H |\varphi - \overline{\varphi}|^2 d\mu$$

$$= e^{-\omega t}\left[\int_H |\varphi|^2 d\mu - (\overline{\varphi})^2\right] \le e^{-\omega t} \int_H |\varphi|^2 d\mu.$$

The proof is complete. $\square$

Necessary and sufficient conditions in order that Poincaré and log-Sobolev inequalities hold can be found in A. Chojnowska-Michalik and B. Goldys [51]. We shall only consider two special situations in the following two subsections.

10.5.1 The case when $M = 1$ and $Q = I$

We are going to prove, following J. D. Deuschel and D. Stroock [106], the following results.

Proposition 10.5.2 *Assume that (10.1.1) holds with $M = 1$ and $Q = I$. Then, for all $\varphi \in W^{1,2}(H,\mu)$ we have*

$$\int_H |\varphi - \overline{\varphi}|^2 d\mu \le \frac{1}{2\omega} \int_H |D\varphi|^2 d\mu. \tag{10.5.9}$$

Proposition 10.5.3 *Assume that (10.1.1) holds with $M = 1$ and $Q = I$. Then, for all $\varphi \in W^{1,2}(H,\mu)$ we have*

$$\int_H \varphi^2 \log(\varphi^2) d\mu \le \frac{1}{\omega} \int_H |D\varphi|^2 d\mu + \|\varphi\|_2^2 \log(\|\varphi\|_2^2). \tag{10.5.10}$$

As we shall see, the proofs of inequalities (10.5.9) and (10.5.10) are similar. The main ingredients are the two following lemmas.

Lemma 10.5.4 *We have*

$$|(DR_t\varphi)(x)|^2 \le e^{-2\omega t} R_t(|D\varphi|^2)(x), \quad \varphi \in W^{1,2}(H,\mu), \quad t > 0, \ x \in H, \tag{10.5.11}$$

Proof. We recall the identity

$$\langle DR_t\varphi(x), h\rangle = \int_H \langle D\varphi(e^{tA}x + y), e^{tA}h\rangle N_{Q_t}(dy), \quad t > 0, \; x, h \in H.$$

$$(10.5.12)$$

By the Hölder inequality it follows that

$$|\langle DR_t\varphi(x), h\rangle|^2 \leq |h|^2 e^{-2\omega t} \int_H |D\varphi(e^{tA}x + y)|^2 N_{Q_t}(dy)$$

$$= |h|^2 e^{-2\omega t} R_t(|D\varphi|^2)(x), \quad t > 0, \; x, h \in H.$$

Now the conclusion follows from the arbitrariness of h. $\square$

Lemma 10.5.5 *For any $g \in C^2(\mathbb{R})$ we have*

$$L_2(g(\varphi)) = g'(\varphi)L_2\varphi + \frac{1}{2} g''(\varphi)|D\varphi|^2, \quad \varphi \in \mathcal{E}_A(H),$$

$$(10.5.13)$$

and

$$\int_H (L_2\varphi)g'(\varphi)d\mu = -\frac{1}{2}\int_H g''(\varphi)|D\varphi|^2 d\mu.$$

$$(10.5.14)$$

Proof. Let $\varphi \in \mathcal{E}_A(H)$. Since

$$Dg(\varphi) = g'(\varphi)D\varphi, \quad D^2g(\varphi) = g''(\varphi)D\varphi \otimes D\varphi + g'(\varphi)D^2\varphi,$$

(10.5.13) follows. Finally integrating (10.5.13) yields (10.5.14) since

$$\int_H L_2(g(\varphi))d\mu = 0,$$

by the invariance of μ. $\square$

Proof of Proposition 10.5.2. Let $\varphi \in W^{1,2}(H, \mu)$, Then by (10.2.3) we have

$$\frac{1}{2}\frac{d}{dt}\int_H |R_t\varphi|^2 d\mu = \int_H L_2 R_t\varphi \; R_t\varphi \; d\mu = -\frac{1}{2}\int_H |DR_t\varphi|^2 d\mu.$$

It follows, taking into account (10.5.11), that

$$\frac{1}{2}\frac{d}{dt}\int_H |R_t\varphi|^2 d\mu \geq -\frac{1}{2}e^{-2t\omega}\int_H R_t(|D\varphi|^2)d\mu = -\frac{1}{2}e^{-2t\omega}\int_H |D\varphi|^2 d\mu,$$

by the invariance of μ. Integrating over t gives

$$\int_H |R_t\varphi|^2 d\mu - \int_H \varphi^2 d\mu \geq -\frac{1}{2\omega}(1 - e^{-2t\omega}) \int_H |D\varphi|^2 d\mu.$$

Finally, letting t tend to $+\infty$ and recalling (10.1.4) gives

$$(\overline{\varphi})^2 - \int_H \varphi^2 d\mu \geq -\frac{1}{2\omega} \int_H |D\varphi|^2 d\mu$$

and the conclusion follows, since

$$\int_H |\varphi - \overline{\varphi}|^2 d\mu = \int_H \varphi^2 d\mu - (\overline{\varphi})^2. \quad \square$$

Proof of Proposition 10.5.3. It is enough to prove the result when $\varphi \in W^{1,2}(H,\mu)$ is such that $\varphi(x) \geq \varepsilon > 0$, $x \in H$. In this case we have

$$\frac{d}{dt} \int_H (R_t(\varphi^2)) \log(R_t(\varphi^2)) d\mu = \int_H L_2 R_t(\varphi^2) \log(R_t(\varphi^2)) d\mu$$

$$+ \int_H L_2 R_t(\varphi^2) d\mu.$$

Now the second term vanishes, due to the invariance of μ. For the first term we use (10.5.14) with $g'(\xi) = \log \xi$. Therefore we have

$$\frac{d}{dt} \int_H R_t(\varphi^2) \log(R_t(\varphi^2)) d\mu = -\frac{1}{2} \int_H \frac{1}{R_t(\varphi^2)} |DR_t(\varphi^2)|^2 d\mu. \quad (10.5.15)$$

By (10.5.12) we have, for any $h \in H$,

$$\langle DR_t(\varphi^2)(x), h \rangle = 2 \int_H \varphi(e^{tA}x + y)\langle D\varphi(e^{tA}x + y), e^{tA}h \rangle N_{Q_t}(dy).$$

It follows by the Hölder inequality that

$$|\langle DR_t(\varphi^2)(x), h \rangle|^2 \leq 4e^{-2t\omega} \int_H \varphi^2(e^{tA}x + y) N_{Q_t}(dy)$$

$$\times \int_H |D\varphi(e^{tA}x + y)|^2 N_{Q_t}(dy),$$

which yields

$$|DR_t(\varphi^2)|^2 \leq 4e^{-2t\omega} R_t(\varphi^2) R_t(|D\varphi|^2).$$

Substituting in (10.5.15) yields

$$\frac{d}{dt} \int_H R_t(\varphi^2) \log(R_t(\varphi^2)) d\mu \geq -2e^{-2t\omega} \int_H R_t(|D\varphi|^2) d\mu$$

$$= -2e^{-2t\omega} \int_H |D\varphi|^2 d\mu,$$

due to the invariance of μ. Integrating over t yields

$$\int_H R_t(\varphi^2) \log(R_t(\varphi^2)) d\mu - \int_H \varphi^2 \log(\varphi^2) d\mu$$

$$\geq \frac{1}{2\omega} (1 - e^{-2t\omega}) \int_H |D\varphi|^2 d\mu.$$

Finally, letting t tend to $+\infty$ and recalling (10.1.4) gives

$$\|\varphi\|_2^2 \log(\|\varphi\|_2^2) - \int_H \varphi^2 \log(\varphi^2) d\mu \geq -\frac{1}{\omega} \int_H |D\varphi|^2 d\mu$$

and the conclusion follows. $\square$

10.5.2   A generalization

Here we assume, besides (10.1.1), that

There exist M_1, $\omega_1 > 0$, *such that* $\|Q^{1/2} e^{tA} Q^{-1/2}\| \leq M_1 e^{-\omega_1 t}$, $t \geq 0$.

$$(10.5.16)$$

This assumption is obviously fulfilled when Q^{-1} is bounded.

Proposition 10.5.6 *Assume that* (10.1.1) *and* (10.5.16) *hold, and let* $\mu = N_{Q_\infty}$. *Then for all* $\varphi \in W_Q^{1,2}(H, \mu)$ *we have*

$$\int_H |\varphi - \overline{\varphi}|^2 d\mu \leq \frac{M_1}{\omega_1} \int_H |Q^{1/2} D\varphi|^2 d\mu. \tag{10.5.17}$$

Proof. For any $h \in H$, $t > 0$, $x \in H$, we have

$$\langle Q^{1/2} D R_t \varphi(x), h \rangle = \int_H \langle D\varphi(e^{tA}x + y), Q^{1/2} e^{tA} Q^{-1/2} h \rangle N_{Q_t}(dy).$$

Consequently for $t > 0$ and $x \in H$ we have

$$|(Q^{1/2} D R_t \varphi)(x)|^2 \leq e^{-2\omega_1 t} R_t(|Q^{1/2} D\varphi|^2)(x), \quad \varphi \in W_Q^{1,2}(H, \mu). \tag{10.5.18}$$

Now the conclusion follows arguing as in the proof of Proposition 10.5.2. $\square$

10.6 Some additional regularity results when Q and A commute

In this section we consider the case when A and Q commute and A is self-adjoint.

We assume that

$$
\begin{cases}
(i) & A : D(A){\subset}H \to H \text{ is self-adjoint, and there exists } \omega > 0 \\
& \text{such that } \langle Ax, x\rangle \le -\omega|x|^2, \; x \in H, \\
(ii) & Q \in L^+(H), \; \text{Ker } Q = \{0\}, \; Q \text{ commutes with } A, \\
(iii) & Q_\infty := -\tfrac{1}{2}\, QA^{-1} \text{ is of trace class}, \\
(iv) & \text{there exist an orthonormal basis } (e_k) \text{ on } H \text{ and sequences of} \\
& \text{positive numbers } (\alpha_k),(q_k),(\lambda_k), \text{ such that}
\end{cases}
$$

$$
Ae_k = -\alpha_k e_k, \; Qe_k = q_k e_k, \; Q_\infty e_k = \lambda_k e_k, \;\; k \in \mathbb{N}.
$$

$$(10.6.1)$$

Obviously $\lambda_k = \tfrac{1}{2}\tfrac{q_k}{\alpha_k}$, $k \in \mathbb{N}$. We still denote by μ the Gaussian measure N_{Q_∞}.

When (R_t) is strong Feller, we have seen that $R_t f$ belongs to $W^{1,p}(H,\mu)$ for all $t > 0$ and $f \in L^p(H,\mu)$. We shall prove that in the present case a smoothing property holds on the directions of Q. We have in fact the following result.

Proposition 10.6.1 *(i) Let $f \in L^p(H,\mu)$ and $t > 0$. Then we have $R_t f \in W_Q^{1,2}(H,\mu)$ and*

$$
\int_H |Q^{1/2}DR_t f|^p d\mu \le (et)^{-p/2} \int_H |f|^p d\mu.
$$

(ii) Let $f \in L^p(H,\mu)$, and $\lambda > 0$. Then $R(\lambda,L)f \in W_Q^{1,p}(H,\mu)$ and

$$
\int_H |Q^{1/2}DR(\lambda,L)f|^p d\mu \le (\pi/(e\lambda))^{-p/2} \int_H |f|^p d\mu.
$$

Proof. Let $h \in H$ and $f \in UC_b(H)$. Then, by using the Cameron-Martin formula, we can prove that there exists the derivative of $R_t f$ in the direction $Q^{1/2}h$, and we have

$$
\langle DR_t f(x), Q^{1/2}h\rangle = \int_H \langle \Lambda_t Q^{1/2}h, Q_t^{-1/2}y\rangle f(e^{tA}x + y) N_{Q_t}(dy)
$$

$$
= \sqrt{2} \int_H \langle (-A)^{1/2}e^{tA}(1 - e^{2tA})^{-1/2}h, Q_t^{-1/2}y\rangle f(e^{tA}x + y) N_{Q_t}(dy).
$$

By the Hölder inequality we find moreover

$$|\langle Q^{1/2}DR_tf(x), h\rangle|^2$$

$$\leq 2\int_H |\langle(-A)^{1/2}e^{tA}(1 - e^{2tA})^{-1/2}h, Q_t^{-1/2}y\rangle|^2 N_{Q_t}(dy)\, R_t(f^2)(x)$$

$$= 2|(-A)^{1/2}e^{tA}(1 - e^{2tA})^{-1/2}h|^2\, R_t(f^2)(x) \leq (te)^{-1}\,|h|^2 R_t(f^2)(x).$$

From the arbitrariness of h it follows that

$$|Q^{1/2}DR_t\varphi(x)|^2 \leq (te)^{-1/2}\, R_t(f^2)(x), \qquad (10.6.2)$$

which yields

$$\|Q^{1/2}DR_t\varphi\|_0 \leq (te)^{-1/2}\, \|f\|_0. \qquad (10.6.3)$$

Moreover, integrating (10.6.2) with respect to μ and recalling that μ is invariant for (R_t), (i) follows for $p = 2$, and, by interpolating with (10.6.3), for any $p \geq 2$. Finally (ii) follows using Laplace transform. $\square$

Chapter 11

Perturbations of Ornstein-Uhlenbeck semigroups

In this chapter we consider the following differential operator in the separable Hilbert space H:

$$N_0\varphi(x) = \frac{1}{2}\mathrm{Tr}[D^2\varphi(x)] + \langle x, AD\varphi(x)\rangle + \langle F(x), D\varphi\rangle, \quad x \in H,$$

where A is a self-adjoint operator and $F : H \to H$ a nonlinear mapping.

Several results can be proved for a more general operator of the form

$$N_0\varphi(x) = \frac{1}{2}\mathrm{Tr}[CD^2\varphi(x)] + \langle x, AD\varphi(x)\rangle + \langle F(x), D\varphi\rangle, \quad x \in H,$$

where $C \in L^+(H)$, but we shall limit ourselves to operators of the first form for the sake of simplicity. Thus we shall assume that assumptions (10.1.8) hold. We shall denote by μ the Gaussian measure N_Q [1] where $Q = -\frac{1}{2} A^{-1}$, and by (R_t) the Ornstein-Uhlenbeck semigroup in $L^2(H, \mu)$:

$$R_t\varphi(x) = \int_H \varphi(e^{tA}x + y)N_{Q_t}(dy), \quad x \in H, \ \varphi \in L^2(H, \mu).$$

L_2 will represent its infinitesimal generator.

Here, §11.1 is devoted to the case when F is Borel and bounded whereas §11.2 concerns the case when F is Lipschitz continuous and dissipative.

---

[1]Warning: Q plays here the rôle of Q_∞ in Chapter 10.

More general situations, concerning reaction-diffusion equations and stochastic quantization , are studied in [79], [99] and [96], [97]. Also, some Kolmogorov equations coming from Burgers and Navier-Stokes equations are studied in [86], [113] and [87].

11.1   Bounded perturbations

Here we shall assume, besides (10.1.8), that F is bounded, that is

$$
\left\{
\begin{array}{l}
(i) \quad A : D(A) {\subset} H \to H \text{ is self-adjoint and there exists } \omega > 0 \text{ such that} \\
\qquad \langle Ax, x \rangle \leq -\omega |x|^2, \ x \in D(A), \\
(ii) \ A^{-1} \in L_1(H), \\
(iii) \ F \in B_b(H; H).
\end{array}
\right.
$$

$$(11.1.1)$$

We are concerned with the differential operator

$$N_2\varphi = L_2\varphi + \langle F(\cdot), D\varphi \rangle, \quad \varphi \in D(L_2).$$

Proposition 11.1.1 *N_2 is the infinitesimal generator of a strongly continuous semigroup (P_t^2) on $L^2(H, \mu)$. Moreover*

(i) the resolvent $R(\lambda, N_2)$ of N_2 is given by

$$R(\lambda, N_2) = R(\lambda, L_2)(1 - T_\lambda)^{-1}, \ \lambda > \lambda_0, \qquad (11.1.2)$$

where

$$T_\lambda\psi(x) = \langle F(x), DR(\lambda, L_2)\psi(x) \rangle, \ \psi \in L^2(H, \mu), \ x \in H, \quad (11.1.3)$$

and

$$\lambda_0 = \pi \|F\|_0^2 = \pi \sup_{x \in H} |F(x)|^2, \qquad (11.1.4)$$

(ii) $\mathcal{E}_A(H)$ is a core for N_2,

(iii) if $\varphi \in L^2(H, \mu)$ is nonnegative then $R(\lambda, N)\varphi$ is nonnegative for all $\lambda > 0$.

Proof. (i) Let $\lambda > \lambda_0$, $f \in L^2(H, \mu)$. Consider the equation

$$\lambda\varphi - L_2\varphi - \langle F(\cdot), D\varphi \rangle = f. \qquad (11.1.5)$$

Setting $\psi = \lambda\varphi - L_2\varphi$, equation (11.1.5) becomes

$$\psi - T_\lambda\psi = f, \tag{11.1.6}$$

where T_λ is defined by (11.1.3). Taking into account (10.3.3) and that we have $\|\Lambda_t\| \le t^{-1/2}$, $t > 0$, we find

$$\|T_\lambda\psi\|_{L^2(H,\mu)} \le \sqrt{\frac{\pi}{\lambda}} \, \|F\|_0 \, \|\psi\|_{L^2(H,\mu)}.$$

Therefore if $\lambda > \lambda_0$ equation (11.1.6) has a unique solution and (11.1.2) follows.

(ii) Let $\varphi \in D(N_2) = D(L_2)$. Since $\mathcal{E}_A(H)$ is a core for L_2, there exists a sequence $(\varphi_n) \subset \mathcal{E}_A(H)$ such that

$$\varphi_n \to \varphi, \quad L_2\varphi_n \to L_2\varphi \quad \text{in } L^2(H,\mu).$$

By Proposition 10.3.1 it follows that $\varphi \in W^{1,2}(H,\mu)$ and

$$\lim_{n\to\infty} \int_H |D\varphi - D\varphi_n|^2 d\mu = 0.$$

Consequently

$$\lim_{n\to\infty} N_2\varphi_n = L_2\varphi + \langle F, D\varphi\rangle = N_2\varphi,$$

and the conclusion follows.

Finally, (iii) follows from Proposition 6.6.4 when $F \in UC_b(H;H)$ and, then, when $F \in B_b(H;H)$, by approximating F by a sequence of functions $(F_n) \subset UC_b(H;H)$, such that $\|F_n\|_0 \le \|F\|_0$, and $F_n(x) \to F(x)$ as $n \to \infty$ μ-a.e. $\square$

We now consider the adjoint semigroup $(P_t^2)^*$ of (P_t^2); we denote by N_2^* its infinitesimal generator and by Σ^* the set of all its stationary points:

$$\Sigma^* = \left\{\varphi \in L^2(H,\mu) : (P_t^2)^*\varphi = \varphi, \, t \ge 0\right\}.$$

Lemma 11.1.2 $(P_t^2)^*$ *has the following properties.*

(i) *For all* $\varphi \ge 0$, μ-*a.e, one has* $(P_t^2)^*\varphi \ge 0$ μ-*a.e.*

(ii) Σ^* *is a lattice, that is if* $\varphi \in \Sigma^*$ *then* $|\varphi| \in \Sigma^*$.

Proof. Let $\varphi \ge 0$, μ-a.e. Then for all $\psi \ge 0$, μ-a.e, and all $t > 0$ we have

$$\int_H \psi \, (P_t^2)^*\varphi \, d\mu = \int_H P_t\psi \, \varphi \, d\mu \ge 0,$$

by Proposition 11.1.1(iii). By the arbitrariness of ψ this implies that $(P_t^2)^* \varphi \geq 0$ μ-a.e, and (i) is proved.

Let us prove (ii). Assume that $\varphi \in \Sigma^*$ so that $\varphi(x) = (P_t^2)^* \varphi(x)$. Then we have

$$|\varphi(x)| = |(P_t^2)^* \varphi(x)| \leq (P_t^2)^* (|\varphi|)(x). \tag{11.1.7}$$

We claim that
$$|\varphi(x)| = (P_t^2)^* (|\varphi|)(x), \quad \mu\text{-a.s.}$$

Assume in contradiction that there is a Borel subset $I \subset H$ such that $\mu(I) > 0$ and
$$|\varphi(x)| < (P_t^2)^* (|\varphi|)(x), \ x \in I.$$

Then we have

$$\int_H |\varphi(x)| \mu(dx) < \int_H (P_t^2)^* (|\varphi|)(x) \mu(dx). \tag{11.1.8}$$

On the other hand

$$\int_H (P_t^2)^* (|\varphi|) d\mu = \langle (P_t^2)^* (|\varphi|), 1 \rangle_{L^2(H,\mu)} = \langle |\varphi|, 1 \rangle_{L^2(H,\mu)} = \int_H |\varphi| d\mu,$$

which is in contradiction with (11.1.8). $\square$

We prove now a regularity result for the domain $D(N_2^*)$ of the infinitesimal generator of the adjoint semigroup $(P_t^2)^*$.

Proposition 11.1.3 *We have* $D(N_2^*) \subset W^{1,2}(H, \mu)$.

Proof. Let $\lambda > \pi \|F\|_0$ and $f \in L^2(H, \mu)$. Let us consider the bilinear form $b : W^{1,2}(H, \mu) \times W^{1,2}(H, \mu) \to \mathbb{R}$ defined as

$$b(\psi, v) = \lambda \int_H \psi v \, d\mu - \int_H \langle AQD\psi, Dv \rangle d\mu - \int_H \langle F, Dv \rangle \psi \, d\mu.$$

Clearly b is continuous since $AQ \in L(H)$, see [72], and

$$|b(\psi, v)| \leq \lambda \|\psi\|_{L^2(H,\mu)} \|v\|_{L^2(H,\mu)} + \|AQ\| \|D\psi\|_{L^2(H,\mu)} \|Dv\|_{L^2(H,\mu)}$$

$$+ \|F\|_0 \|Dv\|_{L^2(H,\mu)} \|\psi\|_{L^2(H,\mu)}.$$

Moreover b is coercive since, recalling the Lyapunov equation (10.1.5), we have

$$b(\psi, \psi) = \lambda \int_H \psi^2 d\mu + \frac{1}{2} \int_H |D\psi|^2 d\mu - \int_H \langle F, Dv \rangle \psi \, d\mu.$$

By the Lax-Milgram theorem there exists $\psi \in W^{1,2}(H, \mu)$ such that

$$b(\psi, v) = \int_H f v d\mu, \quad v \in W^{1,2}(H, \mu).$$

Choosing $v \in D(L_2)$ we have, due to (10.2.2)

$$\lambda \int_H \psi v d\mu - \int_H (N_2 v) \psi d\mu = \int_H f v d\mu.$$

Consequently $\psi \in D(N_2^*)$ and $\lambda \psi - N_2^* \psi = f$. The conclusion follows from the arbitrariness of f. $\square$

The following result is proved in G. Da Prato and J. Zabczyk [102].

Proposition 11.1.4 *There exists an invariant measure ν for (P_t^2) absolutely continuous with respect to μ. If ν_1 is another invariant measure for (P_t^2) absolutely continuous with respect to μ, then $\nu_1 = \nu$.*

Proof. Let $\lambda > 0$ be fixed and let φ_0 be the function identically equal to 1. Clearly $\varphi_0 \in D(N_2)$ and $N_2 \varphi_0 = 0$. Consequently $1/\lambda$ is an eigenvalue of $R(\lambda, N_2)$ since

$$R(\lambda, N_2) \varphi_0 = \frac{1}{\lambda} \varphi_0.$$

Moreover $1/\lambda$ is a simple eigenvalue because μ is ergodic. Since the embedding $W^{1,2}(H, \mu) \subset L^2(H, \mu)$ is compact and $D(L_2) \subset W^{1,2}(H, \mu)$ by Proposition 10.3.1, it follows that $R(\lambda, N_2)$ is compact as well for any $\lambda > 0$. Therefore $R(\lambda, N_2^*)$ is compact and $1/\lambda$ is a simple eigenvalue for $R(\lambda, N_2^*)$. Consequently there exists $\rho \in L^2(H, \mu)$ such that

$$R(\lambda, N_2^*) \rho = \frac{1}{\lambda} \rho. \tag{11.1.9}$$

It follows that $\rho \in D(N_2^*)$ and $N_2^* \rho = 0$. Since Σ^* is a lattice, ρ can be chosen to be nonnegative and such that $\int_H \rho d\mu = 1$.

Now set

$$\nu(dx) = \rho(x) \mu(dx), \quad x \in H.$$

We claim that ν is an invariant measure for (P_t^2). In fact taking the inverse Laplace transform in (11.1.9) we find

$$(P_t^2)^* \rho = \rho, \quad t \geq 0,$$

which implies for any $\varphi \in L^2(H, \mu)$

$$\int_H P_t^2 \varphi d\nu = \int_H P_t^2 \varphi \, \rho d\mu = \int_H \varphi (P_t^2)^* \rho d\mu = \int_H \varphi d\nu.$$

It remains to show uniqueness. Let ν_1 be another invariant measure of P_t^2, and assume that $\nu_1 << \mu$ and $\rho_1 = \frac{d\nu_1}{d\mu}$. Then we have $P_t^2 \rho_1 = \rho_1$, $t \geq 0$, and consequently $R(\lambda, N_2)\rho_1 = \frac{1}{\lambda} \rho_1$. Therefore $\rho = \rho_1$ since $1/\lambda$ is a simple eigenvalue of $R(\lambda, N_2^*)$. $\square$

Remark 11.1.5 If $(P_t^2)^*1 = 1$ then $\nu = \mu$.

Remark 11.1.6 One can show that P_t is irreducible and strong Feller, so that the invariant measure ν is unique, see S. Peszat and J. Zabczyk [184].

We now study the regularity of the density ρ. First notice that, since $\rho \in D(N_2^*)$, by Proposition 11.1.3 we have that $\rho \in W^{1,2}(H, \mu)$.

The following result was proved in V. Bogachev, G. Da Prato and M. Röckner [19].

Proposition 11.1.7 *We have*

$$\frac{1}{2} \int_H |D\rho|^2 d\mu = \int_H \langle F, D\rho \rangle \rho \, d\mu. \qquad (11.1.10)$$

Moreover $\sqrt{\rho} \in W^{1,2}(H, \mu)$ *and we have*

$$2 \int_H |D\sqrt{\rho}|^2 \, d\mu \leq \int_H |\langle F, D\rho \rangle| d\mu. \qquad (11.1.11)$$

Proof. Since ν is an invariant measure for (P_t^2) we have

$$0 = \int_H N_2 \varphi \, \rho \, d\mu = \int_H L_2 \varphi \, \rho \, d\mu + \int_H \langle F, D\varphi \rangle \rho \, d\mu, \quad \varphi \in D(L).$$

Thus, since $\rho \in W^{1,2}(H, \mu)$ we have by Proposition 10.2.3

$$\frac{1}{2} \int_H \langle D\rho, D\varphi \rangle d\mu = \int_H \langle F, D\varphi \rangle \rho d\mu, \qquad (11.1.12)$$

for all $\varphi \in D(L_2)$. Since $D(L_2)$ is dense in $W^{1,2}(H, \mu)$ we can conclude that (11.1.12) holds for all $\varphi \in W^{1,2}(H, \mu)$. Finally, setting $\varphi = \rho$ we obtain (11.1.10).

Moreover, setting in (11.1.12) $\varphi = \log(\rho + \varepsilon)$, with $\varepsilon > 0$, and again using the Lyapunov equation , we find

$$2 \int_H |D\sqrt{\rho + \varepsilon}|^2 d\mu = \int_H \langle F, D\rho \rangle \frac{\rho}{\rho + \varepsilon} d\mu \leq \int_H |\langle F, D\rho \rangle| d\mu.$$

Now (11.1.11) follows letting ε tend to zero. $\square$

Corollary 11.1.8 *We have*

$$\int_H |D\sqrt{\rho}|^2 \, d\mu \le \int_H |F|^2 \rho d\mu. \tag{11.1.13}$$

Proof. Since $D\rho = 2\sqrt{\rho}\, D(\sqrt{\rho})$, we have

$$\int_H |\langle F, D\rho \rangle| d\mu = 2 \int_H |\langle F, D(\sqrt{\rho}) \rangle| |D(\sqrt{\rho}) d\mu.$$

Consequently, by the Hölder inequality we have

$$\left(\int_H |\langle F, D\rho \rangle| d\mu \right)^2 \le 4 \int_H |F|^2 \rho d\mu \int_H |D(\sqrt{\rho})|^2 d\mu.$$

Now the conclusion follows from (11.1.11). $\square$

When F is sufficiently regular we can give an explicit expression for N_2^*. Let $F \in C_b(H;H)$. We say that F has finite *divergence* if for any $x \in H$ the series

$$\text{div } F(x) := \sum_{k=1}^{\infty} D_k F_k(x), \ x \in H,$$

where $F_k(x) = \langle F(x), e_k \rangle$, is convergent and moreover $\text{div } F \in C_b(H)$. If in addition the function

$$Q(X) \to H, \ x \to \langle Q^{-1}x, F(x) \rangle,$$

is extendible to a uniformly continuous and bounded function, we say that F has finite divergence with respect to μ, and we set

$$\text{div}_\mu F(x) = \text{div } F(x) - \langle Q^{-1}x, F(x) \rangle, \ x \in H.$$

The following result is an easy consequence of the integration by parts formula (9.2.1)

Lemma 11.1.9 *Assume that $F \in C_b^1(H;H)$ has finite divergence with respect to μ. Then for any $\varphi, \psi \in W^{1,2}(H, \mu)$ we have*

$$\int_H \langle F, D\varphi \rangle \psi d\mu = - \int_H \varphi \langle F, D\psi \rangle d\mu - \int_H \text{div}_\mu F(x) \, \varphi\psi d\mu. \tag{11.1.14}$$

Proof. It is enough to consider $\varphi, \psi \in \mathcal{E}_A(H)$. In this case taking into account (9.2.1) we find

$$\int_H \langle F, D\varphi \rangle \psi d\mu = \sum_{k=1}^{\infty} \int_H F_k D_k \varphi \, \psi d\mu$$

$$= -\sum_{k=1}^{\infty} \int_H \varphi \left(D_k F_k \psi + F_k D_k \psi \right) d\mu + \sum_{k=1}^{\infty} \frac{1}{\lambda_k} \int_H x_k F_k \varphi \psi d\mu$$

$$= -\int_H \varphi \psi \operatorname{div} F \, d\mu + \int_H \langle F, D\psi \rangle \varphi d\mu - \int_H \langle Q^{-1} x, F \rangle \varphi \psi d\mu,$$

and the conclusion follows. $\square$

Proposition 11.1.10 *Assume that $F \in C_b^1(H; H)$ has finite divergence with respect to μ. Then $D(N_2^*) = D(L_2)$ and, for any $\psi \in D(N_2^*) = D(L_2)$, we have*

$$N_2^* \psi = L_2 \psi - \langle F(\cdot), D\psi \rangle - \operatorname{div}_\mu F(\cdot) \, \psi. \qquad (11.1.15)$$

Proof. Let $\varphi, \psi \in D(L_2)$. Then, taking into account (11.1.14), we find

$$\int_H N_2 \varphi \, \psi d\mu = \int_H L_2 \varphi \, \psi d\mu + \int_H \langle F, D\varphi \rangle \psi d\mu$$

$$= \int_H \varphi \, L_2 \psi d\mu - \int_H \varphi \langle F, D\psi \rangle d\mu - \int_H \operatorname{div}_\mu F \varphi \psi d\mu. \square$$

Now, recalling the characterization of the domain of L_2 given in §10.2.1, we obtain the result

Corollary 11.1.11 *Under the assumptions of Proposition 11.1.10 we have $\rho \in D(L_2)$. Moreover $\rho \in W^{2,2}(H, \mu)$ and $|(-A)^{1/2} D\rho| \in L^2(H, \mu)$.*

11.2 Lipschitz perturbations

We assume here that

$$\begin{cases} (i) \quad A : D(A) \subset H \to H \text{ is self-adjoint and there exists } \omega > 0 \text{ such that} \\ \qquad\qquad \langle Ax, x \rangle \leq -\omega |x|^2, \ x \in D(A), \\ (ii) \quad A^{-1} \in L_1(H), \\ (iii) \quad F \text{ is Lipschitz continuous and dissipative} \\ \qquad\qquad \langle F(x) - F(y), x - y \rangle \leq 0, \ x, y \in H. \end{cases}$$

$$(11.2.1)$$

Let us consider the differential operator

$$N_0\varphi(x) = \frac{1}{2}\mathrm{Tr}[D^2\varphi(x)] + \langle x, AD\varphi(x)\rangle + \langle F(x), D\varphi(x)\rangle, \quad x \in H, \varphi \in \mathcal{E}_A(H),$$

and let us introduce a transition semigroup (P_t) that will be naturally related to N_0. Specifically we set

$$P_t\varphi(x) = \mathbb{E}[\varphi(X(t,x))], \quad \varphi \in UC_b(H), \tag{11.2.2}$$

where $X(t,x)$ is the solution of the differential stochastic equation

$$X(t) = e^{tA}x + \int_0^t e^{(t-s)A}F(X(s))ds + \int_0^t e^{(t-s)A}dW(s), \tag{11.2.3}$$

studied in Chapter 7. It is well known that equation (11.2.3) can be reduced to a family of deterministic equations by the change of variable $Y(t) = X(t,x) - W_A(t)$, where

$$W_A(t) = \int_0^t e^{(t-s)A}dW(s), \ t \geq 0.$$

In this way we obtain

$$Y'(t) = AY(t) + F(Y(t) + W_A(t)), \quad Y(0) = x. \tag{11.2.4}$$

Lemma 11.2.1 *For any $m \in \mathbb{N}$ there is $C_m > 0$, depending only on A and $\|F\|_1$, such that*

$$\mathbb{E}|X(t,x)|^{2m} \leq C_m(1 + e^{-m\omega t}|x|^{2m}), \quad x \in H, \ t \geq 0. \tag{11.2.5}$$

Proof. Multiplying (11.2.4) by $|Y(t)|^{2m-2}Y(t)$ and taking into account assumptions (11.2.1) gives, for a suitable constant C_m^1,

$$\frac{1}{2m}\frac{d}{dt}|Y(t)|^{2m} \leq -\omega|Y(t)|^{2m} + \langle F(W_A), Y(t)\rangle|Y(t)|^{2m-2}$$

$$+\langle F(Y(t) + W_A(t)) - F(W_A(t)), Y(t)\rangle|Y(t)|^{2m-2}$$

$$\leq -\omega|Y(t)|^{2m} + \langle F(W_A), Y(t)\rangle|Y(t)|^{2m-2}$$

$$\leq -\frac{\omega}{2}|Y(t)|^{2m} + C_m^1|F(W_A)|^{2m}.$$

By the Gronwall lemma it follows that

$$|Y(t)|^{2m} \leq e^{-m\omega t}|x|^{2m} + 2mC_m^1 \int_0^t e^{-m\omega(t-s)}|F(W_A(s))|^{2m}ds,$$

and finally, for some constant C_m^2

$$|X(t,x)|^{2m} \leq C_m^2 e^{-m\omega t}|x|^{2m}$$

$$+ C_m^3 \left(\int_0^t e^{-m\omega(t-s)}|F(W_A(s))|^{2m}ds + |W_A(t)|^{2m} \right).$$

$$(11.2.6)$$

Now the conclusion follows taking expectation and recalling that F is sublinear and

$$\sup_{t \geq 0} \mathbb{E}|W_A(t)|^{2m} < +\infty. \quad \square$$

We now show, following G. Da Prato and J. Zabczyk [102], the existence of an invariant measure ν for the semigroup (P_t). Then we study properties of (P_t) on $L^2(H, \nu)$.

To prove the existence of an invariant measure it is convenient to introduce the solution $X(t, s, x)$, $t \geq s$, $x \in H$, of the equation

$$X(t,s,x) = e^{(t-s)A}x + \int_s^t e^{(t-u)A}F(X(u,s,x))du + W_{A,s}(t), \quad (11.2.7)$$

where

$$W_{A,s}(t) = \int_s^t e^{(t-u)A}dW(u). \quad (11.2.8)$$

As in [102] we will show that there exists the limit in $L^2(\Omega, \mathcal{F}, \mathbb{P}; H)$

$$\zeta = \lim_{s \to +\infty} X(0, -s, x),$$

and that the law of ζ is the required invariant measure.

Lemma 11.2.2 *There exists $\zeta \in L^2(\Omega, \mathcal{F}, \mathbb{P}; H)$ such that*

$$\lim_{s \to +\infty} X(0, -s, x) = \eta \quad in \quad L^2(\Omega, \mathcal{F}, \mathbb{P}; H). \quad (11.2.9)$$

Proof. Let $x \in H$ be fixed, and set $X_s(t) = X(t, -s, x)$, $t \geq -s$, and $Y_s(t) = X_s(t) - W_{A,-s}(t)$. Then we find

$$Y_s(t) = e^{(t+s)A} x + \int_{-s}^t e^{(t-\sigma)A} F(Y_s(\sigma)) + W_{A,-s}(\sigma)) \, d\sigma.$$

Therefore $Y_s(t)$ is the mild solution of the following initial value problem:

$$\begin{cases} D_t Y_s(t) = A Y_s(t) + F(Y_s(t) + W_{A,-s}(t)), \\ Y(-s) = x. \end{cases} \qquad (11.2.10)$$

In the following we assume that $Y_s(t)$ is a strict solution to (11.2.10); otherwise we use an approximation of $Y_s(t)$ by strict solutions. We divide the remainder of the proof into three steps.

Step 1. There exists $C_1 > 0$ such that $\mathbb{E}\left(|Y_s(t)|^2\right) \leq C_1$ for any $t > -s$.

We have in fact

$$\begin{aligned}
\frac{1}{2}\frac{d}{dt}|Y_s(t)|^2 &= \langle A Y_s(t), Y_s(t) \rangle + \langle F(Y_s(t) + W_{A,-s}(t)), Y_s(t) \rangle \\[2mm]
&= \langle A Y_s(t), Y_s(t) \rangle + \langle F(Y_s(t) + W_{A,-s}(t)) \\[2mm]
&\quad -F(W_{A,-s}(t)), Y_s(t) \rangle + \langle F(W_{A,-s}(t)), Y_s(t) \rangle \\[2mm]
&\leq -\omega |Y_s(t)|^2 + |F(W_{A,-s}(t))| \, |Y_s(t)| \\[2mm]
&\leq -\omega |Y_s(t)|^2 + \frac{|\omega|}{2} |Y_s(t)|^2 + \frac{2}{|\omega|} |F(W_{A,-s}(t))|^2 \\[2mm]
&\leq -\frac{\omega}{2} |Y_s(t)|^2 + \frac{2}{|\omega|} (a + b|W_{A,-s}(t)|^2),
\end{aligned}$$

where a, b have been chosen such that $|F(x)|^2 \leq a + b|x|^2$, $x \in H$. It follows that

$$|Y_s(t)|^2 \leq e^{-\omega(s+t)}|x|^2 + \frac{4}{|\omega|} \int_{-s}^t e^{-\omega(t-\sigma)}(a + b|W_{A,-s}(\sigma)|^2) d\sigma.$$

Now the claim follows from the estimate

$$\mathbb{E}|W_{A,-s}(t)|^2 = \int_{-s}^t \mathrm{Tr}[e^{2\sigma A}] d\sigma \leq \mathrm{Tr}\, Q.$$

Step 2. There exists $\eta_x \in L^2(\Omega; H)$ such that $\lim_{s \to +\infty} X(0, -s, x) = \eta_x$ in $L^2(\Omega; H)$.

Let $s > s_1$, then we have

$$\begin{cases} D_t(X_s - X_{s_1}) = A(X_s - X_{s_1}) + F(X_s) - F(X_{s_1}), \\ (X_s - X_{s_1})(-s_1) = X_s(-s_1) - x. \end{cases}$$

It follows that

$$\frac{1}{2}\frac{d}{dt}|X_s - X_{s_1}|^2 \leq -\omega|X_s - X_{s_1}|^2,$$

and so

$$|X_s(0) - X_{s_1}(0)|^2 \leq e^{-2\omega s}|X_s(-s_1) - x|^2.$$

Recalling Step 1, we see that there exists a constant $C_2 > 0$ such that

$$\mathbb{E}\left(|X_s(0) - X_{s_1}(0)|^2\right) \leq C_2 e^{-2\omega s}.$$

Consequently $(X_s(0))$ is a Cauchy sequence and Step 2 is proved.

Step 3. We have that η_x is independent of x.

Let $x, y \in H$, and set

$$\rho_s(t) = X_s(0, -s, x) - X_s(0, -s, y).$$

Then we have

$$\begin{cases} D_t\rho_s(t) = A\rho_s(t) + F(X(0, -s, x)) - F(X(0, -s, y)), \\ \rho_s(-s) = x - y. \end{cases}$$

It follows that

$$\frac{1}{2}\frac{d}{dt}|\rho_s(t)|^2 \leq -\omega|\rho_s(t)|^2,$$

which yields

$$|\rho_s(t)| \leq |x - y|e^{-\omega(t+s)}.$$

As $s \to +\infty$ we obtain $\eta_x = \eta_y$ as required. $\square$

We can now prove the following result.

Theorem 11.2.3 *There exists a unique invariant measure ν for (P_t) and for any $\varphi \in C_b(H)$ and $x \in H$ there exists the limit*

$$\lim_{t \to +\infty} P_t\varphi(x) = \int_H \varphi(y)\mu(dy). \tag{11.2.11}$$

Thus ν is ergodic and strongly mixing.

Proof. Denote by ν the law of ζ. Given $\varphi \in C_b(H)$ we have

$$P_t\varphi(x) = \mathbb{E}[\varphi(X(t,0,x))] = \mathbb{E}[\varphi(X(0,-t,x))].$$

Consequently, by the dominated convergence theorem, it follows that

$$\lim_{t\to+\infty} P_t\varphi(x) = \mathbb{E}[\varphi(\zeta)] = \int_H \varphi(y)\nu(dy). \tag{11.2.12}$$

If $s > 0$ and $x_0 \in H$ it follows that

$$\int_H P_s\varphi(y)\mu(dy) = \lim_{t\to+\infty} P_{t+s}\varphi(x_0) = \int_H \varphi(y)\mu(dy),$$

so that μ is invariant. The last statement is left to the reader. $\square$

By Theorem 11.2.3 and Lemma 11.2.1 we have the following result.

Proposition 11.2.4 *Assume that assumptions (11.2.1) hold. Then for any $m \in \mathbb{N}$ there exists $c_m > 0$, depending only on A and $\|F\|_1$, such that*

$$\int_H |x|^{2m}\nu(dx) \le c_m. \tag{11.2.13}$$

Proof. Denote by $\nu_{t,x}$ the law of $X(t,x)$. Then by Lemma 11.2.1 it follows that for any $\beta > 0$,

$$\int_H \frac{|y|^{2m}}{1+\beta|y|^{2m}}\,\nu_{t,x}(dy) \le \int_H |y|^{2m}\,\nu_{t,x}(dy) \le C_m(1+e^{-m\omega t}|x|^{2m}), \quad x \in H.$$

Consequently, letting t tend to ∞ we find, taking into account (11.2.11),

$$\int_H \frac{|y|^{2m}}{1+\beta|y|^{2m}}\,\nu(dy) \le C_m,$$

which yields (11.2.13). $\square$

With the same proof as for Theorem 10.1.5 we see that the semigroup (P_t) can be uniquely extended to a strongly continuous semigroup of contractions in $L^p(H,\nu)$, $p \ge 1$. We shall still denote by (P_t) the extension and by N_p the corresponding infinitesimal generator.

11.2.1 Some additional results on the Ornstein-Uhlenbeck semigroup

In this subsection we prove, for further use, some properties of the Ornstein-Uhlenbeck semigroup

$$R_t\varphi(x) = \int_H \varphi(e^{tA}x + y)N_{Q_t}(dy), \quad t \geq 0,\ x \in H,$$

where N_{Q_t} is the Gaussian measure on H with mean 0 and covariance operator Q_t, and φ is a continuous function from H into $\mathbb{R}$ having *polynomial growth*.

Let us introduce the space $C_{b,2}(H)$ of all mappings $\varphi : H \to \mathbb{R}$ such that the function $H \to \mathbb{R}$, $x \to \frac{\varphi(x)}{1+|x|^2}$, belongs to $C_b(H)$. $C_{b,2}(H)$, endowed with the norm

$$\|\varphi\|_{b,2} := \sup_{x \in H} \frac{|\varphi(x)|}{1 + |x|^2},$$

is a Banach space.

We define $C_{b,2}^1(H)$ as the space of all continuously differentiable functions of $C_{b,2}(H)$ such that

$$\|D\varphi\|_{b,2} := \sup_{x \in H} \frac{|D\varphi(x)|}{1 + |x|^2} < +\infty.$$

The following result has been proved by S. Cerrai; see [36].

Proposition 11.2.5 R_t *maps* $C_{b,2}(H)$ *into itself for all* $t \geq 0$ *and*

$$\|R_t\varphi\|_{b,2} \leq (1 + \operatorname{Tr} Q_\infty)\,\|\varphi\|_{b,2}. \tag{11.2.14}$$

Moreover for all $\varphi \in C_{b,2}(H)$ *and any* $t > 0$ *we have* $R_t\varphi \in C_{b,2}^1(H)$ *and*

$$\|DR_t\varphi\|_{b,2} \leq 8t^{-1/2} \left(1 + 2\operatorname{Tr} Q_\infty^2 + [\operatorname{Tr} Q_\infty]^2\right)^{1/2} \|\varphi\|_{b,2}. \tag{11.2.15}$$

Proof. For any $x \in H$ we have

$$\frac{|R_t\varphi(x)|}{1+|x|^2} \leq \|\varphi\|_{b,2} \int_H \frac{1+|e^{tA}x+y|^2}{1+|x|^2} N_{Q_t}(dy)$$

$$= \|\varphi\|_{b,2} \int_H \frac{1+|e^{tA}x|^2+|y|^2}{1+|x|^2} N_{Q_t}(dy)$$

$$\leq \|\varphi\|_{b,2} \int_H \frac{1+|x|^2+|y|^2}{1+|x|^2} N_{Q_t}(dy)$$

$$\leq \|\varphi\|_{b,2} \left(1 + \int_H |y|^2 N_{Q_t}(dy)\right),$$

$$\leq \|\varphi\|_{b,2} (1 + \operatorname{Tr} Q_t),$$

and (11.2.14) follows. To prove (11.2.15) we notice that, from the Cameron-Martin formula, we have

$$\langle DR_t\varphi(x), h\rangle = \int_H \langle \Lambda_t h, Q_t^{-1/2}y\rangle \varphi(e^{tA}x+y)N_{Q_t}(dy), \quad h \in H,$$

where $\Lambda_t = e^{tA}Q_t^{-1/2}$ fulfills $\|\Lambda_t\| \leq t^{-1/2}$, $t > 0$. By the Hölder inequality it follows that

$$|\langle DR_t\varphi(x), h\rangle|^2 \leq \int_H |\langle \Lambda_t h, Q_t^{-1/2}y\rangle|^2 N_{Q_t}(dy)$$

$$\times \int_H |\varphi(e^{tA}x+y)|^2 N_{Q_t}(dy)$$

$$= |\Lambda_t h|^2 \int_H |\varphi(e^{tA}x+y)|^2 N_{Q_t}(dy).$$

Consequently

$$|DR_t\varphi(x)|^2 \leq \|\Lambda_t\|^2 \int_H |\varphi(e^{tA}x+y)|^2 N_{Q_t}(dy), \quad x \in H.$$

It follows that

$$\frac{|DR_t\varphi(x)|^2}{(1+|x|^2)^2} \leq \|\Lambda_t\|^2\|\varphi\|_{b,2}^2 \int_H \frac{(1+|e^{tA}x+y|^2)^2}{(1+|x|^2)^2}N_{Q_t}(dy)$$

$$\leq 8\|\Lambda_t\|^2\|\varphi\|_{b,2}^2 \int_H \frac{(1+|x|^2)^2+|y|^4}{(1+|x|^2)^2}N_{Q_t}(dy)$$

$$\leq 8\|\Lambda_t\|^2\|\varphi\|_{b,2}^2 \int_H (1+|y|^4)N_{Q_t}(dy),$$

and the conclusion follows. □

R_t is not a strongly continuous semigroup on $C_{b,2}(H)$. Its infinitesimal generator can be defined through its Laplace transform as in S. Cerrai [35]:

$$F(\lambda)f(x) = \int_0^{+\infty} e^{-\lambda t}R_t f(x)dt, \ f \in C_{b,2}(H), \ \lambda > 0.$$

It is easy to check that $F(\lambda)$ maps $C_{b,2}(H)$ into itself for all $\lambda > 0$ and that $F(\lambda)$ is a pseudo-resolvent. Consequently there exists a unique closed operator $L_{b,2}$ in $C_{b,2}(H)$ such that

$$R(\lambda, L_{b,2}) = (\lambda - L_{b,2})^{-1} = F(\lambda), \ \lambda > 0.$$

$L_{b,2}$ is called the *infinitesimal generator* of R_t on $C_{b,2}(H)$.

Remark 11.2.6 By E. Priola [188] it follows that $\varphi \in D(L_{b,2})$ and $L_{b,2}\varphi = \psi$ if and only if $\psi \in C_{b,2}(H)$ and

(i) we have

$$\lim_{h\to 0}\frac{1}{h}(R_h\varphi(x) - \varphi(x)) = \psi(x), \ x \in H,$$

(ii) $\sup_{h>0}\frac{1}{h}\|R_h\varphi - \varphi\|_{C_{b,2}(H)} < +\infty.$

The following result is an immediate consequence of (11.2.15).

Proposition 11.2.7 *We have*

$$D(L_{b,2}) \subset C_{b,2}^1(H), \tag{11.2.16}$$

with continuous embedding. Moreover if $f \in C_{b,2}(H)$, $\lambda > 0$ *and* $\varphi = R(\lambda, L_{b,2})f$, *we have*

$$\|D\varphi\|_{b,2} \leq \sqrt{\frac{\pi}{\lambda}}\left(1 + 2\,\mathrm{Tr}\,Q_\infty^2 + [\mathrm{Tr}\,Q_\infty]^2\right)^{1/2}\|f\|_{b,2}. \tag{11.2.17}$$

By an explicit verification one can check that $\mathcal{E}_A(H) \subset D(L_{b,2})$ and

$$L_{b,2}\varphi = \frac{1}{2}\text{Tr}[D^2\varphi] + \langle x, A^*D\varphi\rangle, \quad \varphi \in \mathcal{E}_A(H). \tag{11.2.18}$$

Remark 11.2.8 It is easy to check that, for any $\varphi \in \mathcal{E}_A(H)$, we have

$$\lim_{t\to 0} R_t\varphi = \varphi \text{ in } C_{b,2}(H), \quad \lim_{t\to 0} R_t L_{b,2}\varphi = L_{b,2}\varphi \text{ in } C_{b,2}(H). \tag{11.2.19}$$

This fact will be used in the proof of Proposition 11.2.10 below. This is the reason for working in the space $C_{b,2}(H)$.

We need some results on approximation by functions in $\mathcal{E}_A(H)$, collected in Lemma 11.2.9 and Proposition 11.2.10, see G. Da Prato and L. Tubaro [99].

Lemma 11.2.9 *For any* $\varphi \in C_{b,2}(H)$ *there exists a multi-sequence* $(\varphi_n) = (\varphi_{n_1,n_2,n_3}) \subset \mathcal{E}_A(H)$ *such that*

$$\lim_{n\to\infty} \varphi_n(x) := \lim_{n_1\to\infty} \lim_{n_2\to\infty} \lim_{n_3\to\infty} \varphi_{n_1,n_2,n_3}(x) = \varphi(x), \quad x \in H, \tag{11.2.20}$$

and

$$\|\varphi_n\|_{b,2} \leq \|\varphi\|_{b,2}. \tag{11.2.21}$$

Proof. Let $\varphi \in C_{b,2}(H)$ and let $(P_{n_1})_{n_1\in\mathbb{N}}$ be a sequence of finite dimensional projection operators on H strongly convergent to the identity. Then, for each $n_1 \in N$ there exists a sequence $(\varphi_{n_1,n_2})_{n_2\in\mathbb{N}} \subset \mathcal{E}(H)$ such that

$$\lim_{n_2\to\infty} \varphi_{n_1,n_2}(x) = \varphi(P_{n_1}x), \quad x \in H,$$

and

$$\frac{|\varphi_{n_1,n_2}(x)|}{1 + |P_{n_1}x|^2} \leq \frac{|\varphi(P_{n_1}x)|}{1 + |P_{n_1}x|^2} \leq \|\varphi\|_{b,2}.$$

Hence we have

$$\frac{|\varphi_{n_1,n_2}(x)|}{1 + |x|^2} \leq \|\varphi\|_{b,2}.$$

Set finally

$$\varphi_{n_1,n_2,n_3}(x) = \varphi_{n_1,n_2}(n_3(n_3 - A^*)^{-1}x), \quad x \in H.$$

Then $\varphi_n = \varphi_{n_1,n_2,n_3} \subset \mathcal{E}_A(H)$, $\lim_{n\to\infty} \varphi_n(x) = \varphi(x), \quad x \in H$, and

$$\frac{|\varphi_{n_1,n_2,n_3}(x)|}{1 + |x|^2} = \frac{|\varphi_{n_1,n_2}(n_3(n_3 - A^*)^{-1}x)|^2}{1 + |n_3(n_3 - A^*)^{-1}x|^2} \frac{1 + |n_3(n_3 - A^*)^{-1}x|^2}{1 + |x|^2}$$

$$\leq \|\varphi_{n_1,n_2}\|_{b,2} \leq \|\varphi\|_{b,2}.$$

Therefore the multi-sequence (φ_{n_1,n_2,n_3}) fulfills (11.2.20) and (11.2.21) as required. $\square$

Proposition 11.2.10 *For any* $\varphi \in D(L_{b,2})$ *there exists a multi-sequence* $(\varphi_n) = (\varphi_{n_1,n_2,n_3,n_4}) \subset \mathcal{E}_A(H)$ *and* $C_\varphi > 0$ *such that for all* $x \in H$

$$\lim_{n\to\infty} \varphi_n(x) = \varphi(x), \quad \lim_{n\to\infty} L_{b,2}\varphi_n(x) = L_{b,2}\varphi(x), \quad \lim_{n\to\infty} D\varphi_n(x) = D\varphi(x)$$

$$(11.2.22)$$

and

$$\|\varphi_n\|_{b,2} + \|L_{b,2}\varphi_n\|_{b,2} + \|D\varphi_n\|_{b,2} \leq C_\varphi. \qquad (11.2.23)$$

Proof. Set $f = \varphi - L\varphi$ and let $(f_n) = (f_{n_1,n_2,n_3}) \subset \mathcal{E}_A(H)$ be a multi-sequence fulfilling (11.2.20) and (11.2.21) (with φ replaced by f). Setting $\varphi_n = (1 - L_{b,2})^{-1} f_n$, we have, taking into account Proposition 11.2.7,

$$\lim_{n\to\infty} \varphi_n(x) \quad = \quad \varphi(x), \quad x \in H,$$

$$\lim_{n\to\infty} L_{b,2}\varphi_n(x) \quad = \quad L_{b,2}\varphi(x), \quad x \in H,$$

$$\lim_{n\to\infty} D\varphi_n(x) \quad = \quad D\varphi(x), \quad x \in H,$$

and

$$\|\varphi_n\|_{b,2} \quad \leq \quad \|f\|_{b,2} \leq (\|\varphi\|_{b,2} + \|L_{b,2}\varphi\|_{b,2}),$$

$$\|L_{b,2}\varphi_n\|_{b,2} \quad \leq \quad (2\|\varphi\|_{b,2} + \|L_{b,2}\varphi\|_{b,2}).$$

Notice that φ_n does not belong to $\mathcal{E}_A(H)$ in general. So we introduce now a further approximation of φ_n. Set for any $M, N \in \mathbb{N}$

$$\varphi_{n,M,N}(x) = \frac{1}{M} \sum_{h=1}^{N} \sum_{k=1}^{M} e^{-(h+\frac{k}{M})} R_{h+\frac{k}{M}} f_n(x),$$

so that

$$|\varphi_{n,M,N}(x)| \leq \|f\|_{b,2}$$

and

$$L_{b,2}\varphi_{n,M,N}(x) = \frac{1}{M} \sum_{h=0}^{N} \sum_{k=1}^{M} e^{-(h+\frac{k}{M})} R_{h+\frac{k}{M}} L_{b,2} f_n(x).$$

Now, by Remark 11.2.8 it follows that $R_t L_{b,2} f_n$ is continuous on t in $C_{b,2}(H)$. Therefore for any $n = (n_1, n_2, n_3)$ we have

$$\lim_{M,N\to\infty} \left\| \int_0^{+\infty} e^{-t} R_t L_{b,2} f_n dt - \frac{1}{M} \sum_{h=1}^{N} \sum_{k=1}^{M} e^{-(h+\frac{k}{M})} R_{h+\frac{k}{M}} L_{b,2} f_n \right\|_{b,2} = 0.$$

Therefore for any $\varepsilon \in (0,1]$ there exist $M_\varepsilon, N_\varepsilon$ such that

$$\|L_{b,2}\varphi_n - L_{b,2}\varphi_{n,M_\varepsilon,N_\varepsilon}\|_{b,2} \leq \varepsilon.$$

Consequently

$$\lim_{\varepsilon \to 0} L_{b,2}\varphi_{n,M_\varepsilon,N_\varepsilon}(x) = L_{b,2}\varphi_n(x),$$

and

$$\|L_{b,2}\varphi_{n,M_\varepsilon,N_\varepsilon}\| \leq \|L_{b,2}\varphi_n\|_{b,2} + \varepsilon.$$

Now the conclusion follows. $\square$

11.2.2   The semigroup (P_t) in $L^p(H,\nu)$

We are here concerned with the contraction semigroup (P_t) defined by
(11.2.2) on $L^p(H,\nu)$, $p \geq 1$. We denote by N_p its infinitesimal generator.

We are going to show, following G. Da Prato [80], that N_p is the closure
of the linear operator N_0 defined by

$$N_0\varphi = L_0\varphi + \langle F(\cdot), D\varphi\rangle, \quad \varphi \in \mathcal{E}_A(H). \tag{11.2.24}$$

Notice that for $\varphi(x) = \varphi_h(x) = e^{i\langle h,x\rangle}$, $x \in H$, with $h \in D(A^*)$, we have

$$N_0\varphi_h(x) = \left[i\langle x, A^*h\rangle + i\langle F(x), h\rangle - \frac{1}{2}|h|^2\right]\varphi_h, \quad x \in H. \tag{11.2.25}$$

First we show that N_p is an extension of N_0.

Proposition 11.2.11 Let $\varphi \in \mathcal{E}_A(H)$. Then $\varphi \in D(N_p)$ and $N_p\varphi = N_0\varphi$.

Proof. Let $\varphi \in \mathcal{E}_A(H)$ and $p \geq 1$. By the Itô formula (see Chapter 7) we
have

$$\lim_{h \to 0} \frac{1}{h}(P_h\varphi(x) - \varphi(x)) = \lim_{h \to 0} \frac{1}{h}\int_0^h P_s N_0\varphi(x)ds = N_0\varphi(x), \quad x \in H.$$

Moreover, taking into account the invariance of ν, we have

$$\int_H \frac{1}{h^p}|P_h\varphi(x) - \varphi(x)|^p \nu(dx) = \int_H \left|\int_0^h P_s N_0\varphi(x)\frac{ds}{h}\right|^p \nu(dx)$$

$$\leq \int_0^h \int_H |P_s N_0\varphi(x)|^p \frac{ds}{h} \nu(dx) = \int_H |N_0\varphi(x)|^p\nu(dx).$$

Since $N_0\varphi$ has linear growth and (11.2.13) holds, it follows by the dominated convergence theorem that

$$\lim_{h\to 0} \frac{1}{h}\left(P_h\varphi - \varphi\right) = N_0\varphi, \text{ in } L^p(H,\nu). \quad \square$$

Obviously N_0 is dissipative in $L^p(H,\nu)$ and consequently closable, let us denote by $\overline{N}_p$ its closure. The main result of this subsection is that $N_p = \overline{N}_p$. To prove this result we need a lemma.

Lemma 11.2.12 *We have* $D(L_{b,2}) \subset D(\overline{N}_p)$ *and for all* $\varphi \in D(L_{b,2})$

$$\overline{N}_p\varphi(x) = L_{b,2}\varphi(x) + \langle F(x), D\varphi(x)\rangle, \ x \in H. \tag{11.2.26}$$

Proof. Let us choose a multi-sequence $(\varphi_n) \subset \mathcal{E}_A(H)$ such that (11.2.22)-(11.2.23) hold. Then for any $x \in H$ we have

$$N_0\varphi_n(x) = L_{b,2}\varphi_n(x) + \langle F(x), D\varphi_n(x)\rangle \to L_0\varphi(x) + \langle F(x), D\varphi(x)\rangle,$$

as $n \to \infty$. Taking into account (11.2.25), Proposition 11.2.10 and the Lipschitzianity of F, we see that there is a constant $C_1 > 0$ such that

$$|N_0\varphi_n(x)| \le C_1(1 + |x|^3), \ x \in H.$$

Now by (11.2.13) and the dominated convergence theorem it follows that $N_0\varphi_n \to L_0\varphi + \langle F(x), D\varphi\rangle$ in $L^p(H,\nu)$. $\square$

We first consider the case when $F \in C_G^1(H;H)$, the subspace of $C^1(H;H)$ of all Gateaux differentiable functions having weakly continuous Gateaux derivatives.

Proposition 11.2.13 *Assume, besides (11.2.1), that* $F \in C_G^1(H;H)$. *Then* N_p *is the closure of* N_0 *in* $L^p(H,\nu)$.

Proof. Let $f \in C_b^1(H)$, $\lambda > 0$ and set

$$\varphi(x) = \int_0^{+\infty} e^{-\lambda t} P_t f(x)dt, \ x \in H.$$

Since $F \in C_G^1(H;H)$, then there exists the Gateaux derivative $X_x(t,x)$ of $X(t,x)$, see Chapter 7. Moreover, setting $\eta^h(t,x) = X_x(t,x) \cdot h$, for any $h \in H$, we have that $\eta^h(t,x)$ fulfills the following equation:

$$\begin{cases} D_t\eta^h(t,x) = A\eta^h(t,x) + DF(X(t,x)) \cdot \eta^h(t,x), \\ \eta^h(0,x) = h. \end{cases} \tag{11.2.27}$$

Now, multiplying both sides of the first equation in (11.2.27) by $\eta^h(t, x)$, and taking into account the dissipativity of F, we find

$$\frac{d}{dt}|\eta^h(t, x)|^2 \leq -2\omega|\eta^h(t, x)|^2.$$

It follows that $\|X_x(t, x)\| \leq e^{-\omega t}$, $t \geq 0$. Consequently φ is also differentiable in the direction h and we have

$$\langle D\varphi(x), h \rangle = \int_0^{+\infty} e^{-\lambda t} \mathbb{E}[\langle Df(X(t, x)), X_x(t, x) \cdot h \rangle], \ h \in H,$$

and

$$\|D\varphi\|_0 \leq \frac{1}{\lambda}\|Df\|_0. \tag{11.2.28}$$

Let us show now that $\varphi \in D(L_{b,2})$ so that, by Lemma 11.2.12, $\varphi \in D(\overline{N}_p)$. Set

$$Z(t, x) = e^{tA}x + W_A(t), \ t \geq 0, \ x \in H,$$

so that

$$X(t, x) = Z(t, x) + \int_0^t e^{(t-s)A} F(X(s, x))ds.$$

For any $h > 0$ we have

$$\frac{1}{h}(R_h\varphi(x) - \varphi(x)) = \frac{1}{h}\mathbb{E}[\varphi(Z(h, x)) - \varphi(x)]$$

$$= \frac{1}{h}\mathbb{E}\left[\varphi(X(h, x)) - \int_0^h e^{(h-s)A} F(X(s, x))ds - \varphi(x)\right]$$

$$= \frac{1}{h}\mathbb{E}(\varphi(X(h, x)) - \varphi(x))$$

$$-\frac{1}{h}\mathbb{E}\left[\left\langle D\varphi(X(h, x)), \int_0^h e^{(h-s)A} F(X(s, x))ds \right\rangle\right] + o(h),$$

where $\lim_{h\to 0} \frac{o(h)}{h} = 0$. By taking the limit $h \to 0$ we find, recalling Remark 11.2.6, that $\varphi \in D(L_{b,2})$ and

$$L_{b,2}\varphi = N_p\varphi - \langle D\varphi, F \rangle.$$

Consequently

$$\lambda\varphi(x) - L_{b,2}\varphi(x) - \langle F(x), D\varphi(x) \rangle = f(x), \ x \in H.$$

Therefore we have proved that $(\lambda - \overline{N}_p)^{-1}(C_b^1(H)) \subset D(\overline{N}_p)$. Since $C_b^1(H)$ is dense in $L^p(H, \nu)$, this implies that $\overline{N}_p = N_p$, by the Lumer-Phillips theorem [161]. $\square$

We now consider the general case.

Theorem 11.2.14 *Assume that assumptions* (11.2.1) *hold. Then N_p is the closure of N_0 in $L^p(H, \nu)$.*

Proof. Let us first define a C^1 approximation of F. For any β we set

$$F_\beta(x) = \int_H e^{\beta A^*} F(e^{\beta A} x + y) N_{Q_\beta}(dy), \ x \in H.$$

Then, by the Cameron-Martin theorem, the mapping $\langle F(\cdot), h \rangle$ is differentiable for any $h \in H$ and for any $k \in H$ we have

$$\langle D \langle F(\cdot), h \rangle, k \rangle = \int_H \langle \Lambda_t k, Q_t^{-1/2} y \rangle \langle F(e^{\beta A} x + y), e^{\beta A} h \rangle N_{Q_\beta}(dy),$$

where $\Lambda_t = Q_t^{-1/2} e^{tA}$. Moreover F_β is dissipative since

$$\langle F_\beta(x) - F_\beta(\overline{x}), x - \overline{x} \rangle$$

$$= \int_H \langle F(e^{\beta A} x + y) - F(e^{\beta A} \overline{x} + y), e^{\beta A}(x - \overline{x}) \rangle N_{Q_\beta}(dy) \leq 0,$$

for all $x, \overline{x} \in H$.

Let $f \in C_b^1(H)$, $\lambda > 0$. Arguing as in the proof of Proposition 11.2.11 we can consider the solution $\varphi_\beta \in D(L_{b,2})$ of the equation

$$\lambda \varphi_\beta - L_{b,2} \varphi_\beta - \langle F_\beta(x), D\varphi_\beta \rangle = f.$$

By Lemma 11.2.12 we have that $\varphi_\beta \in D(\overline{N}_p)$ and

$$\lambda \varphi_\beta - N_p \varphi_\beta = f + \langle F_\beta(x) - F(x), D\varphi_\beta \rangle.$$

Now we have, recalling (11.2.28), and using the dominated convergence theorem,

$$\lim_{\beta \to 0} \langle F_\beta(x) - F(x), D\varphi_\beta \rangle = 0, \text{ in } L^p(H, \nu).$$

Therefore $\lambda - \overline{N}_p$ has dense image in $L^p(H, \nu)$. This concludes the proof again by the Lumer-Phillips theorem [161]. $\square$

11.2.3   The integration by parts formula

Proposition 11.2.15 *For any $\varphi \in \mathcal{E}_A(H)$ we have*

$$\int_H N_0\varphi \, \varphi d\nu = -\frac{1}{2} \int_H |D\varphi|^2 d\nu. \tag{11.2.29}$$

Proof. In fact, by a simple computation, we have

$$N_0(\varphi^2) = 2N_0\varphi \, \varphi + |D\varphi|^2, \quad \varphi \in \mathcal{E}_A(H).$$

Since the measure ν is invariant for N_0 we have $N_0(\varphi^2) = 0$ and the conclusion follows integrating the identity above with respect to ν. $\square$

Since $\mathcal{E}_A(H)$ is a core for N_2, by (11.2.29) it follows that the linear operator

$$D : \mathcal{E}_A(H) \subset D(N_2) \to L^2(H, \nu; H), \quad \varphi \to D\varphi$$

is continuous and consequently it can be extended to all $D(N_2)$ (endowed with the graph norm). We shall still denote by D its extension. By Proposition 11.2.15 we get

Proposition 11.2.16 *For any $\varphi \in D(N_2)$ we have*

$$\int_H N_2\varphi \, \varphi d\nu = -\frac{1}{2} \int_H |D\varphi|^2 d\nu. \tag{11.2.30}$$

A first important consequence of the integration by parts formula is the Poincaré inequality.   For all $\varphi \in L^2(H, \nu)$ we set

$$\overline{\varphi} = \int_H \varphi(x)\nu(dx).$$

Proposition 11.2.17 *Assume that (11.2.1) holds. Then for any $\varphi \in D(N_2)$, we have*

$$\int_H |\varphi - \overline{\varphi}|^2 d\nu \leq \frac{1}{2\omega} \int_H |D\varphi|^2 d\nu. \tag{11.2.31}$$

Proof. The proof is very similar to that of Proposition 10.5.2. Let $\varphi \in D(N_2)$. Then $P_t\varphi \in D(N_2)$ and $DP_t\varphi = N_2 P_t\varphi$. Multiplying both sides of this identity by $P_t\varphi$ and integrating with respect to ν we find, thanks to (11.2.30),

$$\frac{1}{2}\frac{d}{dt} \int_H |P_t\varphi|^2 d\nu = \int_H (N_2 P_t\varphi)P_t\varphi d\nu = -\frac{1}{2}\int_H |DP_t\varphi|^2 d\nu. \tag{11.2.32}$$

Next, we prove that

$$|DP_t\varphi(x)|^2 \le e^{-2\omega t} P_t |D\varphi(x)|^2, \quad x \in H. \tag{11.2.33}$$

For this we need to introduce a weakly differentiable approximation F_β of F as in the proof of Theorem 11.2.14.

We have

$$\langle DP_t^\beta \varphi(x), h \rangle = \mathbb{E}\langle D\varphi(X^\beta(t,x)), X_x^\beta(t,x)h \rangle, \quad x, h \in H,$$

where X^β and P_t^β are defined as X and P_t with F replaced by F_β. It follows that

$$|DP_t^\beta \varphi(x)|^2 \le e^{-2\omega t} P_t^\beta |D\varphi(x)|^2, \quad x \in H. \tag{11.2.34}$$

Now it is easy to see that $P_t^\beta \varphi(x) \to P_t\varphi(x)$ as $\beta \to 0$. Therefore, letting β tend to 0 in (11.2.34) gives (11.2.33).

Consequently by (11.2.32) it follows that

$$\frac{1}{2}\frac{d}{dt}\int_H |P_t\varphi|^2 d\nu \ge -\frac{1}{2}e^{-2t\omega}\int_H P_t(|D\varphi|^2)d\nu = -\frac{1}{2}e^{-2t\omega}\int_H |D\varphi|^2 d\nu,$$

taking into account the invariance of ν. Integrating over t gives

$$\int_H |P_t\varphi|^2 d\nu - \int_H \varphi^2 d\nu \ge -\frac{1}{2\omega}\left(1 - e^{-2t\omega}\right)\int_H |D\varphi|^2 d\nu.$$

Finally, letting t tend to $+\infty$, and recalling (11.2.11), gives

$$(\overline{\varphi})^2 - \int_H \varphi^2 d\nu \ge -\frac{1}{2\omega}\int_H |D\varphi|^2 d\nu$$

and (11.2.31) follows, since

$$\int_H |\varphi - \overline{\varphi}|^2 d\nu = \int_H \varphi^2 d\nu - (\overline{\varphi})^2.$$

We have already seen in §10.5 that the Poincaré inequality yields the *spectral gap* property for N_2.

Proposition 11.2.18 *Let $\sigma(N_2)$ be the spectrum of N_2. Then we have*

$$\sigma(N_2)\backslash\{0\} \subset \{\lambda \in \mathbb{C} : \mathrm{Re}\ \lambda \le -\omega\}. \tag{11.2.35}$$

Let us prove now the logarithmic Sobolev inequality. The proof is similar to that of Proposition 10.5.3. We use the following notation:

$$\|\varphi\|_2^2 = \int_H |\varphi(x)|^2 \nu(dx), \quad \varphi \in L^2(H, \nu).$$

Proposition 11.2.19 *For any* $\varphi \in D(N_2)$ *we have*

$$\int_H \varphi^2 \log(\varphi^2) d\nu \le \frac{1}{\omega} \int_H |D\varphi|^2 d\nu + \|\varphi\|_2^2 \log(\|\varphi\|_2^2). \qquad (11.2.36)$$

Proof. It is enough to prove the result when $\varphi \in \mathcal{E}_A(H)$ is such that $\varphi(x) \ge \varepsilon > 0$, $x \in H$. In this case we have

$$\frac{d}{dt} \int_H (P_t(\varphi^2)) \log(P_t(\varphi^2)) d\nu = \int_H N P_t(\varphi^2) \log(P_t(\varphi^2)) d\nu + \int_H N P_t(\varphi^2) d\nu.$$

Now the second term in the right hand side vanishes, due to the invariance of ν. For the first term we use the identity

$$\int_H N\varphi \, g'(\varphi) d\nu = -\frac{1}{2} \int_H g''(\varphi) |D\varphi|^2 d\nu, \qquad (11.2.37)$$

with $g'(\xi) = \log \xi$. Therefore we have

$$\frac{d}{dt} \int_H P_t(\varphi^2) \log(P_t(\varphi^2)) d\nu = -\frac{1}{2} \int_H \frac{1}{P_t(\varphi^2)} |DP_t(\varphi^2)|^2 d\nu. \qquad (11.2.38)$$

As before, we first consider the C^1 approximation F_β of F. We have for any $h \in H$,

$$\langle DP_t^\beta(\varphi^2)(x), h \rangle = 2\mathbb{E}[\varphi(X^\beta(t,x))\langle D\varphi(X^\beta(t,x)), X_x^\beta(t,x)h \rangle]$$

It follows by the Hölder inequality that

$$|\langle DP_t^\beta(\varphi^2)(x), h \rangle|^2 \le 4e^{-2t\omega} \mathbb{E}[\varphi^2(X^\beta(t,x))]\mathbb{E}[|D\varphi|^2(X^\beta(t,x))],$$

which yields

$$|DP_t^\beta(\varphi^2)|^2 \le 4e^{-2t\omega} P_t^\beta(\varphi^2) \, P_t^\beta(|D\varphi|^2),$$

and, as $\beta \to 0$,

$$|DP_t(\varphi^2)|^2 \le 4e^{-2t\omega} P_t(\varphi^2) \, P_t(|D\varphi|^2).$$

Substituting in (11.2.38) gives

$$\frac{d}{dt} \int_H P_t(\varphi^2) \log(P_t(\varphi^2)) d\nu \geq -2e^{-2t\omega} \int_H P_t(|D\varphi|^2) d\nu$$

$$= -2e^{-2t\omega} \int_H |D\varphi|^2 d\nu,$$

due to the invariance of ν. Integrating over t gives

$$\int_H P_t(\varphi^2) \log(P_t(\varphi^2)) d\nu \geq \frac{1}{\omega} (1 - e^{-2t\omega}) \int_H |D\varphi|^2 d\nu.$$

Finally, letting t tend to $+\infty$, and recalling (11.2.11), gives

$$\|\varphi\|_2^2 \log(\|\varphi\|_2^2) - \int_H \varphi^2 \log(\varphi^2) d\nu \geq -\frac{1}{\omega} \int_H |D\varphi|^2 d\nu$$

and the conclusion follows. $\square$

Remark 11.2.20 In a similar way we can prove that for any $p > 1$ and $\varphi \in D(N_p)$ we have

$$\int_H \varphi^p \log(\varphi^p) d\nu \leq \frac{1}{2\omega} \int_H |D\varphi|^p d\nu + \|\varphi\|_p^p \log(\|\varphi\|_p^p). \qquad (11.2.39)$$

Moreover (P_t) is hypercontractive.

11.2.4 Existence of a density

Here we want to prove that the invariant measure ν of (P_t) is absolutely continuous with respect to the Gaussian measure $\mu = N_Q$ where $Qx = \int_0^{+\infty} e^{tA} e^{tA^*} x dt$, $x \in H$.

Here we assume, besides (11.2.1), that

$$\text{There exists } \delta \in (0, 1/2) \text{ such that } A^{-1+\delta} \in L_1(H). \qquad (11.2.40)$$

Moreover it is convenient to consider the approximating problem

$$dX_n(t) = (AX_n(t) + F_n(X_n(t)))dt + dW_t, \quad X_n(0) = x \in H, \qquad (11.2.41)$$

where F_n are defined by

$$F_n(x) = \frac{nF(x)}{n + |x|}, \quad x \in H, \ n \in \mathbb{N}.$$

We denote by $X_n(\cdot, x)$, be the mild solution of (11.2.41) :

$$X_n(t,x) = e^{tA}x + \int_0^t e^{(t-s)A}F_n(X_n(s,x))dt + \int_0^t e^{(t-s)A}dW_s, x \in H,$$

$$(11.2.42)$$

and by P_t^n be the corresponding transition semigroup

$$P_t^n\varphi(x) = \mathbb{E}[\varphi(X_n(t,x))], \quad \varphi \in C_b(H).$$

By Propositions 11.1.3 and 11.1.4 there exists a unique invariant measure ν_n for problem (11.2.42) which is absolutely continuous with respect to $\mu = N_Q$ with $Q = -\frac{1}{2}A^{-1}$.

We denote by N_p^n the infinitesimal generator of P_t^n in $L^p(H, \nu_n)$. Arguing as before we see that N_p^n is the closure of the linear operator $N_{0,p}^n$ defined by

$$N_{0,p}^n\varphi = L_0\varphi + \langle F_n(x), D\varphi \rangle, \quad \varphi \in \mathcal{E}_A(H),$$

where

$$L_0\varphi = \frac{1}{2}\text{Tr}[D^2\varphi] + \langle x, A^*D\varphi \rangle, \quad \varphi \in \mathcal{E}_A(H).$$

Therefore the following identity holds:

$$\int_H N_2^n\varphi \, \varphi d\nu_n = -\frac{1}{2}\int_H |D\varphi|^2 d\nu_n. \tag{11.2.43}$$

Moreover from (11.1.13) the following estimate holds:

$$\int_H |D(\sqrt{\rho_n})|^2 d\mu \leq \int_H |F(x)|^2 d\nu_n. \tag{11.2.44}$$

Theorem 11.2.21 *Assume that (11.2.1) and (11.2.40) hold and that the embedding $D(A) \subset H$ is compact. Then $\nu \ll \mu$ and the density $\rho = \frac{d\nu}{d\mu}$ is such that $\sqrt{\rho} \in W^{1,2}(H, \mu)$ or, equivalently, $D\log\rho \in W^{1,2}(H, \nu)$.*

Proof.
    Step 1. We have that (ν_n) is tight.

Since the embedding of $D(A)$ into H is compact, it is enough to show that for any $\beta \in (0, \delta)$ there exists $c_\beta > 0$ such that we have

$$\int_H |(-A)^\beta x|^2 \nu_n(dx) \leq c_\beta. \tag{11.2.45}$$

Let $\kappa_\beta > 0$ be such that

$$|(-A)^\beta e^{tA}x| \leq \kappa_\beta t^{-\beta}|x|, \ x \in H, \ t > 0.$$

Now by (11.2.42) it follows that

$$|(-A)^\beta X_n(t,x)| \leq \kappa_\beta t^{-\beta}|x| + \kappa_\beta \int_0^t (t-s)^{-\beta}|F_n(X_n(s,x))|ds$$

$$+ \left| \int_0^t (-A)^\beta e^{(t-s)A} dW(s) \right|.$$

Consequently

$$\mathbb{E}|(-A)^\beta X_n(t,x)|^2 \leq 3\kappa_\beta^2 t^{-2\beta}|x|^2$$

$$+ 3\kappa_\beta^2 \frac{t^{1-2\beta}}{1-2\beta} \int_0^t \mathbb{E}|F_n(X_n(s,x))|^2 ds$$

$$+ 3\kappa_\beta^2 \int_0^t s^{-2\beta} \text{Tr}[e^{sA}e^{sA^*}]ds.$$

Setting $t = 1$ and integrating with respect to ν_n gives

$$\int_H |(-A)^\beta x|^2 d\nu_n \leq 3\kappa_\beta^2 |x|^2 + 3\kappa_\beta^2 \frac{1}{1-2\beta} \int_H |F_n(x)|^2 d\nu_n$$

$$+ 3\kappa_\beta^2 \int_0^t s^{-2\beta} \text{Tr}[e^{sA}e^{sA^*}]ds.$$

Now (11.2.45) follows, taking into account Proposition 11.2.4, since $|F_n(x)| \leq |F(x)|$, $x \in H$, $n \in \mathbb{N}$.

Step 2. Conclusion.

Let $\rho_n = \frac{d\nu_n}{d\mu}$; then by (11.2.44) the sequence $(\sqrt{\rho_n})$ lies in a bounded subset of $W^{1,2}(H, \nu)$. Since the embedding

$$W^{1,2}(H, \mu) \subset L^2(H, \mu)$$

is compact, see Chapter 8, there exists a subsequence (ρ_{n_k}) convergent to some ρ in $L^1(H, \mu)$. Since the sequence (ν_n) is tight we have, as $n \to \infty$,

$$\int_H \varphi d\nu = \lim_{k \to \infty} \int_H \varphi d\nu_{n_k} = \lim_{k \to \infty} \int_H \varphi \rho_{n_k} d\mu = \int_H \varphi \rho d\mu, \ \varphi \in C_b(H),$$

so that $\nu << \mu$. $\square$

Corollary 11.2.22 *The embedding*

$$W^{1,2}(H, \nu) \subset L^2(H, \nu)$$

is compact. Thus P_t is a compact operator on $L^2(H, \nu)$ for any $t > 0$.

Proof. This follows from G. Da Prato, A. Debussche and B. Goldys [88]. $\square$

Chapter 12

Gradient systems

We are given on a separable Hilbert space H a self-adjoint operator $A : D(A) \subset H \to H$ strictly negative and such that A^{-1} is of trace class and a mapping $U : H \to (-\infty, +\infty]$.

We are concerned with the differential operator

$$N_0 \varphi(x) = \frac{1}{2} \mathrm{Tr}[D^2 \varphi(x)] + \langle x, A^* D\varphi(x) \rangle - \langle DU(x), D\varphi(x) \rangle$$

$$:= L\varphi(x) - \langle DU(x), D\varphi(x) \rangle, \quad \varphi \in \mathcal{E}_A(H), \ x \in H,$$

where L is the Ornstein-Uhlenbeck generator introduced in Chapter 6 and $\mathcal{E}_A(H)$ is, as before, the set of real parts of all exponential functions of the form $e^{i\langle h, x \rangle}$ where $h \in D(A^*)$.

Let us introduce a measure ν in H by setting

$$\nu(dx) = \rho(x) N_Q(dx), \quad x \in H,$$

where $Q = -\frac{1}{2} A^{-1}$, $\rho(x) = Z^{-1} e^{-2U(x)}$, $x \in H$, and $Z = \int_H e^{-2U(x)} \mu(dx)$.

Our goal is to show that N_0 is dissipative in $L^p(H, \nu)$, $p \geq 1$, and its closure N_p is m-dissipative. For $p = 2$ this amounts to saying that N_0 is essentially self-adjoint.

Several papers have been devoted to this problem, we quote in particular

- the Dirichlet forms approach, see S. Albeverio and M. Röckner [5], V. Liskevich and M. Röckner [159], etc.,

- the papers by J. Zabczyk [216], G. Da Prato [77], [78], [81], G. Da Prato and L. Tubaro [97], [99].

We notice that similar results can be proved for more general operators of the form

$$N_0\varphi(x) = \frac{1}{2}\text{Tr}[QD^2\varphi(x)] + \langle x, A^*D\varphi(x)\rangle - \langle QDU(x), D\varphi(x)\rangle, \quad x \in H,$$

where $Q \in L^+(H)$ commutes with A and $\mu = N_{Q_\infty}$ with $Q_\infty = -\frac{1}{2}QA^{-1}$, see [159] and [97] for application to *stochastic quantization* . But we will restrict to the first case for the sake of brevity.

Here, §12.2 is devoted to general results, §12.3 to the case when U is a convex lower semicontinuous function.

12.1   General results

12.1.1   Assumptions and setting of the problem

Concerning A we shall assume

$$\begin{cases} A \text{ is self-adjoint and there exists } \omega > 0 \text{ such that} \\ \quad\quad \langle Ax, x\rangle \le -\omega|x|^2, \ x \in D(A), \\ \text{moreover } A^{-1} \text{ is of trace class.} \end{cases} \quad (12.1.1)$$

We shall denote by μ the Gaussian measure N_Q of mean 0 and covariance operator $Q = -\frac{1}{2}A^{-1}$. Since Q is of trace class, there exist a complete orthonormal system (e_k) in H and a sequence of positive numbers (λ_k) such that $Qe_k = \lambda_k e_k$, $k \in \mathbb{N}$. For any $x \in H$ we set $x_h = \langle x, e_h\rangle$, $h \in \mathbb{N}$.

Concerning U we shall assume that

$$U : H \to (-\infty, +\infty] \text{ is Borel and } Z := \int_H e^{-2U(x)}\mu(dx) \in (0, +\infty).$$
$$(12.1.2)$$

Then we set $\rho(x) = Z^{-1}e^{-2U(x)}$, $x \in H$, and assume that

$$\rho, \sqrt{\rho} \in W^{1,2}(H, \mu). \quad (12.1.3)$$

Now we consider the differential operator

$$N_0\varphi = L\varphi + \langle DU, D\varphi\rangle, \quad \varphi \in \mathcal{E}_A(H),$$

where $DU = -\frac{1}{2}D\log\rho$. We notice that

$$D\log\rho \in L^2(H, \nu; H), \quad (12.1.4)$$

since

$$\int_H |D \log \rho|^2 d\nu = \int_H \frac{|D\rho|^2}{\rho} d\mu = 4 \int_H |D\sqrt{\rho}|^2 d\mu < +\infty,$$

by (12.1.3).

Lemma 12.1.1 *For any* $n \in \mathbb{N}$ *we have*

$$\int_H |x|^n d\nu < +\infty. \tag{12.1.5}$$

Proof. We have in fact by the Hölder inequality,

$$\left(\int_H |x|^n d\nu \right)^2 \leq \int_H |x|^{2n} d\mu \int_H \rho^2 d\mu. \quad \square$$

Obviously the measure ν is concentrated on the set $\{U < +\infty\}$ which has positive μ-measure in view of (12.1.2) We notice that there are important cases when $\mu(\{U < +\infty\}) = 0$, see L. Zambotti [224].

We end this subsection by giving two examples.

Example 12.1.2 Let $H = L^2(0,1)$,

$$Ax = D_\xi^2 x, \quad x \in D(A) = H^2(0,1) \cap H_0^1(0,1),$$

and

$$U(x) = \begin{cases} \dfrac{1}{4} \displaystyle\int_0^1 x^4(\xi) d\xi & \text{if } x \in L^4(0,1), \\[3mm] +\infty & \text{if } x \notin L^4(0,1). \end{cases}$$

Then $\{U < +\infty\} = L^4(0,1)$ and $\mu(L^4(0,1)) = 1$.

Moreover U is differentiable in the directions of $L^6(0,1)$ and

$$DU(x) = -x^3, \quad x \in L^6(0,1).$$

Finally we claim that

$$\int_H |DU(x)|^2 \mu(dx) = \int_0^1 d\xi \int_H x^6(\xi) \mu(dx) < +\infty, \tag{12.1.6}$$

so that assumptions (12.1.1)-(12.1.3) are fulfilled.

To prove (12.1.6) notice that, since formally for any $\xi \in [0,1]$,

$$x(\xi) = \sum_{k=1}^{\infty} \langle x, e_k \rangle e_k(\xi) = \left\langle x, \sum_{k=1}^{\infty} e_k(\xi) e_k \right\rangle$$

$$= \frac{1}{\pi} \left\langle Q^{-1/2} x, \sum_{k=1}^{\infty} \frac{e_k(\xi)}{k} e_k \right\rangle,$$

we can write $x(\xi) = W_{\eta_\xi}(x)$, where η_ξ, defined as

$$\eta_\xi = \frac{1}{\pi} \sum_{k=1}^{\infty} \frac{e_k(\xi)}{k} e_k,$$

belongs to $L^2(0,1)$.

Now the integral in 12.1.6 is equal to $\int_H |W_{\eta_\xi}(x)|^6 \mu(dx)$, and, by an explicit straightforward computation, one can easily see that it is finite.

Example 12.1.3 Let U be the convex lower continuous function $H \to \mathbb{R}$ defined by

$$U(x) = \begin{cases} -\log(1 - |x|^2) & \text{if } |x| < 1, \\ +\infty & \text{if } |x| \geq 1. \end{cases}$$

We have

$$\rho(x) = \begin{cases} \dfrac{(1 - |x|^2)^2}{\int_{|x|<1}(1 - |y|^2)^2 \mu(dy)} & \text{if } |x| < 1, \\ 0 & \text{if } |x| \geq 1, \end{cases}$$

Moreover U is differentiable on $B(0,1) = \{x \in H : |x| < 1\}$ and

$$DU(x) = \frac{2x}{1 - |x|^2}, \quad |x| < 1.$$

Finally,

$$\int_{B(0,1)} |DU(x)|^2 \nu(dx) = 4Z^{-1} \int_{B(0,1)} |x|^2 \mu(dx).$$

So assumptions (12.1.1)-(12.1.3) are fulfilled.

12.1.2 The Sobolev space $W^{1,2}(H,\nu)$

We proceed here as we did in §10.2 for the definition of the Sobolev space $W^{1,2}(H,\mu)$. So we first prove an integration by parts formula.

Lemma 12.1.4 *Let* $\varphi, \psi \in \mathcal{E}_A(H)$ *and* $k \in \mathbb{N}$. *Then we have*

$$\int_H D_k\varphi \, \psi d\nu = -\int_H \varphi \, D_k\psi d\nu + \int_H \left[\frac{x_k}{\lambda_k} - D_k \log \rho\right] \varphi\psi d\nu. \qquad (12.1.7)$$

Proof. Let us start from the identity

$$\int_H D_k\varphi \, \psi d\nu = \int_H D_k\varphi \, \psi\rho \, d\mu.$$

Since $\psi\rho \in W^{1,2}(H,\mu)$, we can apply the integration by parts formula (9.2.1) and we find

$$\int_H D_k\varphi \, \psi d\nu = -\int_H \varphi D_k(\psi\rho)d\mu + \frac{1}{\lambda_k} \int_H x_k\varphi\psi d\nu$$

$$= -\int_H \varphi D_k\psi \, d\nu - \int_H \varphi\psi D_k \log \rho \, d\nu + \frac{1}{\lambda_k} \int_H x_k\varphi\psi d\nu. \qquad \square$$

Proposition 12.1.5 *For any* $k \in \mathbb{N}$ *the operator* D_k *is closable on* $L^2(H,\nu)$.

Proof. Let (φ_n) be a sequence in $\mathcal{E}_A(H)$ and $g \in L^2(H,\nu)$ such that

$$\varphi_n \to 0, \quad D_k\varphi_n \to g, \quad \text{in } L^2(H,\nu).$$

We have to show that $g = 0$. Let $\psi \in \mathcal{E}_A(H)$. Then by Lemma 12.1.4 we have

$$\int_H D_k\varphi_n \, \psi d\nu = -\int_H \varphi_n \, D_k\psi \, d\nu + \int_H \frac{x_k}{\lambda_k} \varphi_n\psi d\nu - \int_H D_k \log \rho \, \varphi_n\psi \, d\nu.$$

$$(12.1.8)$$

Clearly the first and the third term on the right hand side tend to 0 as $n \to \infty$. Concerning the second we have

$$\lim_{n\to\infty} \int_H x_k\varphi_n\psi d\nu = 0. \qquad (12.1.9)$$

In fact

$$\left| \int_H x_k \varphi_n \psi d\nu \right|^2 \le \int_H \varphi_n^2 d\nu \int_H x_k^2 \psi^2 \rho d\mu$$

$$\le \int_H \varphi_n^2 d\nu \left(\int_H \rho^2 d\mu \right)^{1/2} \left(\int_H x_k^4 \psi^4 d\mu \right)^{1/2} \to 0.$$

Now, letting $n \to \infty$ in (12.1.8) yields

$$\int_H g\psi d\nu = 0,$$

so that the conclusion follows from the arbitrariness of g. $\square$

If φ belongs to the domain of the closure of D_k in $L^2(H, \nu)$ (which we shall still denote by D_k) we shall write $D_k \varphi \in L^2(H, \nu)$.

Now we consider the derivative operator

$$D : \ \mathcal{E}_A(H) \subset L^2(H, \nu) \to L^2(H, \nu; H).$$

The proof of the following result is similar to that of Proposition 9.2.4 and it is left to the reader.

Proposition 12.1.6 *The operator D is closable.*

If φ belongs to the domain of the closure of D (which we shall still denote by D) in $L^2(H, \nu)$ we shall write $D\varphi \in L^2(H, \nu; H)$.

Now we define $W^{1,2}(H, \nu)$ as the space of all functions $\varphi \in L^2(H, \nu)$ such that $D\varphi \in L^2(H, \nu; H)$. Endowed with the inner product

$$\langle \varphi, \psi \rangle_{W^{1,2}(H,\nu)} = \int_H \varphi \psi d\nu + \int_H \langle D\varphi, D\psi \rangle d\nu$$

$W^{1,2}(H, \nu)$ is a Hilbert space. In an analogous way we can define $W^{k,2}(H, \nu)$, $k \ge 2$.

12.1.3   Symmetry of the operator N_0

Let us consider the linear operator

$$N_0 \varphi = L\varphi - \langle DU, D\varphi \rangle, \quad \varphi \in \mathcal{E}_A(H), \tag{12.1.10}$$

where L is the infinitesimal generator of the Ornstein-Uhlenbeck semigroup (R_t) introduced in Chapter 6.

We shall prove first that N_0 is well defined in $L^2(H, \nu)$, secondly that it is symmetric.

Lemma 12.1.7 *For all $\varphi \in \mathcal{E}_A(H)$ we have $N_0\varphi \in L^2(H,\nu)$.*

Proof. We have in fact

$$\int_H |N_0\varphi|^2 d\nu = \int_H |L\varphi|^2 d\nu + \int_H |\langle DU, D\varphi \rangle|^2 d\nu$$
$$-2\int_H L\varphi \langle DU, D\varphi \rangle d\nu.$$

Since $\varphi \in \mathcal{E}_A(H)$ there exist $a, b > 0$ such that

$$|L\varphi(x)| \le a + b|x|, \quad x \in H.$$

Then we have

$$\int_H |L\varphi|^2 d\nu \le 2a^2 + 2b^2 \int_H |x|^2 \rho d\mu$$
$$\le 2a^2 + 2b^2 \left(\int_H \rho^2 d\mu\right)^{1/2} \left(\int_H |x|^4 d\mu\right)^{1/2} < +\infty.$$

Moreover, recalling (12.1.4), as $DU = -\frac{1}{2} D\log\rho$, we find

$$\int_H |\langle DU, D\varphi \rangle|^2 d\nu \le \|\varphi\|_1^2 \int_H |DU|^2 d\nu$$
$$= \frac{1}{4} \|\varphi\|_1^2 \int_H |D\log\rho|^2 d\nu < +\infty.$$

Finally,

$$\left| \int_H L\varphi \langle DU, D\varphi \rangle d\nu \right| \le \|\varphi\|_1 \left(\int_H |L\varphi|^2 d\nu\right)^{1/2} \left(\int_H |DU| d\nu\right)^{1/2}.$$

and the conclusion follows. $\square$

Proposition 12.1.8 *For all $\varphi, \psi \in \mathcal{E}_A(H)$ we have*

$$\int_H N_0\varphi\psi d\nu = -\frac{1}{2}\int_H \langle D\varphi, D\psi \rangle d\nu. \qquad (12.1.11)$$

Therefore N_0 is symmetric in $L^2(H,\nu)$.

Proof. Let $\varphi, \psi \in \mathcal{E}_A(H)$. Then we have

$$\int_H N_0 \varphi \psi d\nu = \int_H L\varphi \psi \rho d\mu - \int_H \langle DU, D\varphi \rangle \psi d\nu.$$

Since $\rho \in W^{1,2}(H, \mu)$ we have $\psi \rho \in W^{1,2}(H, \mu)$ and

$$\begin{aligned}
\int_H L\varphi \, \psi \rho d\mu &= -\frac{1}{2} \int_H \langle D\varphi, D(\psi \rho) \rangle d\mu \\
&= -\frac{1}{2} \int_H \langle D\varphi, D\psi \rangle d\nu - \frac{1}{2} \int_H \langle D\varphi, D \log \rho \rangle \psi d\nu \\
&= -\frac{1}{2} \int_H \langle D\varphi, D\psi \rangle d\nu + \int_H \langle D\varphi, DU \rangle \psi d\nu,
\end{aligned}$$

and the conclusion follows. $\square$

Remark 12.1.9 Arguing as in Eberle [109], we see that N_0 is dissipative in $L^p(H, \nu)$ for all $p \geq 1$. In what follows we shall denote by N_p the closure of N_0 in $L^p(H, \nu)$.

12.1.4   The m-dissipativity of N_1 on $L^1(H, \nu)$.

Here we follow G. Da Prato and L. Tubaro [99].

We first introduce an approximating problem. Let $(F_n) \subset C_b^\infty(H; H)$ be such that

$$\lim_{n \to \infty} F_n = DU \quad \text{in } L^2(H, \nu; H).$$

Since $\mathcal{E}_A(H)$ is dense in $L^2(H, \nu)$, such a sequence can be easily constructed. Then consider the approximating equation

$$\lambda \varphi_n - L\varphi_n + \langle F_n(\cdot), D\varphi_n \rangle = f. \tag{12.1.12}$$

Lemma 12.1.10 *Let $f \in C_b(H)$ and $\lambda > 0$. Then equation (12.1.12) has a unique solution $\varphi_n \in D(L) \cap C_b^1(H)$ and*

$$\|\varphi_n\|_0 \leq \frac{1}{\lambda} \|f\|_0. \tag{12.1.13}$$

If in addition $f \in C_b^k(H)$, $k \in \mathbb{N}$, we have $\varphi_n \in C_b^{k+1}(H)$.

Proof. Let us recall that, in view of Proposition 6.6.4, the linear operator

$$N_n\varphi = L\varphi + \langle F_n, D\varphi \rangle, \quad \varphi \in D(L),$$

is m-dissipative on $C_b(H)$. Consequently for any $\lambda > 0$ and any $\varphi \in C_b(H)$ there exists a unique solution $\varphi \in D(L)$ of equation (12.1.12) and (12.1.13) holds. Moreover, by Corollary 6.4.3 we have $\varphi \in C_b^1(H)$. The last statement follows easily arguing as in Proposition 6.6.5. $\square$

Lemma 12.1.11 *Let $\varphi \in D(L)$. Then $\varphi \in D(N_1)$ and*

$$N_1\varphi = L\varphi - \langle DU, D\varphi \rangle. \tag{12.1.14}$$

Proof. By Proposition 11.2.10 there exist a multi-sequence $(\varphi_k) = (\varphi_{k_1 k_2 k_3 k_4}) \subset \mathcal{E}_A(H)$ and $M > 0$ such that

$$\varphi_k(x) \to \varphi(x), \ D\varphi_k(x) \to D\varphi(x), \ L\varphi_k(x) \to L\varphi(x), \ x \in H,$$

and

$$|\varphi_k(x)| + |D\varphi_k(x)| + |L\varphi_k(x)| \le M(1 + |x|^2), \ x \in H.$$

It follows that

$$N_1\varphi_n(x) \to L\varphi(x) - \langle DU(\cdot), D\varphi(x) \rangle, \ x \in H,$$

and that

$$|N_1\varphi_n(x)| \le M(1 + |x|^2) + M|DU(x)|, \ x \in H.$$

Now the conclusion follows from (12.1.5) and the dominated convergence theorem. $\square$

Lemma 12.1.12 *Let $f \in C_b^1(H)$ and $\lambda > 0$. Then the solution φ_n to (12.1.12) belongs to $D(N_1)$ and we have*

$$N_1\varphi_n = L\varphi_n - \langle F_n(\cdot), D\varphi_n \rangle. \tag{12.1.15}$$

Proof. By Lemma 12.1.10 we have $\varphi_n \in D(L)$ and by Lemma 12.1.11 we know that $\varphi_n \in D(N_1)$. Thus the conclusion follows. $\square$

Theorem 12.1.13 *Under the assumptions (12.1.1), (12.1.2) and (12.1.3), N_1 is m-dissipative on $L^1(H, \nu)$.*

Proof. Let $f \in C_b(H)$ and let φ_n be the solution to (12.1.12):

$$\lambda \varphi_n - L\varphi_n + \langle F_n(\cdot), D\varphi_n \rangle = f.$$

Then by Lemma 12.1.11 $\varphi_n \in D(N_1) \cap C_b^1(H)$ and

$$N_1 \varphi_n = L\varphi_n - \langle DU(\cdot), D\varphi_n \rangle.$$

Therefore

$$\lambda \varphi_n - N_1 \varphi_n = f - \langle F_n(\cdot) - DU(\cdot), D\varphi_n \rangle. \tag{12.1.16}$$

It follows, taking into account (12.1.11), that

$$\lambda \int_H \varphi_n^2 d\nu + \frac{1}{2} \int_H |D\varphi_n|^2 d\nu = \int_H f\varphi_n d\nu - \int_H \varphi_n \langle F_n - DU, D\varphi_n \rangle d\nu.$$

Moreover, in view of Lemma 12.1.10, $\|\varphi_n\|_0 \leq \frac{1}{\lambda} \|f\|_0$, and consequently

$$\lambda \int_H \varphi_n^2 d\nu + \frac{1}{2} \int_H |D\varphi_n|^2 d\nu$$

$$\leq \frac{1}{\lambda} \|f\|_0^2 + \frac{1}{\lambda} \|f\|_0 \int_H |F_n - DU| \, |D\varphi_n| d\nu$$

$$\leq \frac{1}{\lambda} \|f\|_0^2 + \frac{1}{4} \int_H |D\varphi_n|^2 d\nu + \frac{4}{\lambda^2} \|f\|_0^2 \int_H |F_n - DU|^2 d\nu.$$

Therefore there exists a constant M_1, independent of n, such that

$$\int_H |D\varphi_n|^2 d\nu \leq M_1. \tag{12.1.17}$$

It follows that

$$\lim_{n \to \infty} \langle F_n(x) - DU(x), D\varphi_n \rangle = 0 \quad \text{in } L^1(H, \nu), \tag{12.1.18}$$

and so

$$\lim_{n \to \infty} (\lambda \varphi_n - N_1 \varphi_n) = f \quad \text{in } L^1(H, \nu).$$

Therefore the closure of the range of $\lambda - N_1$ contains $C_b(H)$ and so it is dense in $L^1(H, \nu)$. Now the conclusion follows from a classical result due to Lumer and Phillips [161] . $\square$

12.2 The m-dissipativity of N_2 on $L^2(H, \nu)$

In this section we are going to prove, following G. Da Prato and L. Tubaro [97], the following result.

Theorem 12.2.1 *Assume that, besides the assumptions* (12.1.1),(12.1.2) *and* (12.1.3), *we have* $|DU| \in L^4(H, \nu)$. *Then* N_2 *is self-adjoint on* $L^2(H, \nu)$.

Proof. The proof is similar to that of Theorem 12.1.13. However, we need to show, instead of (12.1.18), that

$$\lim_{n \to \infty} \langle F_n(\cdot) - DU(\cdot), D\varphi_n \rangle = 0 \text{ in } L^2(H, \nu). \tag{12.2.1}$$

To this purpose we need an estimate

$$\int_H |D\varphi_n|^4 d\nu \le M_2, \tag{12.2.2}$$

for the solution φ_n of the approximating equation (12.1.12) where $\lambda > 0$ and $f \in C_b^2(H)$.

To find (12.2.2) we first estimate $D_h \varphi_n$, $h \in \mathbb{N}$. Differentiating (12.1.12) with respect to x_h gives

$$\lambda D_h \varphi_n - N_0 D_h \varphi_n + \langle F_n - DU, DD_h \varphi_n \rangle$$

$$+ a_h D_h \varphi_n + \langle D_h F_n, D\varphi_n \rangle = D_h f, \tag{12.2.3}$$

where $a_h = \frac{1}{2\lambda_h}$. Multiplying both sides of (12.2.3) by $D_h \varphi_n$, integrating with respect to ν and taking into account (12.1.11), yields

$$\lambda \int_H |D_h \varphi_n|^2 d\nu + \frac{1}{2} \int_H |DD_h \varphi_n|^2 d\nu + \int_H a_h |D_h \varphi_n|^2 d\nu$$

$$+ \frac{1}{2} \int_H \langle F_n - DU, D(D_h \varphi_n)^2 \rangle d\nu + \int_H \langle D_h F_n, D\varphi_n \rangle D_h \varphi_n d\nu$$

$$= \int_H D_h f \, D_h \varphi_n d\nu.$$

Summing up over h gives

$$\lambda \int_H |D\varphi_n|^2 d\nu + \frac{1}{2} \int_H \text{Tr}[(D^2 \varphi_n)^2] d\nu + \int_H |(-A)^{1/2} D\varphi_n|^2 d\nu$$

$$+ \int_H \langle D^2 \varphi_n \, D\varphi_n, F_n - DU \rangle d\nu + I = -2 \int_H N_0 \varphi_n \, f \, d\nu, \tag{12.2.4}$$

where

$$I = \sum_{h,k=1}^{\infty} \int_{H} D_h F_{n,k} \, D_k \varphi_n \, D_h \varphi_n \, d\nu,$$

and $F_{n,k} = \langle F_n, e_k \rangle$. Notice that for the factor $D_h F_{n,k}$ appearing in I there is no chance to have an estimate independent of n (in fact only a uniform estimate of F_n on $L^4(H, \nu)$ is available). For this reason we transform I by an integration by parts. Using (12.1.7) we obtain

$$I = - \sum_{h,k=1}^{\infty} \int_{H} F_{n,k} \, D_h D_k \varphi_n \, D_h \varphi_n \, d\nu$$

$$- \sum_{h,k=1}^{\infty} \int_{H} F_{n,k} \, D_k \varphi_n \, D_h^2 \varphi_n \, d\nu$$

$$+ \sum_{h,k=1}^{\infty} \frac{1}{\lambda_h} \int_{H} x_h F_{n,k} \, D_k \varphi_n \, D_h \varphi_n \, d\nu$$

$$+2 \sum_{h,k=1}^{\infty} \int_{H} D_h U \, F_{n,k} \, D_k \varphi_n \, D_h \varphi_n \, d\nu.$$

Consequently

$$I = - \int_{H} \langle D^2 \varphi_n \, D\varphi_n, F_n \rangle d\nu - 2 \int_{H} N_0 \varphi_n \langle F_n, D\varphi_n \rangle d\nu.$$

Substituting in (12.2.4) gives

$$\lambda \int_{H} |D\varphi_n|^2 d\nu + \frac{1}{2} \int_{H} \mathrm{Tr}[(D^2 \varphi_n)^2] d\nu + \int_{H} |(-A)^{1/2} D\varphi_n|^2 d\nu$$

$$= \int_{H} \langle D^2 \varphi_n \, D\varphi_n, DU \rangle d\nu + 2 \int_{H} N_0 \varphi_n \langle F_n, D\varphi_n \rangle d\nu - 2 \int_{H} N_0 \varphi_n \, f \, d\nu.$$
$$(12.2.5)$$

Let us estimate the first two terms on the left hand side of (12.2.5). Recalling the elementary inequality

$$|\langle Gx, y \rangle|^2 \le \mathrm{Tr}[G^2]|x|^2|y|^2, \quad x, y \in H, \ G \in L^+(H),$$

we obtain for the first term,

$$\left| \int_H \langle D^2\varphi_n \, D\varphi_n, DU \rangle d\nu \right|^2 \le \int_H \text{Tr}[(D^2\varphi_n)^2]d\nu \int_H |D\varphi_n|^2 |DU|^2 d\nu$$

$$\le \int_H \text{Tr}[(D^2\varphi_n)^2]d\nu \left(\int_H |D\varphi_n|^4 d\nu \right)^{1/2} \left(\int_H |DU|^4 d\nu \right)^{1/2}.$$

$$(12.2.6)$$

Concerning the second we have

$$\left| \int_H N_0\varphi_n \langle F_n, D\varphi_n \rangle d\nu \right|^2$$

$$\le \int_H |N_0\varphi_n|^2 d\nu \left(\int_H |D\varphi_n|^4 d\nu \right)^{1/2} \left(\int_H |F_n|^4 d\nu \right)^{1/2}. \quad (12.2.7)$$

Now, substituting (12.2.6) and (12.2.7) in (12.2.5) yields

$$\lambda \int_H |D\varphi_n|^2 d\nu + \frac{1}{2} \int_H \text{Tr}[(D^2\varphi_n)^2]d\nu + \int_H |(-A)^{1/2} D\varphi_n|^2 d\nu$$

$$\le \left(\int_H \text{Tr}[(D^2\varphi_n)^2]d\nu \right)^{1/2} \left(\int_H |D\varphi_n|^4 d\nu \right)^{1/4} \left(\int_H |DU|^4 d\nu \right)^{1/4}$$

$$+2 \left(\int_H |N_0\varphi_n|^2 d\nu \right)^{1/2} \left(\int_H |D\varphi_n|^4 d\nu \right)^{1/2} \left(\int_H |F_n|^4 d\nu \right)^{1/2}$$

$$-2 \int_H N_0\varphi_n \, f \, d\nu. \quad (12.2.8)$$

To conclude, we need another independent estimate for $\int_H |D\varphi_n|^4$. We have

$$\int_H |D\varphi_n|^4 d\nu = \sum_{h,k=1}^{\infty} \int_H (D_h\varphi_n)^2 (D_k\varphi_n)^2 d\nu$$

$$= \sum_{h,k=1}^{\infty} \int_H D_h\varphi_n \left(D_h\varphi_n \left(D_k\varphi_n \right)^2 \right) d\nu$$

$$= -\sum_{h,k=1}^{\infty} \int_H \varphi_n D_h^2\varphi_n \left(D_k\varphi_n \right)^2 d\nu$$

$$-2 \sum_{h,k=1}^{\infty} \int_H \varphi_n \, D_h\varphi_n \, D_k\varphi_n \, D_h D_k\varphi_n \, d\nu$$

$$+2 \sum_{h,k=1}^{\infty} \int_H \varphi_n \, D_h U \, D_h\varphi_n \left(D_k\varphi_n \right)^2 d\nu$$

$$+ \sum_{h,k=1}^{\infty} \frac{1}{\lambda_h} \int_H x_h\varphi_n \, D_h\varphi_n \left(D_k\varphi_n \right)^2 d\nu.$$

Therefore

$$\int_H |D\varphi_n|^4 d\nu = -\int_H \varphi_n \, \mathrm{Tr}[D^2\varphi_n] \, |D\varphi_n|^2 d\nu$$

$$-2 \int_H \varphi_n \langle D^2\varphi_n \, D\varphi_n, D\varphi_n \rangle d\nu$$

$$+2 \int_H \varphi_n \langle DU, D\varphi_n \rangle |D\varphi_n|^2 d\nu$$

$$-2 \int_H \varphi_n \langle Ax, D\varphi_n \rangle |D\varphi_n|^2 d\nu,$$

and hence

$$\int_H |D\varphi_n|^4 d\nu = -2 \int_H \varphi_n \, N_0\varphi_n |D\varphi_n|^2 d\nu$$

$$-2 \int_H \varphi_n \langle D^2\varphi_n \, D\varphi_n, D\varphi_n \rangle d\nu.$$

This yields

$$\int_H |D\varphi_n|^4 d\nu \leq 2\frac{\|f\|_0}{\lambda} \left(\int_H |N_0\varphi_n|^2 d\nu \right)^{1/2} \left(\int_H |D\varphi_n|^4 d\nu \right)^{1/2}$$

$$+ 2\frac{\|f\|_0}{\lambda} \int_H (\mathrm{Tr}[(D^2\varphi_n)^2] d\nu)^{1/2} \left(\int_H |D\varphi_n|^4 d\nu \right)^{1/2}.$$

$$(12.2.9)$$

Now, comparing (12.2.8) with (12.2.9), we see that there exists $M > 0$ such that

$$\frac{1}{2} \int_H \mathrm{Tr}[(D^2\varphi_n)^2] d\nu + \int_H |(-A)^{1/2} D\varphi_n|^2 d\nu + \int_H |D\varphi_n|^4 d\nu \leq M,$$

$$(12.2.10)$$

and the conclusion follows. $\square$

12.3 The case when U is convex

Here we assume, besides (12.1.1)-(12.1.3), that

$$\begin{cases} (i) & U \text{ is convex, nonnegative and lower semicontinuous,} \\ (ii) & \exists \varepsilon > 0 \text{ such that } \int_H |DU|^{2+\varepsilon} d\nu < +\infty. \end{cases} \quad (12.3.1)$$

Our goal is to show that N_2 is self-adjoint and that $P_t = e^{tN_2}$ is strong Feller. We follow here [78] and [81], see also [96].

It is convenient to introduce some approximations of U that are convex and regular, namely the Yosida approximations U_α of U:

$$U_\alpha(x) = \inf \left\{ U(y) + \frac{|x-y|^2}{2\alpha} : y \in H \right\}, \quad x \in H, \ \alpha > 0.$$

The following properties are well known, see e.g. [64]:

$$\lim_{\alpha \to 0} U_\alpha(x) = U(x), \ x \in H, \ \alpha > 0. \quad (12.3.2)$$

$$\lim_{\alpha \to 0} DU_\alpha(x) = DU(x), \ x \in H \ \mu\text{-a.e.}, \ \alpha > 0. \quad (12.3.3)$$

$$|DU_\alpha(x)| \leq |DU(x)|, \ x \in H \ \mu\text{-a.e.}, \ \alpha > 0. \quad (12.3.4)$$

Moreover $U_\alpha \in C^1(H)$ and DU_α is Lipschitz continuous.

Let us introduce as before an approximating equation,

$$\lambda \varphi_\alpha - L_p \varphi_\alpha + \langle DU_\alpha, D\varphi_\alpha \rangle = f, \quad \lambda > 0, \tag{12.3.5}$$

with $p \geq 1$. Here by L_p we mean the infinitesimal generator of the Ornstein-Uhlenbeck semigroup (R_t) in $L^p(H, \mu)$.

By a *solution* of (12.3.5) we mean a function $\varphi_\alpha \in C_b^{0,1}(H) \cap D(L_p)$, $p \geq 1$ which fulfills (12.3.5). ([1])

Notice that this definition is meaningful. In fact if $\varphi_\alpha \in C_b^{0,1} \cap D(L_p)$, we have, in view of (12.3.5), $D\varphi_\alpha \in L^p(H, \mu) \cap L^\infty(H, \mu)$; therefore $\langle DU_\alpha, D\varphi_\alpha \rangle \in L^p(H, \mu)$ since DU_α is Lipschitz continuous.

Lemma 12.3.1 *Let $f \in C_b^1(H)$, $\lambda, \alpha > 0$. Then equation (12.3.5) has a unique solution $\varphi_\alpha \in C_b^{0,1} \cap D(L_p)$, $p \geq 1$. Moreover the following estimate holds:*

$$\|\varphi_\alpha\|_1 \leq \frac{1}{\lambda} \|Df\|_0, \quad \lambda, \alpha > 0. \tag{12.3.6}$$

Proof. Since DU_α is Lipschitz continuous, we know by Chapter 7 that the stochastic differential equation

$$dX = (AX - DU_\alpha(X))dt + dW_t, \quad X(0) = x, \tag{12.3.7}$$

has a unique solution $X_\alpha(\cdot, x)$. Set

$$\varphi_\alpha(x) = \int_0^{+\infty} e^{-\lambda t} \mathbb{E}\left[f(X_\alpha(t, x))\right] dt, \quad x \in H.$$

We claim that φ_α belongs to $D(L_p) \cap C_b^{0,1}(H)$ and is a solution of (12.3.5). In fact $\varphi_\alpha \in C_b^{0,1}(H)$ since $f \in C_b^1(H)$ and for all $x, y \in H$,

$$|\mathbb{E}\left[f(X_\alpha(t, x))\right] - \mathbb{E}\left[f(X_\alpha(t, y))\right]| \leq \|f\|_1 \mathbb{E}|X_\alpha(t, x) - X_\alpha(t, y)|$$

$$\leq e^{-\omega t} \|f\|_1 |x - y|.$$

Moreover, since $f - \langle DU_\alpha, D\varphi_\alpha \rangle$ is Lipschitz continuous, it is not difficult to show that $\varphi_\alpha \in D(L_p)$ and (12.3.5) holds. $\square$

Now we are ready to prove the following result.

---

[1]We recall that $C_b^{0,1}(H)$ is the subspace of $C_b(H)$ of all Lipschitz continuous mappings.

Theorem 12.3.2 *Assume, besides* (12.1.1)-(12.1.3), *that* (12.3.1) *holds. Then N_0 is essentially self-adjoint in $L^2(H,\nu)$. Moreover the following identity holds for the closure N_2 of N_0:*

$$\int_H N_2\varphi\,\psi d\nu = -\frac{1}{2}\int_H \langle D\varphi, D\psi\rangle d\nu, \quad \varphi,\psi \in D(N_2). \tag{12.3.8}$$

Proof. We are going to prove that for $\lambda > 0$, the closure of $(\lambda - N_0)(\mathcal{E}_A(H))$ contains $C_b^1(H)$. Since $C_b^1(H)$ is dense in $L^2(H,\nu)$, this will prove the result by the Lumer-Phillips theorem [161] .

Let in fact $f \in C_b^1(H)$, $\lambda, \alpha > 0$ and let φ_α be the solution to (12.3.5), with $p \geq 4$. Since $\mathcal{E}_A(H)$ is a core for L_p (Proposition 10.2.1), there exists a sequence $(\varphi_\alpha^{(n)}) \subset \mathcal{E}_A(H)$ such that

$$\varphi_\alpha^{(n)} \to \varphi_\alpha, \quad L_p\varphi_\alpha^{(n)} \to L_p\varphi_\alpha, \quad \text{in } L^p(H,\mu), \tag{12.3.9}$$

as n tends to ∞. Since, by Proposition 10.6.1, $D(L) \subset W^{1,p}(H,\mu)$ with continuous inclusion, we have

$$D\varphi_\alpha^{(n)} \to D\varphi_\alpha, \quad \text{in } L^p(H,\mu;H), \tag{12.3.10}$$

as n tends to ∞. We claim that

$$\lim_{n\to\infty} \langle DU_\alpha, D\varphi_\alpha^{(n)}\rangle = \langle DU_\alpha, D\varphi_\alpha\rangle \quad \text{in } L^2(H,\nu). \tag{12.3.11}$$

In fact, since DU_α is Lipschitz continuous, there exists a constant $C_\alpha > 0$ such that $|DU_\alpha(x)| \leq C_\alpha(1+|x|)$, $x \in H$. It follows that

$$\int_H |\langle DU_\alpha, D\varphi_\alpha^{(n)} - D\varphi_\alpha\rangle|^2 d\nu$$

$$\leq Z^{-1}C_\alpha^2 \int_H (1+|x|)^2|D\varphi_\alpha^{(n)} - D\varphi_\alpha|^2 e^{-2U} d\mu$$

$$\leq Z^{-1}C_\alpha^2 \left(\int_H (1+|x|)^4 e^{-4U} d\mu\right)^{1/2}\left(\int_H |D\varphi_\alpha^{(n)} - D\varphi_\alpha|^4 d\mu\right)^{1/2},$$

and so (12.3.11) follows.

Set now

$$f_\alpha^{(n)} = \lambda\varphi_\alpha^{(n)} - L_p\varphi_\alpha^{(n)} + \langle DU_\alpha, D\varphi_\alpha^{(n)}\rangle.$$

Then, since $L^4(H,\mu) \subset L^2(H,\nu)$, with continuous embedding, we have, taking into account (12.3.11), that $f_\alpha^{(n)} \to f$ in $L^2(H,\nu)$ as $n \to \infty$. But we have

$$\lambda\varphi_\alpha^{(n)} - N_0\varphi_\alpha^{(n)} = f_\alpha^{(n)} + \langle (DU - DU_\alpha), D\varphi_\alpha^{(n)}\rangle.$$

Then by (12.3.1) it follows that

$$\lim_{n \to \infty} \langle (DU - DU_\alpha), D\varphi_\alpha^{(n)} \rangle = \langle (DU - DU_\alpha), D\varphi_\alpha \rangle \quad \text{in } L^2(H, \nu).$$

Consequently

$$\lim_{n \to \infty} (\lambda - N_0)(\varphi_\alpha^{(n)}) = f + \langle DU - DU_\alpha, D\varphi_\alpha \rangle,$$

in $L^2(H, \nu)$. This implies

$$f + \langle DU - DU_\alpha, D\varphi_\alpha \rangle \in \overline{(\lambda - N_0)(\mathcal{E}_A(H))}. \tag{12.3.12}$$

We claim that

$$\lim_{\alpha \to 0} \langle DU - DU_\alpha, D\varphi_\alpha \rangle = 0 \quad \text{in } L^2(H, \nu). \tag{12.3.13}$$

We have in fact by (12.3.6),

$$\int_H |\langle DU - DU_\alpha, D\varphi_\alpha \rangle|^2 d\nu \le \frac{\|f\|_1}{\lambda^2} \int_H |DU - DU_\alpha|^2 d\nu.$$

Now (12.3.13) follows from (12.3.3). Therefore, letting α tend to 0 in (12.3.12), we have $f \in \overline{(\lambda - N_0)(\mathcal{E}_A(H))}$. This means that $\overline{(\lambda - N_0)(\mathcal{E}_A(H))}$ contains $C_b^1(H)$ and, consequently, coincides with $L^2(H, \nu)$.

Finally the identity (12.3.8) follows from (12.1.11). $\square$

Remark 12.3.3 From identity (12.3.8) it follows that, setting $\psi = 1$, ν is an invariant measure for the semigroup $P_t = e^{tN_2}$, $t \ge 0$. Moreover it implies that $D((-N_2)^{1/2}) = W^{1,2}(H, \nu)$.

Let us denote by (P_t^α) the transition semigroup

$$P_t^\alpha \varphi(x) = \mathbb{E}\left[\varphi(X_\alpha(t, x))\right], \quad \varphi \in C_b(H),$$

where X_α is the solution to (12.3.7). Clearly the probability measure ν_α defined by

$$\nu_\alpha(dx) = Z_\alpha^{-1} e^{-2U_\alpha(x)} \mu(dx), \quad Z_\alpha = \int_H e^{-2U_\alpha(y)} \mu(dy),$$

is invariant for (P_t^α) and ν_α is weakly convergent to ν as $\alpha \to 0$. The unique extension of (P_t^α) to $L^2(H, \nu_\alpha)$ will be denoted by the same symbol. Moreover we shall denote by N_α the infinitesimal generator of (P_t^α).

Proposition 12.3.4 *Let* $f \in C_b^1(H)$ *and* $\lambda > 0$. *Then the following statements hold.*

(i) We have $R(\lambda, N_2)f \in C_b^{0,1}(H)$ *and*

$$\|R(\lambda, N_2)f\|_1 \le \frac{1}{\lambda}\,\|f\|_1. \qquad (12.3.14)$$

(ii) For any $\alpha > 0$ *we have*

$$\|R(\lambda, N_2)f - R(\lambda, N_\alpha)f\|_{L^2(H,\nu)} \le \frac{1}{\lambda^2}\,\|f\|_1\|DU - DU_\alpha\|_{L^2(H,\nu)}. \qquad (12.3.15)$$

(iii) We have

$$\lim_{\alpha \to 0} R(\lambda, N_\alpha)f = R(\lambda, N_2)f \quad in\ L^2(H, \nu).$$

Proof. We first prove that $\varphi_\alpha = R(\lambda, N_\alpha)f$ belongs to the domain of N_2. Thanks to Proposition 11.2.10 there exists a sequence $(\varphi_\alpha^{(n)}) \subset \mathcal{E}_A(H)$ such that for $p \ge 4$,

$$\varphi_\alpha^{(n)} \quad \to \quad \varphi_\alpha, \ \ L_p\varphi_\alpha^{(n)} \to L_p\varphi_\alpha \ \ in\ L^p(H, \mu)$$

$$D\varphi_\alpha^{(n)} \quad \to \quad D\varphi_\alpha \ \ in\ L^p(H, \mu; H),$$

as n tends to ∞. Then the sequence $(N_0\varphi_\alpha^{(n)})$ is convergent in $L^2(H, \nu)$, so that $\varphi_\alpha \in D(N)$.

Now from the identity

$$\lambda\varphi_\alpha - N_2\varphi_\alpha = f + \langle DU_\alpha - DU, D\varphi_\alpha\rangle,$$

and (12.3.6) it follows that

$$\|R(\lambda, N_2)f - R(\lambda, N_\alpha)f\|_{L^2(H,\nu)} \ \le \ \frac{1}{\lambda}\,\|\langle DU - DU_\alpha, D\varphi_\alpha\rangle\|_{L^2(H,\nu)}$$

$$\le \ \frac{1}{\lambda^2}\,\|f\|_1\|DU - DU_\alpha\|_{L^2(H,\nu)}.$$

Thus (ii) and (iii) are proved. Finally (i) follows from the estimate

$$\|R(\lambda, N_\alpha)f\|_1 \le \frac{1}{\lambda}\,\|f\|_1. \qquad \square$$

Theorem 12.3.5 *For any $f \in \mathcal{E}_A(H)$ we have*

$$\lim_{\alpha \to 0} P_t^\alpha f = e^{tN_2} f \quad \text{in } L^2(H,\nu). \tag{12.3.16}$$

Proof. We recall that, by the Hille-Yosida theorem, for any $\alpha > 0$ we have

$$P_t^\alpha f = \lim_{n \to \infty} \left(1 - \frac{t}{n} N_\alpha\right)^{-n} f \quad \text{in } L^2(H,\nu_\alpha),$$

and

$$P_t f = \lim_{n \to \infty} \left(1 - \frac{t}{n} N_2\right)^{-n} f \quad \text{in } L^2(H,\nu),$$

Moreover

$$\left|P_t^\alpha f - \left(1 - \frac{t}{n} N_\alpha\right)^{-n} f\right|_{L^2(H,\nu_\alpha)} \leq \frac{1}{n} \|N_\alpha f\|_{L^2(H,\nu_\alpha)}, \tag{12.3.17}$$

and

$$\left|P_t f - \left(1 - \frac{t}{n} N_2\right)^{-n} f\right|_{L^2(H,\nu)} \leq \frac{1}{n} \|N_2 f\|_{L^2(H,\nu)}. \tag{12.3.18}$$

Taking into account the obvious inequality

$$\|\psi\|_{L^2(H,\nu)} \leq Z_\alpha^{1/2} Z^{-1/2} \|\psi\|_{L^2(H,\nu_\alpha)}, \tag{12.3.19}$$

it follows that

$$|P_t f - P_t^\alpha f|_{L^2(H,\nu)} \leq \left|P_t f - \left(1 - \frac{t}{n} N_2\right)^{-n} f\right|_{L^2(H,\nu)}$$

$$+ \left|\left(1 - \frac{t}{n} N\right)^{-n} f - \left(1 - \frac{t}{n} N_\alpha\right)^{-n} f\right|_{L^2(H,\nu)}$$

$$+ \left|P_t^\alpha f - \left(1 - \frac{t}{n} N_\alpha\right)^{-n} f\right|_{L^2(H,\nu)}$$

$$\leq \frac{1}{n}\left(\|N_2 f\|_{L^2(H,\nu)} + Z_\alpha^{1/2} Z^{-1/2} \|N_\alpha f\|_{L^2(H,\nu_\alpha)}\right)$$

$$+ \left|\left(1 - \frac{t}{n} N_2\right)^{-n} f - \left(1 - \frac{t}{n} N_\alpha\right)^{-n} f\right|_{L^2(H,\nu)}.$$

Moreover the following identity holds.

$$
\left(1 - \frac{t}{n} N_2\right)^{-n} f - \left(1 - \frac{t}{n} N_\alpha\right)^{-n} f
$$

$$
= \sum_{h=0}^{n-1} \left(1 - \frac{t}{n} N_2\right)^{-h} \left[\left(1 - \frac{t}{n} N_2\right)^{-1} - \left(1 - \frac{t}{n} N_\alpha\right)^{-1}\right]
$$

$$
\times \left(1 - \frac{t}{n} N_\alpha\right)^{-n+1+h} f.
$$

By Proposition 12.3.4 it follows that

$$
\left|\left(1 - \frac{t}{n} N_2\right)^{-n} f - \left(1 - \frac{t}{n} N_\alpha\right)^{-n} f\right|_{L^2(H,\nu)}
$$

$$
\leq n \|f\|_1 \|DU - DU_\alpha\|_{L^2(H,\nu)}.
$$

Finally, it follows that

$$
|P_t f - P_t^\alpha f|_{L^2(H,\nu)} \leq \frac{1}{n}\left(\|N_2 f\|_{L^2(H,\nu)} + Z_\alpha^{1/2} Z^{-1/2}\|N_\alpha f\|_{L^2(H,\nu_\alpha)}\right)
$$

$$
+ n\|f\|_1 \|DU - DU_\alpha\|_{L^2(H,\nu)},
$$

and the conclusion follows. □

We end the introductory part of this section by proving the strong Feller property of P_t.

Proposition 12.3.6 *Assume* (12.1.1)-(12.1.3) *and* (12.3.1). *Then the semigroup* $P_t = e^{tN_2}$, $t \geq 0$, *is strong Feller* .

Proof. By Theorem 7.7.1 we know that the semigroup (P_t^α) is strong Feller for any $\alpha > 0$. Moreover, for any f Borel and bounded the following estimate holds:

$$
|P_t^\alpha f(x) - P_t^\alpha f(y)| \leq t^{-1/2}\|f\|_0 |x - y|, \quad x, y \in H, \ t > 0. \qquad (12.3.20)
$$

Now the conclusion follows, letting α tend to 0. □

12.3.1   Poincaré and log-Sobolev inequalities

We first prove the strongly mixing property for the measure ν.

Proposition 12.3.7 *For any $\varphi \in L^2(H, \nu)$ we have*

$$\lim_{t \to +\infty} e^{tN_2}\varphi = \int_H \varphi d\nu = \overline{\varphi}. \tag{12.3.21}$$

Proof. It is enough to show (12.3.21) for $\varphi \in \mathcal{E}(H)$. In this case we have

$$|e^{tN_2}\varphi(x) - \overline{\varphi}| \le |e^{tN_2}\varphi(x) - P_t^\alpha \varphi(x)| + |P_t^\alpha \varphi(x) - \overline{\varphi}_\alpha| + |\overline{\varphi}_\alpha - \overline{\varphi}|, \tag{12.3.22}$$

where

$$\overline{\varphi}_\alpha = \int_H \varphi d\nu_\alpha.$$

By Theorem 12.3.5 for any $\varepsilon > 0$ there exists $\alpha_\varepsilon > 0$ such that for any $\alpha < \alpha_\varepsilon$

$$|\overline{\varphi}_\alpha - \overline{\varphi}_\alpha| \le \varepsilon, \quad \|e^{tN_2}\varphi - P_t^\alpha \varphi\|_{L^2(H,\nu)} \le \varepsilon.$$

Taking into account (12.3.19), we find from (12.3.22)

$$\|e^{tN_2}\varphi(x) - \overline{\varphi}\|_{L^2(H,\nu)} \le 2\varepsilon + \|P_t^\alpha \varphi - \overline{\varphi}_\alpha\|_{L^2(H,\nu)}$$

$$\le \varepsilon + Z_\alpha^{1/2} Z^{-1/2} \|P_t^\alpha \varphi - \overline{\varphi}_\alpha\|_{L^2(H,\nu_\alpha)}.$$

Recalling that P_t^α is strongly mixing, see G. Da Prato and J. Zabczyk [102], the conclusion follows. $\square$

Theorem 12.3.8 *For all $\varphi \in W^{1,2}(H, \nu)$ we have*

$$\int_H |\varphi - \overline{\varphi}|^2 d\nu \le \frac{1}{2\omega} \int_H |D\varphi|^2 d\nu. \tag{12.3.23}$$

Proof. The Poincaré inequality for the measure ν_α follows from Proposition 11.2.17:

$$\int_H |\varphi - \overline{\varphi}_\alpha|^2 d\nu_\alpha \le \frac{1}{2\omega} \int_H |D\varphi|^2 d\nu_\alpha. \tag{12.3.24}$$

Now it is enough to let α tend to 0. $\square$

In a similar way, see Proposition 11.2.19, we can prove the log-Sobolev inequality.

Theorem 12.3.9 *For all $\varphi \in W^{1,2}(H,\nu)$ we have*

$$\int_H \varphi^2 \log(\varphi^2) d\nu \le \frac{1}{2\omega} \int_H |D\varphi|^2 d\nu + \|\varphi\|_2^2 \log(\|\varphi\|_2^2), \qquad (12.3.25)$$

where

$$\|\varphi\|_2^2 = \int_H |\varphi(x)|^2 \nu(dx).$$

Part III

APPLICATIONS TO CONTROL THEORY

Chapter 13

Second order Hamilton-Jacobi equations

This chapter is devoted to the study of Hamilton-Jacobi equations in a separable Hilbert space H, and to their relationship with stochastic optimal control problems . We shall be essentially concerned with the case of semilinear equations, which correspond to stochastic optimal control problems driven by additive noise.

More precisely let us consider the following controlled system on H :

$$\begin{cases} dX(t) = (AX(t) + G(X(t)) + z(t))dt + Q^{1/2}dW(t), \ t \in [0,T], \\ X(0) = x \in H, \end{cases}$$

(13.0.1)

where $A : D(A) \subset H \to H$ is a linear operator, $G : H \to H$ is a continuous mapping, Q is a symmetric nonnegative operator on H and W is a cylindrical Wiener process on a probability space $(\Omega, \mathcal{F}, \mathbb{P})$, see Chapter 7. X represents the *state*, z the *control* and $T > 0$ is fixed.

Under suitable assumptions, which we will make precise later, problem (13.0.1) has a unique mild solution which we shall denote by $X(t, x; z)$ or, for short, by $X(t, x)$.

Given $g, \varphi \in UC_b(H)$ and a convex lower semicontinuous function $h : H \to [0, +\infty)$, we want to minimize the cost functional

$$J(x, z) = \mathbb{E} \left(\int_0^T [g(X(t, x; z)) + h(z(t))] \, dt + \varphi(X(T, x; z)) \right), \quad (13.0.2)$$

over all $z \in L^2_W(0, T; L^2(\Omega, H))$, the Hilbert space of all square integrable processes adapted to W defined on $[0, T]$ and with values in H.

We denote by

$$J^*(x) = \inf\{J(x, z) : z \in L^2_W(0, T; L^2(\Omega, H))\}$$

the *value function* of the problem.

Using the dynamic programming approach, we look for a regular solution of the Hamilton-Jacobi equation ([1])

$$\begin{cases} D_t u = \frac{1}{2}\mathrm{Tr}[QD^2 u] + \langle Ax + G(x), Du \rangle - F(Du) + g, \\ u(0) = \varphi, \end{cases} \qquad (13.0.3)$$

where the *Hamiltonian* F is given by the Legendre transform of h:

$$F(x) = \sup_{y \in H} \{\langle x, y \rangle - h(y)\}, \quad x \in H. \qquad (13.0.4)$$

Then we show that the *optimal control* z^* is related to the *optimal state* X^* by the *feedback* formula

$$z^*(t) = -DF(Du(T - t, X^*(t, x))), \ t \in [0, T], \qquad (13.0.5)$$

where X^* is the solution of the *closed loop equation*

$$\begin{cases} dX = (AX + G(X) - DF(Du(T - t, X)))dt + Q^{1/2}dW(t), \ t \in [0, T], \\ X(0) = x \in H. \end{cases}$$

$$(13.0.6)$$

Finally, the *optimal cost* is given by

$$J^*(x) = u(T, x), \quad x \in H. \qquad (13.0.7)$$

Possible choices of h are the following.

(i) Let $h(x) = \frac{1}{2}|x|^2$, $x \in H$. Then the Hamiltonian F is given by $F(x) = \frac{1}{2}|x|^2$, $x \in H$.

(ii) Let $R > 0$ and $h(x) = \begin{cases} \frac{1}{2}|x|^2, \text{ if } |x| \le R, \\ +\infty, \text{ if } |x| > R. \end{cases}$

Then $F(x) = \begin{cases} \frac{1}{2}|x|^2, \text{ if } |x| \le R, \\ R|x| - \frac{R^2}{2}, \text{ if } |x| > R. \end{cases}$

---

[1]We set in the following $D = D_x$, $D^2 = D_x^2$.

Note that in case (a) F is locally Lipschitz continuous, whereas in case (b) it is Lipschitz continuous.

We shall also consider the problem with *infinite horizon* that consists in minimizing the cost functional

$$J_\infty(x, z) = \mathbb{E}\left(\int_0^{+\infty} e^{-\lambda t}\left[g(X(t, x; z)) + h(z(t))\right] dt\right), \qquad (13.0.8)$$

over all $z \in L_W^2(0, +\infty; H)$, where $\lambda > 0$ is given. We denote by

$$J_\infty^*(x) = \inf\{J(x, z) : z \in L_W^2(0, +\infty; H)\}$$

the *optimal cost*.

In this case, we look for a regular solution of the stationary Hamilton-Jacobi equation

$$\lambda\varphi - \frac{1}{2}\text{Tr}[QD^2\varphi] - \langle Ax + G(x), D\varphi\rangle + F(D\varphi) = g. \qquad (13.0.9)$$

We show that the *optimal control* z^* is related to the *optimal state* X^* by the *feedback* formula

$$z^*(t) = -DF(D\varphi(X^*(t, x))), \ t \geq 0, \qquad (13.0.10)$$

where X^* is the solution of the *closed loop equation*

$$\begin{cases} dX = (AX + G(X) - DF(D\varphi(X)))dt + Q^{1/2}dW(t), \ t \geq 0, \\ X(0) = x \in H. \end{cases} \qquad (13.0.11)$$

Second order Hamilton-Jacobi equations are a classical subject in the finite dimensional case, see the monograph by W. Fleming and M. Soner [114] and references therein.

In infinite dimensions they were studied first, when the cost functional is convex, by V. Barbu and G. Da Prato [9] and then by M. G. Crandall and P. L. Lions, see [59], using the approach of viscosity solutions . For a presentation of this method see M. G. Crandall, H. Ishii and P. L. Lions [58]. The viscosity approach applies to fully nonlinear equations but does not allow us to prove existence of regular solutions , and so the feedback formula (13.0.5) can be only intendended in a weak sense. Also it requires, with the exception of the paper by F. Gozzi, E. Rouy and A. Swiech, see [136], that the operator Q is of trace class.

The approach presented here consists in solving (under suitable assumptions) the Hamilton-Jacobi equation (13.0.3) by generalizing the classical finite dimensional theory of semilinear evolution equations to the infinite dimensional case. This will make meaningful the feedback formula (13.0.5).

This approach was introduced in G. Da Prato [65], then it was developped by P. Cannarsa and G. Da Prato in [28], [30], by F. Gozzi in [133], [134], by F. Gozzi and E. Rouy in [135]. Recently Hamilton-Jacobi equations have been also considered in spaces $L^2(H, \mu)$ where μ is a suitable probability measure on H, see [55], [56], [57], [128].

For the sake of simplicity we shall present here only basic facts concerning a simple situation when the nonlinear mapping G is Lipschitz or locally Lipschitz continuous. However, these results can be considered as a first step in studying more general situations.

Some more general Hamilton-Jacobi equations, arising in optimal control problems for the reaction-diffusions equation, have been studied by S. Cerrai in [39], [42] and in the monograph [43]. Moreover control of turbulence for Burgers and Navier-Stokes equations has been considered by G. Da Prato and A. Debussche in [83], [84] and [85].

We quote also some recent results about the *ergodic control* by G. Tessitore [208] and B. Goldys and B. Maslowski; see [131].

Finally, very few results seem to be available for controlled equations with multiplicative noise. A systematic study of this case has been started by M. Fuhrman and M. Tessitore, by using backward equations; see [122], [123]. We also recall an approach based on the Hopf transformation due to G. Da Prato and J. Zabczyk [103].

Here, §13.1 is devoted to assumptions and some preliminaries. We solve the Hamilton-Jacobi equation when the Hamiltonian H is Lipschitz continuous in §13.2 and in §13.3 when it is locally Lipschitz. Applications to the control problem are presented in §13.4.

13.1 Assumptions and setting of the problem

We shall assume that

$$
\begin{cases}
(i) & A \text{ is the infinitesimal generator of a } C_0 \text{ semigroup } e^{tA} \text{ on } H, \\
(ii) & Q \in L^+(H) \text{ and } \int_0^t \mathrm{Tr}[e^{sA}Qe^{sA^*}]ds < +\infty, \ t > 0, \ x \in H, \\
(iii) & G \in C_b^1(H; H), \\
(iv) & e^{tA}(H) \subset Q^{1/2}(H) \text{ and there is } \beta \in (0,1) \text{ such that} \\
& \quad \|Q^{-1/2}e^{tA}\| \leq Ct^{-\frac{\beta}{2}}, \ t \in (0,1].
\end{cases}
$$

$$(13.1.1)$$

By assumptions (13.1.1)(i), (iii) it follows, recalling results of Chapter 7, that the differential stochastic equation

$$\begin{cases} dX(t) = (AX(t) + G(X(t)) + z(t))ds + Q^{1/2}dW(t), \ t \in [0, T], \\ X(0) = x \in H, \end{cases}$$

$$(13.1.2)$$

has a unique *mild* solution $X(\cdot) = X(\cdot, x)$ on $[0, T]$. That is $X(\cdot)$ is an adapted stochastic process such that $X(0) = x$ and for $t \in [0, T]$,

$$X(t, x) = e^{tA}x + \int_0^t e^{(t-s)A}z(s)ds + \int_0^t e^{(t-s)A}G(X(s, x))ds + W_A(t),$$

where

$$W_A(t) = \int_0^t e^{(t-s)A}Q^{1/2}dW(s) \qquad (13.1.3)$$

is the *stochastic convolution*.

Remark 13.1.1 The assumption (13.1.1)(iv) implies that

$$\|\Lambda_t\| = \|Q_t^{-1/2}e^{tA}\| \le Kt^{-\frac{1+\beta}{2}}, \ t \in [0, 1], \qquad (13.1.4)$$

where

$$Q_t x = \int_0^t e^{sA}Qe^{sA^*}x\,dx, \quad t \ge 0, \ x \in H.$$

Therefore, the Ornstein-Uhlenbeck semigroup (R_t) introduced in Chapter 6,

$$R_t\varphi(x) = \int_H \varphi(e^{tA}x + y)N_{Q_t}(dy), \ \varphi \in UC_b(H),$$

is *strong Feller*.

To prove (13.1.4), let us consider the controlled system

$$\xi' = A\xi + Q^{1/2}u, \quad \xi(0) = x,$$

and notice that, for fixed $T > 0$, the control

$$u(t) = -\frac{1}{T}\,Q^{-1/2}e^{tA}x, \ t \in [0, T],$$

drives x to 0 in time T. Since, in view of (13.1.1)(iv) (see Appendix B),

$$|\Lambda_T x|^2 \le \int_0^T |u(t)|^2 dt \le \frac{c^2}{T^{1+\beta}}|x|^2,$$

the conclusion follows.

Notice that in hypothesis 13.1.1(iv) one could replace the condition $\|Q^{-1/2}e^{tA}\| \le Ct^{-\frac{\beta}{2}}$, $t \in (0, 1]$, with (13.1.4).

Example 13.1.2 Let $d \in \mathbb{N}$, $H = L^2([0, \pi]^d)$, and let A be the Laplace operator on $\mathcal{O} = [0, \pi]^d$ with Dirichlet boundary conditions:

$$Ax = \Delta x = \sum_{k=1}^{d} D_k^2 x, \ x \in D(A) = H^2(\mathcal{O}) \cap H_0^1(\mathcal{O}). \qquad (13.1.5)$$

As is well known, A is self-adjoint and

$$Ae_k = -|k|^2 e_k, \ k = (k_1, \dots, k_d) \in \mathbb{N}^d,$$

where $e_k(\xi) = 2^{d/2} \sin k_1 \xi_1 \dots \sin k_d \xi_d$.

Let us choose $Q = (-A)^{-\beta}$ with $\beta \geq 0$. Then we have

$$Qe_k = |k|^{-2\beta} e_k, \ k = (k_1, \dots, k_d) \in \mathbb{N}^d, \qquad (13.1.6)$$

where $\beta \geq 0$. As is easily seen, the operator Q_t is of trace class if and only if $2 + 2\beta > d$.

Thus we can take $Q = I$ if and only if $d = 1$, and $\beta < 1$ if and only if $d < 4$.

Finally, if $d < 4$ and $Q = (-A)^{-\beta}$ with $d/2 - 1 < \beta < 1$ we see which (13.1.1) holds.

It is useful to consider Galerkin approximations of problem (13.1.1). They will require additional assumptions (which are however often fulfilled), but they will allow simplifications in several proofs.

Let us assume

$$\begin{cases} (i) \quad \text{there are an orthonormal basis } (e_k) \subset D(A) \text{ and a sequence } (\lambda_k) \text{ of} \\ \text{positive numbers such that } Qe_k = \lambda_k e_k, \ k \in \mathbb{N}, \\ (ii) \quad \text{we have } \lim_{n \to \infty} e^{tA_n} x = e^{tA} x, \ x \in H, \text{ where } A_n = \Pi_n A \Pi_n \text{ and} \\ \Pi_n x = \sum_{k=1}^{n} \langle x, e_k \rangle e_k, \ x \in H, \\ (iii) \text{ there is } \beta \in [0, 1) \text{ such that } Q_n^{-1/2}(-A_n)^{-\beta/2} \in L(H), \text{ where} \\ Q_n = \Pi_n Q, \ A_n = \Pi_n A \Pi_n, \\ \text{moreover there is } C_1 > 0 \text{ such that } \|Q_n^{-1/2} e^{tA_n}\| \leq C_1 t^{-\beta/2}, \ t \in [0, 1]. \end{cases}$$
$$(13.1.7)$$

Now we consider the approximating problem

$$\begin{cases} dX_n(t) = (A_n X_n(t) + G_n(X_n(t)) + \Pi_n z(t))dt + \Pi_n Q^{1/2} dW(t), \\ X_n(0) = \Pi_n x, \end{cases}$$
$$(13.1.8)$$

where $G_n(x) = \Pi_n G(\Pi_n x)$, $x \in H$.

The mild solution X_n of problem (13.1.8) is given by the solution of the integral equation

$$X_n(t) = e^{tA_n}x + \int_0^t e^{(t-s)A_n}[G_n(X_n(s)) + \Pi_n z(s)]ds + W_A^n(t), \ t \geq 0,$$

(13.1.9)

where

$$W_{A,n}(t) = \int_0^t e^{(t-s)A_n}\Pi_n Q^{1/2}dW(s), \ t \geq 0.$$

(13.1.10)

Notice that $X_n(t) \in \Pi_n(H)$, and so problem (13.1.9) is in fact finite dimensional.

The following result was proved in G. Da Prato and J. Zabczyk [101].

Proposition 13.1.3 *Assume that (13.1.1) and (13.1.7) hold. Let $x \in H$, $n \in \mathbb{N}$. Then we have*

$$\lim_{n \to \infty} X_n(t) = X(t) \quad in \ L^2(\Omega; H),$$

(13.1.11)

uniformly on $t \in [0, T]$.

Proposition 13.1.4 *Assume that (13.1.1) and (13.1.7), hold. Let*

$$R_t^n \varphi(x) = \int_H \varphi(e^{tA_n}x + y)N_{Q_t^n}(dy), \quad \varphi \in UC_b(H),$$

where

$$Q_{n,t} = \int_0^t e^{sA_n}Qe^{sA_n^*}ds, \ t \geq 0.$$

Then there exists $C_2 > 0$, such that

$$|DR_t^n \varphi(x)| \leq C_2 t^{-\frac{1+\beta}{2}}\|\varphi\|_0, \quad x \in H, \varphi \in UC_b(H).$$

(13.1.12)

Moreover, for all $\varphi \in UC_b(H)$, we have

$$\lim_{n \to \infty} R_t^n \varphi(x) = R_t\varphi(x), \quad t \geq 0, x \in H,$$

(13.1.13)

and

$$\lim_{n \to \infty} DR_t^n \varphi(x) = DR_t\varphi(x), \quad t > 0, \ x \in H.$$

(13.1.14)

Proof. We first notice that (13.1.12) follows from Remark 13.1.1. We have moreover

$$\langle DR_t^n \varphi(x), h\rangle = \int_H \langle \Lambda_{n,t}h, Q_{n,t}^{-1/2}y\rangle \varphi(e^{tA_n}x + y)N_{Q_{n,t}}(dy),$$

where $\Lambda_{n,t} = Q_{n,t}^{-1/2}e^{tA_n}$. The conclusion follows letting n tend to ∞. $\square$

13.2 Hamilton-Jacobi equations with a Lipschitz Hamiltonian

We assume here that (13.1.1) and (13.1.7) hold and consider the problem

$$\begin{cases} D_t u = Lu + \langle G, Du \rangle + F(Du) + g, \\ u(0) = \varphi, \end{cases} \tag{13.2.1}$$

where L is the infinitesimal generator of the Ornstein-Uhlenbeck semigroup (R_t) introduced in §6.3.3 and $F : H \to \mathbb{R}$ is Lipschitz continuous. Moreover $\varphi, g \in UC_b(H)$ are fixed.

We write problem (13.2.1) in the integral (or *mild*) form

$$u(t, \cdot) = R_t \varphi + \int_0^t R_{t-s}[\langle G, Du(s, \cdot)\rangle + F(D(u(s, \cdot))) + g]ds. \tag{13.2.2}$$

We will look for a solution of (13.2.2) in the space Z_T consisting of all continuous bounded functions $u : [0, T] \times H \to \mathbb{R}$ such that $u(t, \cdot) \in UC_b^1(H)$ for all $t > 0$ and the mapping

$$(0, T] \times H \to H, \ (t, x) \to t^{\frac{1+\beta}{2}} Du(t, x),$$

is measurable and bounded.

It is easy to check that Z_T, endowed with the norm

$$\|u\|_{Z_T} = \sup_{t \in [0,T]} \|u(t, \cdot)\|_0 + \sup_{t \in [0,T]} t^{\frac{1+\beta}{2}} \|Du(t, \cdot)\|_0,$$

is a Banach space.

The following result was proved in P. Cannarsa and G. Da Prato [27] when $\beta = 0$, and in F. Gozzi [133] if $\beta \in (0, 1)$. For the linear case see §6.5

Theorem 13.2.1 *Assume that (13.1.1) hold, that $F : H \to \mathbb{R}$ is Lipschitz continuous and that $\varphi, g \in UC_b(H)$. Then equation (13.2.2) has a unique solution $u \in Z_T$.*

Proof. We write equation (13.2.2) as

$$u(t, \cdot) = \gamma(u)(t) + R_t \varphi + \int_0^t R_{t-s} g \, ds,$$

where

$$\gamma(u)(t) = \int_0^t R_{t-s}[\langle G, Du(s, \cdot)\rangle + F(D(u(s, \cdot)))]ds, \ t \in [0, T].$$

Thanks to (13.1.12) and (13.1.14) it is easy to see that γ maps Z_T into itself. Let us show that γ is a contraction on Z_T. In fact, if $u, \bar{u} \in Z_T$ we have ([2])

$$\|\gamma(u)(t, \cdot) - \gamma(\bar{u})(t, \cdot)\|_0 \leq (\|F\|_1 + \|G\|_1) \int_0^t \|Du(s, \cdot) - D\bar{u}(s, \cdot)\|_0 \, ds$$

$$\leq (\|F\|_1 + \|G\|_1) \int_0^t s^{-\frac{1+\beta}{2}} ds \, \|u - \bar{u}\|_{Z_T}.$$

$$= \frac{2}{1-\beta} t^{\frac{1-\beta}{2}} (\|F\|_1 + \|G\|_1) \|u - \bar{u}\|_{Z_T}.$$

Moreover, taking into account formulae (13.1.12) and (13.1.14),

$$\|D\gamma(u)(t, \cdot) - D\gamma(\bar{u})(t, \cdot)\|_0$$

$$\leq K(\|F\|_1 + \|G\|_1) \int_0^t (t - s)^{-\frac{1+\beta}{2}} \|Du(s, \cdot) - D\bar{u}(s, \cdot)\|_0 ds$$

$$\leq K(\|F\|_1 + \|G\|_1) \int_0^t (t - s)^{-\frac{1+\beta}{2}} s^{-\frac{1+\beta}{2}} ds \, \|u - \bar{u}\|_{Z_T}$$

$$\leq K(\|F\|_1 + \|G\|_1) t^{-\beta} \int_0^1 (1 - \sigma)^{-\frac{1+\beta}{2}} \sigma^{-\frac{1+\beta}{2}} d\sigma \, \|u - \bar{u}\|_{Z_T}.$$

It follows that

$$t^{\frac{1+\beta}{2}} \|D\gamma(u)(t, \cdot) - D\gamma(\bar{u})(t, \cdot)\|_0$$

$$\leq Kt^{\frac{1-\beta}{2}} (\|F\|_1 + \|G\|_1) \int_0^1 (1 - \sigma)^{-\frac{1+\beta}{2}} \sigma^{-\frac{1+\beta}{2}} d\sigma \, \|u - \bar{u}\|_{Z_T}, \ t \in [0, T].$$

Therefore γ is a contraction on Z_{T_0} provided T_0 is sufficiently small, and so there exists a unique solution of (13.2.2) on $[0, T_0]$. Finally this solution can be continued in the whole interval $[0, T]$ by a standard argument. $\square$

Let now consider the approximating problems

$$u_n(t, \cdot) = R_t^n \varphi \circ \Pi_n + \int_0^t R_{t-s}^n \left(\langle G_n, Du_n(s, \cdot) \rangle + F(Du_n(s, \cdot)) + g_n \right) ds.$$

$$(13.2.3)$$

By Proposition 13.1.4 we obtain, using standard arguments, the following result.

---

[2]We recall that $\|F\|_1$ is the Lipschitz norm of F.

Theorem 13.2.2 *Assume that* (13.1.1) *and* (13.1.7) *hold, and that* $F : H \to \mathbb{R}$ *is Lipschitz continuous and that* $\varphi, g \in UC_b(H)$. *Then equation* (13.2.3) *has a unique solution* u_n. *Moreover*

$$\lim_{n \to \infty} u_n(t, x) = u(t, x), \ t \in [0, T], \ x \in H, \quad (13.2.4)$$

and

$$\lim_{n \to \infty} Du_n(t, x) = Du(t, x), \ t \in [0, T], \ x \in H. \quad (13.2.5)$$

13.2.1   Stationary Hamilton-Jacobi equations

We assume here that (13.1.1) and (13.1.7) hold and that $F : H \to \mathbb{R}$ is Lipschitz continuous.

We are concerned with the stationary equation

$$\lambda\varphi - L\varphi - \langle G, D\varphi \rangle - F(D\varphi) = g, \quad (13.2.6)$$

where $\lambda > 0$ and $g \in UC_b(H)$ are given.

Let us consider the approximating equation

$$\lambda\varphi_n - L_n\varphi_n - \langle G_n, D\varphi_n \rangle - F(\Pi_n D\varphi_n) = g_n, \quad (13.2.7)$$

where $g_n(x) = g(\Pi_n(x))$, $x \in H$. Here L (resp. L_n) is the infinitesimal generator of (R_t) (resp. (R_t^n)).

By Corollary 6.4.3 it follows that $D(L) \subset UC_b^1(H)$. Moreover, in view of (13.1.12) and (13.1.14), there exists a constant $K_1 > 0$ such that for all $\lambda > 0$, and $g \in UC_b(H)$, we have

$$|DR(\lambda, L)g(x)| \le K_1\lambda^{\frac{\beta-1}{2}} \|g\|_0, \ x \in H. \quad (13.2.8)$$

We first prove existence of a solution to (13.2.6) when λ is sufficiently large.

Lemma 13.2.3 *Let* $\lambda_0 = K_1^{\frac{2}{1-\beta}}$, *where* K_1 *is defined in* (13.2.8). *Then for all* $\lambda > \lambda_0$ *and* $g \in UC_b(H)$ *there exist unique solutions* $\varphi \in D(L)$ *and* $\varphi_n \in D(L_n)$ *of* (13.2.6) *and* (13.2.7) *respectively.*

Moreover

$$\lim_{n \to \infty} \varphi_n(x) = \varphi(x), \quad x \in H. \quad (13.2.9)$$

Proof. Let $\lambda > \lambda_0$. Setting $\lambda\varphi - L\varphi = \psi$ equation (13.2.6) becomes

$$\psi - \langle G, DR(\lambda, L)\psi \rangle - F(DR(\lambda, L)\psi) = g.$$

For all $\psi, \overline{\psi} \in UC_b(H)$, we have

$$\|\langle G, DR(\lambda, L)\psi - R(\lambda, L)\overline{\psi} \rangle\|_0 \quad \leq \quad \|G\|_0 K_1 \lambda^{-\frac{1-\beta}{2}} \|\psi - \overline{\psi}\|_0,$$

$$\|F(DR(\lambda, L)\psi) - F(DR(\lambda, L)\overline{\psi})\|_0 \leq \|F\|_1 K_1 \lambda^{-\frac{1-\beta}{2}} \|\psi - \overline{\psi}\|_0.$$

Therefore the mapping

$$\psi \to \langle G, DR(\lambda, L)\psi \rangle - F(DR(\lambda, L)\psi)$$

is a contraction in $UC_b(H)$ and hence equation (13.2.6) has a unique solution φ.

Existence and uniqueness of a solution φ_n of (13.2.7) can be proved analogously. Finally (13.1.14) follows from a straightforward argument. $\square$

We want now to remove the condition $\lambda > \lambda_0$. To this purpose we introduce the operators

$$K\varphi = L\varphi + \langle G, D\varphi \rangle + F(D\varphi), \quad \varphi \in D(L), \qquad (13.2.10)$$

and

$$K_n\varphi = L_n\varphi + \langle G_n, D\varphi \rangle + F(\Pi_n D\varphi), \quad \varphi \in D(L_n). \qquad (13.2.11)$$

Notice that K_n is dissipative on $UC_b(H)$, by the maximum principle in finite dimensions. We recall that this implies

$$\|\varphi - \overline{\varphi}\|_0 \leq \|\varphi - \overline{\varphi} - \lambda(K_n\varphi - K_n\overline{\varphi})\|_0, \quad \lambda > 0, \; \varphi, \overline{\varphi} \in D(L_n).$$

Therefore, by Lemma 13.2.3, K_n is m-dissipative on $UC_b(H)$, see e.g. [64].

Theorem 13.2.4 *Assume that (13.1.1) and (13.1.7) hold and that $F : H \to \mathbb{R}$ is Lipschitz continuous. Then K is m-dissipative on $UC_b(H)$. Consequently for all $\lambda > 0$ and $g \in UC_b(H)$, there exists a unique solution $\varphi \in D(L)$ of (13.2.6).*

Proof. Let $\lambda > \lambda_0$, $\varphi, \overline{\varphi} \in D(L)$. Set

$$g = \lambda\varphi - K\varphi, \quad \overline{g} = \lambda\overline{\varphi} - K\overline{\varphi},$$

and $g_n(x) = g(\Pi_n x)$, $\bar{g}_n(x) = \bar{g}(\Pi_n x)$. Let φ_n, $\bar{\varphi}_n$ be such that

$$\lambda \varphi_n - K_n \varphi_n = g_n, \quad \lambda \bar{\varphi}_n - K_n \bar{\varphi}_n = \bar{g}_n.$$

From the dissipativity of K_n it follows that

$$\|\varphi_n - \bar{\varphi}_n\|_0 \leq \frac{1}{\lambda} \|g_n - \bar{g}_n\|_0.$$

Therefore

$$|\varphi_n(x) - \bar{\varphi}_n(x)| \leq \frac{1}{\lambda} \|g - \bar{g}\|_0, \quad x \in H.$$

Letting n tend to infinity and taking the supremum on x, we find

$$\|\varphi - \bar{\varphi}\|_0 \leq \frac{1}{\lambda} \|g - \bar{g}\|_0, \quad x \in H.$$

This shows that K is dissipative, and therefore is m-dissipative by Lemma 13.2.3. $\square$

We end this section with an estimate on the gradient of the solution φ of (13.2.6), independent of F. This estimate will be useful later for studying Hamilton-Jacobi equations with locally Lipschitz Hamiltonians.

Proposition 13.2.5 *Assume that* (13.1.1) *and* (13.1.7) *hold and that* $F \in C_b^1(H)$. *Assume in addition that* e^{tA} *is a contraction semigroup,* G *is dissipative,* $\lambda > 0$ *and* $g \in C_b^1(H)$. *Let* φ *be the solution to* (13.2.6). *Then we have*

$$\|D\varphi\|_0 \leq \|Dg\|_0. \tag{13.2.12}$$

Proof. In view of (13.1.14), it is enough to prove the result assuming that H is a n-dimensional space. Thus we write equation (13.2.6) as

$$\lambda \varphi - \frac{1}{2} \sum_{h=1}^{n} \lambda_h D_h^2 \varphi - \sum_{h,k=1}^{n} a_{h,k} x_k D_h \varphi - \sum_h^n G_h^n D_h \varphi - F(\Pi_n D\varphi) = g, \tag{13.2.13}$$

where $a_{h,k} = \langle Ae_k, e_h \rangle$ and $G_h^n = \langle G_n, e_h \rangle$, $h, k \in \mathbb{N}$.

Setting $\psi_j = D_j \varphi$, differentiating (13.2.13) with respect to x_j and multiplying both sides by $\bar{\psi}_j$, we obtain

$$\lambda \psi_j^2 - \frac{1}{2} \sum_{h=1}^{n} \lambda_h D_h^2 \psi_j \, \psi_j - \sum_{h,k=1}^{n} a_{h,k} x_k D_h \psi_j \, \psi_j - \sum_{h=1}^{n} G_h^n D_h \psi_j \, \psi_j$$

$$-\sum_{h=1}^{n} (D_j G_h^n) \psi_h \psi_j - \sum_{h=1}^{n} a_{h,j} \psi_h \psi_j - \sum_{h=1}^{n} D_h F(\Pi_n D\varphi) D_h \psi_j \, \psi_j = D_j g \, \psi_j. \tag{13.2.14}$$

Since

$$D_h^2 \psi_j \, \psi_j = \frac{1}{2} \, D_h^2(\psi_j^2) - (D_h \psi_j)^2,$$

(13.2.14) can be written as

$$\lambda \psi_j^2 \quad - \quad \frac{1}{4} \sum_{h=1}^{n} \lambda_h D_h^2(\psi_j^2) + \frac{1}{2} \sum_{h=1}^{n} \lambda_h (D_h \psi_j)^2 - \frac{1}{2} \sum_{h,k=1}^{n} a_{h,k} x_k D_h(\psi_j^2)$$

$$-\frac{1}{2} \sum_{h=1}^{n} G_h^n D_h(\psi_j^2) - \sum_{h=1}^{n} a_{h,j} \psi_h \psi_j - \sum_{h=1}^{n} (D_j G_h^n) \psi_h \psi_j$$

$$-\frac{1}{2} \sum_{h=1}^{n} D_h F(\Pi_n D\varphi) D_h(\psi_j^2) = D_j g \psi_j. \qquad (13.2.15)$$

Summing up over j and setting $z = |D\varphi|^2$, we find

$$\lambda z - \frac{1}{4} \, \mathrm{Tr}[Q D^2 z] + \frac{1}{2} \sum_{h,j=1}^{n} \lambda_h (D_h \psi_j)^2 - \langle Ax, Dz \rangle - \langle G(x), Dz \rangle$$

$$-\langle AD\varphi, D\varphi \rangle - \langle DG \cdot D\varphi, D\varphi \rangle - \frac{1}{2} \langle DF(\Pi_n D\varphi), Dz \rangle = \langle D\varphi, Dg \rangle. \qquad (13.2.16)$$

Therefore

$$\lambda z \quad - \quad \frac{1}{4} \, \mathrm{Tr}[Q D^2 z] - \langle Ax, Dz \rangle - \langle G(x), Dz \rangle - \frac{1}{2} \langle DF(\Pi_n D\varphi), Dz \rangle$$

$$\leq \langle D\varphi, Dg \rangle.$$

By the maximum principle in finite dimensions it follows that $\|D\varphi\|_0^2 \leq \frac{1}{\lambda} \|Dg\|_0 \|D\varphi\|_0$, which yields the conclusion. $\square$

13.3 Hamilton-Jacobi equation with a quadratic Hamiltonian

We assume here that (13.1.1) and (13.1.7) hold. We are concerned with existence and uniqueness of solutions for the problem

$$\begin{cases} D_t u = Lu + \langle G, Du \rangle - \frac{1}{2} |Du|^2 + g, \\ u(0) = \varphi, \end{cases} \qquad (13.3.1)$$

with $\varphi, g \in UC_b(H)$. Again we consider the approximating problem

$$\begin{cases} D_t u_n = L_n u_n + \langle G_n, Du_n \rangle - \frac{1}{2}|\Pi_n Du_n|^2 + g_n, \\ u(0) = \varphi_n, \end{cases} \qquad (13.3.2)$$

For a more general method that does not require Galerkin approximations see F. Gozzi [134].

We write problems (13.3.1) and (13.3.2) in integral form:

$$u(t, \cdot) = R_t \varphi + \int_0^t R_{t-s} \left[\langle G, Du(s, \cdot) \rangle - \frac{1}{2}|Du(s, \cdot)|^2 + g \right] ds, \qquad (13.3.3)$$

$$u_n(t, \cdot) = R_t^n \varphi_n + \int_0^t R_{t-s}^n \left[\langle G_n, Du_n(s, \cdot) \rangle - \frac{1}{2}|\Pi_n Du_n(s, \cdot)|^2 + g_n \right] ds. \qquad (13.3.4)$$

We will look as before for a solution of (13.3.1) in the space Z_T introduced previously.

Theorem 13.3.1 *Assume that (13.1.1) and (13.1.7) hold and in addition that $\|e^{tA}\| \le e^{\omega t}$, $t \ge 0$, for some $\omega \in \mathbb{R}$, and G is dissipative. Then for any $\varphi \in C_b^1(H)$ the equations (13.3.3) and (13.3.4) have unique gs u and u_n respectively. Moreover*

$$\lim_{n \to \infty} u_n(t, x) = u(t, x), \ t \ge 0, \ x \in H, \qquad (13.3.5)$$

and

$$\lim_{n \to \infty} Du_n(t, x) = Du(t, x), \ t > 0, \ x \in H. \qquad (13.3.6)$$

Proof. Since by Proposition 13.1.4 $|DR_t^n \varphi(x)| \le C_1 t^{-\frac{1+\beta}{2}} \|\varphi\|_0$, existence and uniqueness of local solutions in a suitable interval $[0, T_0]$ (as well as the convergence of u_n to u in Z_{T_0}) follows by a standard argument, see P. Cannarsa and G. Da Prato [30] for details. To continue the solution to $[0, T]$ it is enough to find an a priori estimate for the finite dimensional parabolic problem (13.3.2), which we write in the following form (dropping the index

n in u for simplicity):

$$
\begin{cases}
D_t u = \dfrac{1}{2} \sum_{h=1}^{n} \lambda_h D_h^2 u + \sum_{h,k=1}^{n} a_{h,k} x_k D_h u \\[2mm]
\qquad + \sum_{h}^{n} G_h^n D_h u - \dfrac{1}{2} \sum_{h=1}^{n} (D_h u)^2 + g_n, \\[3mm]
u(0,x) = \varphi_n,
\end{cases}
\tag{13.3.7}
$$

where $D_h u = \langle Du, e_h \rangle$, $a_{h,k} = \langle Ae_k, e_h \rangle$, $G_h^n = \langle G_n, e_h \rangle$, $x_k = \langle x, e_k \rangle$, $g_n(x) = g(\Pi_n x)$ and $\varphi_n(x) = \varphi(\Pi_n x)$.

Since the local solution is regular, we can assume from now on that $\varphi, g \in UC_b^1(H)$.

Setting $v_j = D_j u$, and differentiating with respect to x_j, we obtain

$$
\begin{aligned}
D_t v_j =\ & \frac{1}{2} \sum_{h=1}^{n} \lambda_h D_h^2 v_j + \sum_{h,k=1}^{n} a_{h,k} x_k D_h v_j + \sum_{h}^{n} G_h^n D_h v_j \\
& - \sum_{h=1}^{n} D_h u D_h v_j - \sum_{h=1}^{n} a_{h,j} v_h - \sum_{h=1}^{n} D_j G_h^n v_h + D_j g_n.
\end{aligned}
\tag{13.3.8}
$$

Now, multipling both sides of (13.3.8) by v_j and summing up over j, we obtain

$$
\begin{aligned}
\frac{1}{2} D_t |v|^2 =\ & \frac{1}{2} \sum_{j,h=1}^{n} \lambda_h D_h^2 v_j v_j + \sum_{h,k,j=1}^{n} a_{h,k} x_k (D_h v_j) v_j \\
& - \sum_{h,j=1}^{n} (D_h u)(D_h v_j) v_j - \sum_{h,j=1}^{n} G_h^n (D_h v_j) v_j - \sum_{h,j=1}^{n} a_{h,j} v_h v_j \\
& - \sum_{h,j=1}^{n} D_j G_h^n v_h v_j + D_j g_n v_j.
\end{aligned}
\tag{13.3.9}
$$

Notice that

$$\frac{1}{2}\sum_{j,h=1}^{n}\lambda_h D_h^2 v_j v_j \;=\; \frac{1}{4}\sum_{h=1}^{n}\lambda_h D_h^2 |v|^2 - \frac{1}{2}\sum_{j,h=1}^{n}\lambda_h |D_h v_j|^2,$$

$$\sum_{h,k,j=1}^{n} a_{h,k} x_k D_h v_j v_j \;=\; \frac{1}{2}\langle A_n x, D|v|^2\rangle,$$

$$\sum_{h,j=1}^{n} G_h^n (D_h v_j) v_j \;=\; \frac{1}{2}\langle G_n, D|v|^2\rangle,$$ (13.3.10)

$$\sum_{h,j=1}^{n} D_h u D_h v_j v_j \;=\; \frac{1}{2}\langle Du, D|v|^2\rangle.$$

Now, substituting (13.3.10) into (13.3.9), we find

$$D_t |v|^2 \;\le\; \frac{1}{2}\mathrm{Tr}[QD^2(|v|^2)] + \langle A_n x + G_n(x) - Du(t,x), D|v|^2\rangle$$

$$+\omega |v|^2 + \frac{1}{2}|Dg_n|^2 + \frac{1}{2}|v|^2.$$

Finally, the required a priori estimate follows again from the maximum principle. □

13.3.1   Stationary equation

We assume here that (13.1.1) and (13.1.7) hold. We are concerned with the equation

$$\lambda \varphi - L\varphi - \langle G, D\varphi\rangle + \frac{1}{2}|D\varphi|^2 = g,$$ (13.3.11)

where $\lambda > 0$, and $g \in UC_b(H)$. We set

$$K\varphi = L\varphi + \langle G, D\varphi\rangle - \frac{1}{2}|D\varphi|^2, \; \varphi \in D(L).$$

We recall that, in view of Proposition 4.2.1, we have $D(L) \subset UC_b^1(H)$, so that the above definition is meaningful.

Proposition 13.3.2 *K is dissipative.*

Proof. Let $\lambda > 0$, $\varphi_i \in D(L)$, $i = 1, 2$, and set

$$g_i = \lambda\varphi_i - L\varphi_i - \langle G, D\varphi_i \rangle + \frac{1}{2}|D\varphi_i|^2, \ i = 1, 2.$$

Then for any $\varepsilon > 0$ we have

$$\lambda\varphi_i - L\varphi_i - \langle G, D\varphi_i \rangle + \frac{1}{2}\frac{|D\varphi_i|^2}{1 + \varepsilon|D\varphi_i|^2} = g_i - \frac{\varepsilon}{2}\frac{|D\varphi_i|^4}{1 + \varepsilon|D\varphi_i|^2}.$$

By Theorem 13.2.4, applied to the function F defined by

$$F(x) = -\frac{1}{2}\frac{|x|^2}{1 + \varepsilon|x|^2}, \ x \in H,$$

it follows that

$$\|\varphi_1 - \varphi_2\|_0 \le \frac{1}{\lambda}\|g_1 - g_2\|_0 + \frac{\varepsilon}{2\lambda}\sup_{x \in H}\sum_{i=1}^{2}\frac{|D\varphi_i(x)|^4}{1 + \varepsilon|D\varphi_i(x)|^2}.$$

Therefore, for $\varepsilon \to 0$, we have

$$\|\varphi_1 - \varphi_2\|_0 \le \frac{1}{\lambda}\|g_1 - g_2\|_0. \quad \square$$

Theorem 13.3.3 K *is m-dissipative.*

Proof. Let $\lambda > 0$ and let $g \in UC_b^1(H)$. For any $\varepsilon > 0$ denote by φ_ε the solution to

$$\lambda\varphi_\varepsilon - L\varphi_\varepsilon - \langle G, D\varphi_\varepsilon \rangle + \frac{1}{2}\frac{|D\varphi_\varepsilon|^2}{1 + \varepsilon|D\varphi_\varepsilon|^2} = g.$$

It follows that

$$\lambda\varphi_\varepsilon - L\varphi_\varepsilon\langle G, D\varphi_\varepsilon \rangle + \frac{1}{2}|D\varphi_\varepsilon|^2 = g + \frac{\varepsilon}{2}\frac{|D\varphi_\varepsilon|^4}{1 + \varepsilon|D\varphi_\varepsilon|^2}.$$

By Proposition 13.2.5 we have

$$\|D\varphi_\varepsilon\|_0 \le C(\lambda)\|Dg\|_0, \ \varepsilon > 0.$$

Therefore

$$\lim_{\varepsilon \to 0}[\lambda\varphi_\varepsilon - L\varphi_\varepsilon + \frac{1}{2}|D\varphi_\varepsilon|^2] = g.$$

This implies that the closure of the image of $\lambda - K$ contains $UC_b^1(H)$. Since $UC_b^1(H)$ is dense in $UC_b(H)$ (Theorem 2.2.1), this shows that the image of $\lambda - K$ is dense in $UC_b(H)$. This yields the conclusion. $\square$

13.4   Solution of the control problem

We assume here that 13.1.1 and 13.1.7 hold. We study the control problem only in the case of quadratic Hamiltonian.

13.4.1   Finite horizon

We are concerned with the following optimal control problem.
    Given $\varphi, g \in UC_b^1(H)$, minimize

$$J(T, x, z) = \mathbb{E}\left(\int_0^T \left[g(X(t)) + \frac{1}{2}|z(t)|^2\right] dt + \varphi(X(T))\right), \qquad (13.4.1)$$

over all controls $z \in L_W^2(0, T; L^2(\Omega; H))$, subject to the state equation (13.1.2).
    We set

$$J^*(T, x) = \inf\left\{J(z) : z \in L_W^2(0, T; L^2(\Omega; H))\right\}, \qquad (13.4.2)$$

J^* is the *value function* of the problem.
    We say that $z^* \in L_W^2(0, T; L^2(\Omega; H))$ is an *optimal control*, if $J^*(T, x) = J(T, x, z^*)$. Then the solution X^* of (13.1.2) corresponding to z^* is called an *optimal state* and the pair (z^*, X^*) an *optimal pair*.

    We are going to show existence of an optimal control. We first prove two basic identities.

Proposition 13.4.1 *Let $z \in L_W^2(0, T; L^2(\Omega; H))$, $X(\cdot, x)$ (resp. $X_n(\cdot, x)$) be the mild solution of (13.1.2) (resp. (13.1.8)) and let u (resp. u_n) be the mild solutions of Hamilton-Jacobi equation (13.3.1) (resp. (13.3.2)) with $F(x) = \frac{1}{2}|x|^2$. Let moreover $z \in L_W^2(0, T; L^2(\Omega; H))$ and $z_n(t) = \Pi_n z(t)$, $t \in [0, T]$. Then the following identities hold:*

$$u(T, x) + \frac{1}{2}\mathbb{E}\left(\int_0^T |z(s) + Du(T - s, X(s, x))|^2 ds\right) = J(T, x, z), \quad (13.4.3)$$

and

$$u_n(T, x) \; + \; \frac{1}{2}\mathbb{E}\left(\int_0^T |z_n(s) + Du_n(T - s, X_n(s, x))|^2 ds\right)$$

$$= \; \mathbb{E}\left(\int_0^T \left[g(X_n(t)) + \frac{1}{2}|z_n(t)|^2\right] dt + \varphi_n(X_n(T))\right). \quad (13.4.4)$$

Proof. Since u_n is regular, by Itô's formula (see Chapter 7) we have

$$d_s u_n(t - s, X_n(s, x)) = -D_t u_n(t - s, X_n(s, x))ds$$

$$+ \frac{1}{2}\mathrm{Tr}[Q_n D^2 u_n(t - s, X_n(s, x))]ds$$

$$+ \langle A_n X_n(s, x) + G_n(X_n(s, x)) + \Pi_n z(s), Du_n(t - s, X_n(s, x)) \rangle ds$$

$$+ \langle Du_n(t - s, X_n(s, x)), Q_n^{1/2}dW(s) \rangle$$

$$= \left[\frac{1}{2}|Du_n(t - s, X_n(s, x)) + \Pi_n z(s)|^2 - \frac{1}{2}|\Pi_n z(s)|^2 - g(X_n(s, x)) \right] ds$$

$$+ \langle Du_n(t - s, X_n(s, x)), Q_n^{1/2}dW(s) \rangle.$$

Setting $t = T$, integrating over s in $[0, T]$, and taking expectation we find (13.4.4). Now (13.4.3) follows letting n tend to infinity. $\square$

Theorem 13.4.2 *There exists an optimal pair (X^*, z^*) for problem (13.4.1) and the feedback formula holds:*

$$z^*(t) = -Du\left(T - t, X^*(t)\right), \; t \in [0, T]. \tag{13.4.5}$$

Moreover the optimal cost $J^*(T, x)$ is given by

$$J^*(T, x) = u(T, x). \tag{13.4.6}$$

Proof. We first note that, in view of (13.4.3), we have

$$u(T, x) \leq J(T, x, z), \quad z \in L^2_W(0, T; L^2(\Omega; H)). \tag{13.4.7}$$

Let now X_n be the solution to

$$\begin{cases} dX_n(t) = [A_n X_n + G_n(X_n) - Du_n(T - t, X_n(t))]dt + Q_n^{1/2}dW(t) \\ X_n(0) = \Pi_n x, \end{cases}$$

$$\tag{13.4.8}$$

and set $z_n(t) = -Du_n(T - t, X_n)$. Then we have

$$X_n(t, x) = e^{tA_n}x_n + \int_0^t e^{(t-s)A_n} z_n(s)ds + \int_0^t e^{(t-s)A_n} G_n(z_n(s))ds$$

$$+ W_{A,n}(t), \tag{13.4.9}$$

and, in view of (13.4.4),

$$u_n(T,x) = \mathbb{E}\left(\int_0^T \left[g(X_n(t)) + \frac{1}{2}|z_n(t)|^2\right]dt + \varphi_n(X_n(T))\right). \quad (13.4.10)$$

Consequently, there exists a positive constant C such that

$$\frac{1}{2}\mathbb{E}\left(\int_0^T |z_n(t)|^2 dt\right) \le u_n(T,x) \le C. \quad (13.4.11)$$

Therefore, there exists a subsequence (z_{n_k}) of (z_n) which converges weakly in $L^2_W(0,T;L^2(\Omega;H))$ to an element z^*. From (13.4.9), it follows that (X_{n_k}) converges weakly in $L^2_W(0,T;L^2(\Omega;H))$ to an element X^* and we have

$$X(t,x) = e^{tA}x + \int_0^t e^{(t-s)A}z^*(s)ds + W_A(t). \quad (13.4.12)$$

Now from (13.4.3), it follows that $u(T,x) = J(T,x,z^*)$, so that z^* is optimal. $\square$

Remark 13.4.3 The equation

$$\begin{cases} dX = [AX + G(X) - Du(T-t,X(t))]dt + Q^{1/2}dW(t), \ t \in [0,T], \\ X(0) = x \in H, \end{cases}$$

$$(13.4.13)$$

is called the *closed loop equation*. It can be solved directly either when e^{tA}, $t > 0$, is compact, see P. Cannarsa and G. Da Prato [31] and F. Gozzi [133], or when u is of class C^2, see [27], [134]. In the second case the optimal control is unique.

13.4.2 Infinite horizon

We are here concerned with the following optimal control problem.
    Given $\lambda > 0, g \in UC^1_b(H)$ nonnegative, minimize the cost functional

$$J_\infty(x,z) = \mathbb{E}\left(\int_0^{+\infty} e^{-\lambda t}\left[g(X(t)) + \frac{1}{2}|z(t)|^2\right]dt\right), \quad (13.4.14)$$

over all controls $z \in L^2_W(0,+\infty;L^2(\Omega;H))$, subject to the state equation (13.1.2).
    We set

$$J^*_\infty(x) = \inf\left\{J(x,z) : z \in L^2_W(0,+\infty;L^2(\Omega;H))\right\}. \quad (13.4.15)$$

We say that $z^* \in L_W^2(0, +\infty; L^2(\Omega; H))$ is an *optimal control*, if $J_\infty^*(x) = J_\infty(x, z^*)$. Then the corresponding solution X^* of (13.1.2) is called an *optimal state*, and the pair (z^*, X^*) an *optimal pair*.

We are going to show existence of an optimal control. As before we start with an identity.

Proposition 13.4.4 *Let $z \in L_W^2(0, +\infty; L^2(\Omega; H))$, $X(\cdot, x)$ be the solution of (13.1.2) and let φ be the mild solution of the Hamilton-Jacobi equation (13.3.11). Then the following identity holds:*

$$\varphi(x) + \frac{1}{2}\mathbb{E}\left(\int_0^{+\infty} e^{-\lambda t}|z(s) + D\varphi(X(s, x))|^2 ds\right) = J_\infty(x, z). \quad (13.4.16)$$

Proof. We assume that $\varphi \in UC_b^2(H)$, otherwise we proceed using Galerkin approximations . By Itô's formula we have

$$de^{-\lambda t}\varphi(X(t, x)) = e^{-\lambda t}[(L - \lambda)\varphi(X(t, x)) + \langle G(X(t, x)), D\varphi(X(t, x))\rangle]dt$$

$$+ e^{-\lambda t}\left(\langle D\varphi(X(t, x)), z(t)\rangle + \langle D\varphi(X(t, x)), Q^{1/2}dW(t)\rangle\right)$$

$$= e^{-\lambda t}\left(\frac{1}{2}|D\varphi(X(t, x)) + z(t)|^2\right)dt + \langle D\varphi(X(t, x)), Q^{1/2}dW(t)\rangle$$

$$-\frac{1}{2}e^{-\lambda t}|z(t)|^2 - g(X(t, x)).$$

Then (13.4.16) follows taking the expectation and integrating with respect to t from 0 to infinity. $\square$

Arguing as in the proof of Theorem 13.4.2 we find

Theorem 13.4.5 *There exists an optimal pair (z^*, X^*) for problem (13.4.14). Moreover the following feedback formula holds:*

$$z^*(t) = -D\varphi(X^*(t)), \quad t \in [0, T]. \quad (13.4.17)$$

Finally, the optimal cost $J_\infty^*(x)$ is given by

$$J_\infty^*(x) = \varphi(x). \quad (13.4.18)$$

13.4.3   The limit as $\varepsilon \to 0$

We assume here that (13.1.1) and (13.1.7) hold. We fix $\varepsilon \in [0,1]$ and consider the following optimal control problem.

Given $\varphi, g \in C_b^1(H)$ nonnegative, minimize

$$J_\varepsilon(T,x,z) = \mathbb{E}\left(\int_0^T \left[g(X_\varepsilon(t)) + \frac{1}{2}|z(t)|^2\right]dt + \varphi(X_\varepsilon(T))\right), \qquad (13.4.19)$$

over all controls $z \in L_W^2(0,T;L^2(\Omega;H))$, subject to the state equation

$$\begin{cases} dX_\varepsilon(t) = (AX_\varepsilon(t) + G(X_\varepsilon(t)) + z(t))dt + \sqrt{\varepsilon}\,Q^{1/2}dW(t), \ t \in [0,T], \\ X_\varepsilon(0) = x \in H. \end{cases}$$

$$(13.4.20)$$

Obviously for $\varepsilon = 0$ the problem above reduces to a deterministic problem

$$\begin{cases} D_t X_0(t) = (AX_0(t) + G(X_0(t)) + z(t)), \ t \in [0,T], \\ X_0(0) = x \in H. \end{cases} \qquad (13.4.21)$$

We set

$$J_\varepsilon^*(T,x) = \inf\left\{J_\varepsilon(T,x,z) : \ z \in L_W^2(0,T;L^2(\Omega;H))\right\}. \qquad (13.4.22)$$

Moreover, we consider the Hamilton-Jacobi equation

$$\begin{cases} D_t u_\varepsilon = \frac{1}{2}\varepsilon\mathrm{Tr}[QD^2 u_\varepsilon] + \langle Ax + G(x), Du_\varepsilon\rangle - \frac{1}{2}|Du_\varepsilon|^2 + g, \\ u_\varepsilon(0) = \varphi. \end{cases} \qquad (13.4.23)$$

Obviously, for $\varepsilon = 0$, problem (13.4.23) reduces to the first order equation

$$\begin{cases} D_t u_0 = \langle Ax + G(x), Du_0\rangle - \frac{1}{2}|Du_0|^2 + g, \\ u_0(0) = \varphi. \end{cases} \qquad (13.4.24)$$

Proposition 13.4.6 *We have*

$$\lim_{\varepsilon \to 0} u_\varepsilon(T,x) = J_0^*(T,x). \qquad (13.4.25)$$

Proof. For all $n \in \mathbb{N}$ there exists $\zeta_n \in L^2(0,T;H)$ such that

$$J_0^*(T,x) \le J_0(T,x,\zeta_n) < J_0^*(T,x) + \frac{1}{n}.$$

Moreover

$$J_0(T,x,z_\varepsilon^*) \ge J_0^*(T,x),$$

and, due to Proposition 13.4.1,

$$u_\varepsilon(T, x) \leq J_\varepsilon(T, x, \zeta_n). \qquad (13.4.26)$$

It follows that

$$J_\varepsilon(T, x, z_\varepsilon^*) - J_0(T, x, z_\varepsilon^*) \leq u_\varepsilon(T, x) - J_0^*(T, x)$$

$$\leq J_\varepsilon(T, x, \zeta_n) - J_0(T, x, \zeta_n) + \frac{1}{n}. \qquad (13.4.27)$$

Notice now that for any $z \in L^2_W(0, T; L^2(\Omega; H))$ we have

$$J_\varepsilon(T, x, z) - \mathbb{E}(J_0(T, x, z)) = \mathbb{E} \int_0^T \left((g(X_\varepsilon(s, x)) - g(X_0(s, x)) \right) ds$$

$$+ \mathbb{E}(\varphi(X_\varepsilon(s, x)) - \varphi(X_0(s, x))) \to 0$$

$$\text{as } \varepsilon \to 0.$$

Thus the conclusion follows. $\square$

Remark 13.4.7 $J_0^*(T, x)$ can be considered as a *viscosity solution* of (13.4.23).

Chapter 14

Hamilton-Jacobi inclusions

14.1 Introduction

Let H be a separable Hilbert space and $X(t,x)$, $t \geq 0$, $x \in H$, a solution of a stochastic evolution equation on H,

$$\begin{cases} dX(t) = (AX(t) + F(X(t)))dt + G(X(t))dW(t), \\ X(0) = x \in H, \end{cases} \tag{14.1.1}$$

defined on a probability space $(\Omega, \mathcal{F}, \mathbb{P})$ with a filtration $(\mathcal{F}_t)_{t \geq 0}$. Let α, g, h be real functions on H, with $g \geq h$. For arbitrary $t \in [0, +\infty)$ and stopping time $\tau \leq t$ define the functional

$$J_t(\tau, x) = \mathbb{E}\left(e^{\int_0^\tau \alpha(X(\sigma, x))d\sigma} [h(X(\tau, x))1_{\tau < t} + g(X(\tau, x))1_{\tau = t}]\right), \tag{14.1.2}$$

and let V be the corresponding value function

$$V(t, x) = \sup_{\tau \leq t} J_t(\tau, x). \tag{14.1.3}$$

A heuristic dynamic programming argument leads to the following Hamilton-Jacobi inclusion for the function V,

$$\begin{cases} D_t V(t, x) \in \frac{1}{2}\text{Tr}[G(x)D^2V(t,x)G^*(x)] + \langle Ax + F(x), DV(t,x) \rangle \\ \qquad\qquad + \alpha(x)V(t,x) - \partial I_{K_h}(V(t,x)), \\ V(0, x) = g(x), \ x \in H, \ t \geq 0. \end{cases} \tag{14.1.4}$$

The inclusion (14.1.4) will be considered in a Hilbert space $\mathcal{H} = L^2(H, \mu)$ where μ is a properly choosen measure. In (14.1.4)

$$K_h = \{f \in \mathcal{H} : f \geq h\},$$

I_{K_h} is the indicator function of K_h :

$$I_{K_h}(f) = \begin{cases} 0 \text{ if } f \in K_h, \\ +\infty \text{ if } f \notin K_h, \end{cases} \qquad (14.1.5)$$

and ∂I_{K_h} is the subgradient of I_{K_h}.

Let us recall that the subgradient of I_{K_h} of the function I_{K_h} is a set valued mapping defined on the domain $D(\partial I_{K_h}) = K_h$ by the formula

$$\partial I_{K_h}(f) = \{\xi \in \mathcal{H} : \langle \xi, \eta - f \rangle \leq 0, \ \forall \eta \in K_h\}. \qquad (14.1.6)$$

It is easy to see that

$$\partial I_{K_h}(f) = \{\xi \in \mathcal{H} : \xi \leq 0 \text{ and } \xi(x) = 0 \text{ if } f(x) > h(x)\}. \qquad (14.1.7)$$

Our aim in the present chapter is to show that (14.1.4) has a solution which can be interpreted as the value function (14.1.3). Note that if $V(t,x) > h(x)$, for some $t > 0$ and all $x \in H$, then (14.1.4) becomes the usual Kolmogorov equation,

$$\begin{cases} D_t V(t,x) &= \frac{1}{2}\mathrm{Tr}[G(x)D^2 V(t,x)G^*(x)] + \langle Ax + F(x), DV(t,x) \rangle \\ &\quad + a(x)V(t,x), \\ V(0,x) &= g(x). \end{cases}$$
$$(14.1.8)$$

We shall follow J. Zabczyk [218] and [222].

14.2 Excessive weights and an existence result

For $\varphi \in B_b(H)$ and $\Gamma \in \mathcal{B}(H)$ define, as usual,

$$P_t\varphi(x) = \mathbb{E}[\varphi(X(t,x))], \ P_t(x,\Gamma) = P_t 1_\Gamma, \ t \geq 0, \ x \in H.$$

Let $\omega \geq 0$. A measure μ on $(II, \mathcal{B}(II))$, finite on bounded sets, is called ω-*excessive* if

$$P_t^* \mu(\Gamma) \leq e^{\omega t}\mu(\Gamma), \quad t \geq 0, \ \Gamma \in \mathcal{B}(H), \qquad (14.2.1)$$

where $P_t^*\mu$ is given by

$$P_t^*\mu(\Gamma) = \int_H P_t(x,\Gamma)\mu(dx), \quad t \geq 0, \ \Gamma \in \mathcal{B}(H). \qquad (14.2.2)$$

In particular if μ is an invariant measure for (P_t), that is $P_t^*\mu = \mu$ for all $t \geq 0$, finite or not, then μ is 0-excessive or, shortly, excessive for (P_t). It turns out that ω-excessive measures are natural weights in which the formula (14.1.4) should be considered. Unlike invariant measures, ω-excessive measures exist for arbitrary Markovian semigroups (P_t). For instance if ν is a finite measure on H and $\omega > 0$, then

$$\mu = \int_0^{+\infty} e^{-\omega s} P_s^* \nu ds \tag{14.2.3}$$

is ω-excessive. If the formula (14.2.3) defines a measure μ for $\omega = 0$ finite on bounded sets then μ is 0-excessive. Both facts follow from the identity

$$P_t^*\mu = e^{\omega t} \int_t^{+\infty} e^{-\omega s} P_s^* \nu ds.$$

We will require that

$$\begin{cases} \text{For arbitrary } \varphi \in C_b(H), \text{ the function of two variables } P_t\varphi(x), \\ t \geq 0, \ x \in H, \text{ is continuous.} \end{cases}$$
$$\tag{14.2.4}$$

We have the following result.

Proposition 14.2.1 *Assume that a measure μ is ω-excessive, $\omega \geq 0$, for (P_t), and that (14.2.4) holds. Then the semigroup (P_t) has a unique extension from $B_b(H) \cap L^2(H,\mu)$ to a strongly continuous semigroup on $L^2(H,\mu)$ such that*

$$\|P_t\|_{L^2(H,\mu)} \leq e^{\frac{\omega}{2}t}, \quad t \geq 0. \tag{14.2.5}$$

We show first that for arbitrary nonnegative measurable ψ, we have

$$\int_H \left[\int_H P_t(x,dy)\psi(y) \right] \mu(dx) \leq e^{\omega t} \int_H \psi(y)\mu(dy). \tag{14.2.6}$$

Notice that (14.2.6) holds for the characteristic function $\psi = \chi_\Gamma$ and therefore for ψ nonnegative and simple. Since an arbitrary measurable $\psi \geq 0$ is a limit of an increasing sequence of nonnegative simple functions the inequality (14.2.6) holds in general. If now $\varphi \in B_b(H) \cap L^2(H,\mu)$ then, by definition,

$$|P_t\varphi|_{\mathcal{H}}^2 = \int_H \left| \int_H P_t(x,dy)\varphi(y) \right|^2 \mu(dx),$$

Hamilton-Jacobi inclusions

and by the Hölder inequality

$$\left| \int_H P_t(x, dy)\varphi(y) \right| \leq \left(\int_H P_t(x, dy) \right)^{\frac{1}{2}} \left(\int_H R_t(x, dy)|\varphi(y)|^2 \right)^{\frac{1}{2}}.$$

Consequently,

$$|P_t\varphi|^2_{\mathcal{H}} \leq \int_H P_t(x, dy)|\varphi(y)|^2\mu(dx),$$

and taking into account (14.2.6)

$$|P_t\varphi|^2_{\mathcal{H}} \leq e^{\omega t} \int_H |\varphi(y)|^2\mu(dy).$$

This shows that the operator P_t can be extended from $B_b(H) \cap L^2(H, \mu)$ to the whole of $L^2(H, \mu)$ and that (14.2.5) holds. To show that the extended semigroup is strongly continuous note that the set $C_b(H) \cap L^2(H, \mu)$ is dense in $L^2(H, \mu)$. It is therefore enough to show that for $\varphi \in C_b(H) \cap L^2(H, \mu)$,

$$\lim_{t \to 0} \int_H |P_t\varphi(x) - \varphi(x)|^2\mu(dx) = 0. \tag{14.2.7}$$

Assume first that $\varphi \in \mathcal{H}$ is a bounded, continuous function such that $\varphi = 0$ outside of an open set $U \subset H$, $\mu(U) < +\infty$. Then

$$\|P_t\varphi - \varphi\|^2_{\mathcal{H}} = \int_H |P_t\varphi(x) - \varphi(x)|^2\mu(dx) \tag{14.2.8}$$

$$= -2 \int_H P_t\varphi(x)\varphi(x)\mu(dx) + \int_E |P_t\varphi(x)|^2\mu(dx) + \|\varphi\|^2_{\mathcal{H}}.$$

However,

$$\int_U P_t\varphi(x)\varphi(x)\mu(dx) = \int_H P_t\varphi(x)\varphi(x)\mu(dx),$$

and by the dominated convergence theorem

$$\lim_{t \to 0} \int_U P_t\varphi(x) \, \varphi(x)\mu(dx) = \int_U |\varphi(x)|^2\mu(dx) = \|\varphi\|^2_{\mathcal{H}}.$$

Moreover

$$\limsup_{t \to 0} \|P_t\varphi\|^2_{\mathcal{H}} \leq \|\varphi\|^2_{\mathcal{H}},$$

and consequently

$$\limsup_{t \to 0} \|P_t\varphi - \varphi\|^2_{\mathcal{H}} \leq -2\|\varphi\|^2_{\mathcal{H}} + \|\varphi\|^2_{\mathcal{H}} + \|\varphi\|^2_{\mathcal{H}}.$$

Since the functions φ with the imposed properties form a dense set in $\mathcal{H}$, the proof of the proposition is complete. $\square$

The new extended semigroup will be denoted by (P_t) like the initial one. With a similar proof one can demonstrate also the validity of the following generalization of Proposition 14.2.1 concerned with the semigroup

$$T_t\varphi(x) = \mathbb{E}\left[e^{\int_0^t \alpha(X(\sigma,x))d\sigma}\varphi(X(t,x)) : t \geq 0, \ x \in H\right]. \qquad (14.2.9)$$

Proposition 14.2.2 *Assume in addition to the assumptions of Proposition 14.2.1 that α is a continuous function on H bounded from above by a. Then (T_t) has a unique extension to a strongly continuous semigroup on $L^2(H,\mu)$ such that*

$$\|T_t\|_{L^2(H,\mu)} \leq e^{(\frac{\omega}{2}+a)t}, \ \ t \geq 0. \qquad (14.2.10)$$

Let $\mathcal{L}$ be the infinitesimal generator of the semigroup (T_t) considered on $\mathcal{H} = L^2(H,\mu)$ and $D(\mathcal{L})$ its domain. We will consider an abstract version of (14.1.4) in the form

$$\begin{cases} D_t u(t,x) & \in \ \mathcal{L}u(t,x) - \partial I_{K_h}(u(t,x)), \\ u(0) & = \ g. \end{cases} \qquad (14.2.11)$$

A locally Lipschitz $\mathcal{H}$-valued function $u(t)$, $t \geq 0$, is said to be a *strong solution* to (14.2.11) if for all $t \geq 0$, $u(t) \in K_h \cap D(\mathcal{L})$, $u(0) = g$ and the inclusion (14.2.11) holds for almost $t \geq 0$ and μ-almost all $x \in H$. If $h \in \mathcal{H}$ and $g \in K_h$ and there exist functions $g_n \geq h_n$, converging in $\mathcal{H}$ to g and h respectively, for which the inclusion (14.2.11) has a strong solution u_n converging uniformly on bounded intervals of $\mathbb{R}_+$ to a continuous u we call u a generalized solution of (14.2.11).

Theorem 14.2.3 *Assume that (14.2.4) holds and that α is the continuous function from above. If the functions h and g are in $D(\mathcal{L})$ and $g \geq h$ then the inclusion (14.2.11) has a unique strong solution.*

Proof. The theorem will be a consequence of a result on maximal monotone operators , see e.g. [25], and of a lemma.

Let $\mathcal{M}$ be a transformation from a set $D(\mathcal{M}) \subset \mathcal{H}$ into the set of non-empty subsets of a Hilbert space $\mathcal{H}$ and ω a nonnegative number. The transformation $\mathcal{M}$ is said to be ω-*maximal-monotone* if the operator $\mathcal{M}+\omega I$ is maximal monotone, see [25, page 106], and [64, page 82]. If $\mathcal{M}$ is ω-maximal-monotone then, for arbitrary $\lambda \in (0, \frac{1}{\omega})$, the range of $I + \lambda\mathcal{M}$ is the whole of $\mathcal{H}$ and for arbitrary $y \in \mathcal{H}$, there exists a unique $x \in D(\mathcal{M})$

such that $y \in x + \lambda \mathcal{M}(x)$. The unique element y is denoted by $\mathcal{J}_\lambda(x)$ and the family of transformations $\mathcal{J}_\lambda$, $\lambda \in (0, \frac{1}{\omega})$, is called *the resolvent* of $\mathcal{M}$. Operators $\mathcal{M}_\lambda = \frac{1}{\lambda}(I - \mathcal{J}_\lambda)$, $\lambda \in (0, \frac{1}{\omega})$ are called *Yosida approximations* of $\mathcal{M}$. If $\mathcal{N}$ is a maximal monotone operator on $\mathcal{H}$ then the sum $\mathcal{M} + \mathcal{N}$: is ω-monotone but not always ω-maximal-monotone.

The following result is due to Brézis, Crandall and Pazy [26].

Theorem 14.2.4 *Assume that operators $\mathcal{M}$ and $\mathcal{N}$, defined on subsets of a Hilbert space $\mathcal{H}$, are respectively ω-maximal-monotone and maximal monotone. If for arbitrary $y \in \mathcal{H}$, arbitrary $\lambda > 0$ and arbitrary $\delta \in (0, \frac{1}{\omega})$ there exists a solution x_λ^δ of the problem*

$$y \in x + \delta(\mathcal{M}x + \mathcal{N}_\lambda x)$$

such that for each $\delta \in (0, \frac{1}{\omega})$ the functions $\mathcal{N}_\lambda x_\lambda^\delta, \lambda > 0$, are bounded as $\lambda \to 0$, then the operator $\mathcal{M} + \mathcal{N}$, with the domain $D(\mathcal{M}) \cap D(\mathcal{N})$, is ω-maximal-monotone.

First we derive from Theorem 14.2.4 the following crucial proposition.

Proposition 14.2.5 *Assume that an operator $\mathcal{L}$ is an infinitesimal generator of a C_0 semigroup of positive linear operators (T_t) on a Hilbert space $\mathcal{H} = L^2(E, \mu)$ such that for some $\gamma \geq 0$*

$$|T_t \psi|_{\mathcal{H}} \leq e^{\gamma t} |\psi|_{\mathcal{H}}, \ t \geq 0, \ \psi \in \mathcal{H}. \tag{14.2.12}$$

If $\varphi \in D(\mathcal{L})$ then the operator

$$-\mathcal{L} + \partial I_{K_\varphi}, \tag{14.2.13}$$

is γ-maximal-monotone.

Proof. It is well known that a linear operator $-\mathcal{L}$ is γ-maximal-monotone if and only if it generates a C_0 semigroup of linear operators satisfying (14.2.12). Let $\mathcal{M} = -\mathcal{L}$ and $\mathcal{N} = \partial I_{K_\varphi}$. Then $\mathcal{N}$ is maximal monotone, see [25], and

$$\mathcal{N}_\lambda(\eta) = -\frac{1}{\lambda}(\varphi - \eta)^+, \lambda > 0, \eta \in \mathcal{H}.$$

The proposition is now a consequence of Theorem 14.2.4 and of the following lemma.

Lemma 14.2.6 *If $\delta \in \left(0, \frac{1}{\gamma}\right)$, $\lambda > 0$, $\psi \in \mathcal{H}$ then, for arbitray $\psi \in \mathcal{H}$, the equation*

$$\psi = v - \delta \left(\mathcal{L}v + \frac{1}{\lambda}(\varphi - v)^+ \right) \qquad (14.2.14)$$

has a unique solution v_λ such that

$$\left\| \frac{1}{\lambda}(\varphi - v_\lambda)^+ \right\|_{\mathcal{H}} \leq \left\| \left[\frac{1}{\delta}(\varphi - \psi) - \mathcal{L}\varphi \right]^+ \right\|_{\mathcal{H}}.$$

Proof. Define $\mathcal{R}_\sigma = (\sigma I - \mathcal{L})^{-1}$, $\sigma > \gamma^+$. Then the basic equation is equivalent to

$$v = \mathcal{R}_{\frac{1}{\delta}} \left(\frac{1}{\delta}\psi + \frac{1}{\lambda}(\varphi - v)^+ \right),$$

or, by the resolvent identity, to

$$v = \mathcal{R}_{\frac{1}{\delta}+\frac{1}{\lambda}} \left(\frac{1}{\delta}\psi \right) + \mathcal{R}_{\frac{1}{\delta}+\frac{1}{\lambda}} \left(\frac{1}{\lambda}((\varphi - v)^+ + v) \right). \qquad (14.2.15)$$

Since the norm of the operator $\mathcal{R}_{\frac{1}{\delta}+\frac{1}{\lambda}}$ is at most $\left(\frac{1}{\delta} - \gamma^+ \frac{1}{\lambda} \right)^{-1}$ and the real function $z \to \frac{1}{\lambda}((a - z)^+ + z)$ is Lipschitz with constant $\frac{1}{\lambda}$, the transformed equation (14.2.15) has a unique solution by the contraction mapping principle.

Since $\varphi \in D(\mathcal{L})$, there exists a function $\eta \in \mathcal{H}$ such that

$$\varphi = \mathcal{R}_{\frac{1}{\delta}}\eta, \quad \frac{1}{\delta}\varphi - \mathcal{L}\varphi = \eta,$$

or, again by the resolvent identity,

$$\varphi = \mathcal{R}_{\frac{1}{\delta}+\frac{1}{\lambda}} \left(\eta + \frac{1}{\lambda}\varphi \right).$$

Taking into account that v_λ satisfies the transformed equation (14.2.15) one gets, by subtraction,

$$v_\lambda - \varphi = \mathcal{R}_{\frac{1}{\delta}+\frac{1}{\lambda}} \left(\frac{1}{\delta}\psi - \eta \right) + \frac{1}{\lambda}\mathcal{R}_{\frac{1}{\delta}+\frac{1}{\lambda}} \left[(\varphi - v_\lambda)^+ - (\varphi - v_\lambda) \right].$$

Consequently,

$$\varphi - v_\lambda \leq \mathcal{R}_{\frac{1}{\delta}+\frac{1}{\lambda}} \left(\eta - \frac{1}{\delta}\psi \right)^+.$$

In particular,

$$(\varphi - v_\lambda)^+ \le \mathcal{R}_{\frac{1}{\delta} + \frac{1}{\lambda}} \left(\eta - \frac{1}{\delta} \psi \right)^+,$$

and

$$\left\| \frac{1}{\lambda} (\varphi - v_\lambda)^+ \right\| \le \frac{1}{\lambda} \left\| \mathcal{R}_{\frac{1}{\delta} + \frac{1}{\lambda}} \left(\eta - \frac{1}{\delta} \psi \right)^+ \right\|_{\mathcal{H}}$$

$$\le \frac{\frac{1}{\lambda}}{(\frac{1}{\delta} - \omega) + \frac{1}{\lambda}} \left\| \left(\frac{1}{\delta} (\varphi - \psi) - \mathcal{L}\eta \right)^+ \right\|_{\mathcal{H}}$$

$$\le \left\| \left(\frac{1}{\delta} (\varphi - \psi) - \mathcal{L}\eta \right)^+ \right\|_{\mathcal{H}},$$

as required. □

To complete the proof of Theorem 14.2.3 it is enough to use Proposition 14.2.5 and recall, see [25], that if an operator $\mathcal{M} = -\mathcal{L} + \partial I_{K_\varphi}$ is γ-maximal-monotone then the differential inclusion

$$\frac{dz(t)}{dt} + \mathcal{M}z(t) \ni 0, \ z(0) = x \in \overline{D(\mathcal{M})}, \tag{14.2.16}$$

has a unique strong solution $z(t, x)$, $t \ge 0$, and for arbitrary $x \in \overline{D(\mathcal{M})}$, the inclusion (14.2.16) has a weak solution denoted also by $z(t, x)$, $t \ge 0$. Moreover the operators $S(t) : \overline{D(\mathcal{M})} \to \overline{D(\mathcal{M})}$, $t \ge 0$, are given by

$$S(t)x = z(t, x), \ t \ge 0, \ x \in \overline{D(\mathcal{M})}. \quad \square$$

It is of interest to notice that implicitly we have shown existence of a solution to the stationary inclusion

$$0 \in \mathcal{L}V(x) - \partial_{K_\varphi}(V(x)) \tag{14.2.17}$$

on the value function V for a stopping problem on an infinite time interval. In fact we have the following result.

Theorem 14.2.7 *If assumption (14.2.4) is satisfied with $a < -\omega$ and the function φ is in $D(\mathcal{L})$ then the inclusion (14.2.17) has a unique strong solution.*

Proof. Define $\mathcal{G} = \mathcal{L} - (\frac{\omega + a}{2})I$. Then $\mathcal{G}$ generates a C_0 semigroup satisfying the inequality (14.2.12) from Proposition 14.2.5 with $\gamma = 0$. Consequently the operator

$$-\mathcal{L} + \left(\frac{\omega + a}{2}\right) I + \partial I_{K_\varphi},$$

is maximal monotone. In particular for arbitrary $\lambda > 0$ the inclusion

$$0 \in \psi + \lambda - \mathcal{L}\psi + \left(\frac{\omega + a}{2}\right) \psi + \partial I_{K_\varphi}(\psi) \qquad (14.2.18)$$

has a unique solution $\psi \in D(\mathcal{L}) \cup K_\varphi$. However (14.2.18) is equivalent to

$$-\left(\frac{1}{\lambda} + \frac{\omega + a}{2}\right) \psi \in N\psi - \partial_{K_\varphi}(\psi).$$

Taking $\frac{1}{\lambda} = -(\frac{\omega + a}{2})$, the result follows. $\square$

14.3   Weak solutions as value functions

In this section we sketch the proof that weak solutions to (14.2.11) are in fact identical with value functions. For more detils we refer to [218]. On the functions h, g and on the (discount) function α we impose continuity and growth conditions, usually satisfied in applications.

$\left\{\begin{array}{l}
(i) \quad \text{The functions } h, g \text{ are continuous and bounded on bounded sets} \\
\qquad \text{and } h \le g, \\
(ii) \quad \text{for an arbitrary compact set } K \subset E \text{ and arbitrary } T > 0 \\
\qquad \mathbb{E}(\sup_{x \in K} \sup_{t \in [0,T]}(|\varphi(X(t,x))| + |\psi(X(t,x))|) < +\infty, \\
(iii) \text{ for arbitrary } T > 0 \text{ there exists a function } \zeta \in \mathcal{H} \text{ such that} \\
\qquad \mathbb{E}(\sup_{t \in [0,T]}(|\varphi(X(t,x))| + |\psi(X(t,x))|) \le \zeta(x), x \in H, \\
(iv) \quad \text{the function } \alpha \text{ is contiuous and bounded from above by a constant} \\
\qquad a, \\
\qquad\qquad\qquad\quad \alpha(x) \le a, \quad x \in H.
\end{array}\right.$

$$(14.3.1)$$

We can state now the main result of the present section.

Theorem 14.3.1 *Under assumptions (14.2.4) and (14.3.1), the value function $V(t,x), t \ge 0, x \in H$, is continuous and is a weak solution of the Hamilton-Jacobi inclusion (14.2.11). If, in addition, $\varphi, \psi \in D(\mathcal{L})$ then V is the unique strong solution of (14.2.11).*

Proof. The fact that V is continuous has been established, in the present generality, in [218]. The proofs of the remaining parts of the theorem use the penalization technique, see [11] and [204], and will be divided into several steps.

Step 1. The functions h and g are in $C_b(H) \cap D(\mathcal{L})$.

Consider first the so called *penalized Hamilton-Jacobi equation*,

$$\begin{cases} D_t V^\lambda(t,x) & = \ \mathcal{L}V^\lambda(t,x) + \frac{1}{\lambda}(h - V^\lambda(t,x))^+, \\ V^\lambda(0,x) & = \ g(x), \ x \in H, \ t > 0, \end{cases} \tag{14.3.2}$$

in its integral form:

$$V^\lambda(t) = T_t g + \frac{1}{\lambda} \int_0^t T_s (h - V^\lambda(t-s))^+ ds, \quad t \geq 0. \tag{14.3.3}$$

Note that the transformation $\zeta \to (\varphi - \zeta)^+$ satisfies a Lipschitz condition both in $\mathcal{H}$ and in $C_b(H)$. By an easy contraction argument the equation (14.3.2) has a unique solution in $C([0,T]; \mathcal{H})$, first for small T, and then for all $T \geq 0$. By a similar argument the equation (14.3.3) also has a unique solution in $C([0,T]; C_b(H))$ for arbitrary $T \geq 0$. It is clear that if $h \in C_b(H) \cap \mathcal{H}$ then the two solutions coincide.

Let $\lambda > 0$ be an arbitrary positive number and $\psi(t), t \geq 0$ an $\mathcal{F}_t$-adapted process taking values in the interval $[0, \frac{1}{\lambda}]$. Define functionals

$$\tilde{J}_s(\psi x) \ = \ \mathbb{E} \left[\int_0^s e^{\int_0^t (\alpha(X(\sigma,x)-\psi(\sigma))d\sigma} \psi(t) h(X(t,x))dt \right]$$

$$+ \mathbb{E} \left(e^{\int_0^s \alpha(X(\sigma,x)-\psi(\sigma))d\sigma} g(X(s,x)) \right). \tag{14.3.4}$$

and consider the following value function:

$$V_\lambda(s,x) = \sup_{0 \leq u(\cdot) \leq \frac{1}{\lambda}} \tilde{J}_s(u,x) \tag{14.3.5}$$

where the supremum is taken with respect to all processes $u(t), t \geq 0$, $(\mathcal{F}_t)$-adapted, having values in the interval $[0, \frac{1}{\lambda}]$.

We need the following result.

Proposition 14.3.2 *Assume that conditions* (14.2.4) *and* (14.3.1) *hold and that* $h, g \in C_b(H)$, $g \geq h$. *Then*

(i) $V^\lambda(s,x) = V_\lambda(s,x)$, *for all* $\lambda > 0$, $s > 0$, $x \in H$,

(ii) $V_\lambda(s,x) \uparrow V(s,x)$, *as* $\lambda \downarrow 0$, *for* $s > 0$, $x \in H$.

The proof from [204] can be adapted to the present, more general situation. The following lemma is a generalization of the first part of Lemma 1 from [204].

Lemma 14.3.3 *Let* $(x_t)_{t\geq0}$, $(\psi_t)_{t\geq0}$ *and* $(\alpha_t)_{t\geq0}$ *be progressively measurable, real processes such that* x *and* ψ *are bounded and* α *is bounded from above and locally integrable. If* $T \in [0,+\infty]$ *and* $(w_t)_{t\in[0,t]}$ *is a right continuous process such that for each* $t \in [0,T]$,

$$w_t = \mathbb{E}\left(\int_t^T e^{\int_t^s \alpha_r dr} x_s ds \Big| \mathcal{F}_t \right), \quad \mathbb{P}\text{-}a.e.$$

Then, for each $t \in [0,T]$,

$$w_t = \mathbb{E}\left(\int_t^T e^{\int_t^s (\alpha_r - \psi_r)dr} (x_s + \psi_s W_s) ds \Big| \mathcal{F}_t \right), \quad \mathbb{P}\text{-}a.e.$$

Denote $\mathcal{M} = -\mathcal{L}$ and $\mathcal{N} = \partial I_{K_h}$. Then the Yosida approximations $\mathcal{N}_\lambda$ are given by the formula

$$\mathcal{N}_\lambda(\eta) = \frac{1}{\lambda}(h - \eta)^+, \lambda > 0, \eta \in \mathcal{H}.$$

Moreover the penalized equations (14.3.2) are of the form

$$\begin{cases} D_t V^\lambda + \mathcal{M}V^\lambda + \mathcal{N}_\lambda V^\lambda = 0, \\ V^\lambda(0) = g(x), \; x \in H, \end{cases}$$

and the solutions $V^\lambda(t), t \geq 0$, form a continuous semigroup of transformations $g \to S^\lambda(t)g, t \geq 0$, on $\mathcal{H}$. By Theorem 14.2.4 and Theorem 14.2.7, the operator $\mathcal{M} + \mathcal{N}$ is $\frac{\omega+a}{2}$-maximal-monotone and the value function V is identical with the strong solution of the equation

$$\begin{cases} -D_t V & \in & \mathcal{M}V, +\mathcal{N}V, \\ V(0) & = & g(x), \; x \in H. \end{cases} \tag{14.3.6}$$

Let $S(t), t \geq 0$, be the semigroup determined by (14.3.6). We need a version of Bénilan's theorem [10]

Theorem 14.3.4 *Assume that the assumptions of Theorem 14.2.7 hold. Then, for arbitrary* $x \in \overline{D(\mathcal{M} + \mathcal{N})}$, $S^\lambda(t)x \to S(t)x$ *uniformly on bounded subsets of* $[0,+\infty)$.

Proof. One shows, see [25, page 35], that for arbitrary $\delta \in (0, \frac{1}{\omega})$ the limit $x_\infty^\delta = \lim_{\lambda \to 0} x_\lambda^\delta$ is the unique solution of the inclusion

$$y \in x + \delta(\mathcal{M}(x) + \mathcal{N}(x)).$$

This means that for arbitrary $\delta \in (0, \frac{1}{\omega})$ and $y \in \mathcal{H}$,

$$\left(I + \delta(\mathcal{M} + \mathcal{N}_\lambda)\right)^{-1} y \to \left(I + \delta(\mathcal{M} + \mathcal{N})\right)^{-1} y$$

as $\lambda \to 0$. The result follows now from Benilan's theorem, see Theorem 4.2 in [25] or [10]. $\square$

Taking into account Lemma 14.3.3 and Benilan's theorem, $S^\lambda \psi \to S\psi$, uniformly on bounded intervals of $\mathbb{R}_+$, as functions with values in $\mathcal{H}$. Since $S^\lambda \psi = V_\lambda$ and V_λ converges to the value function V pointwise the value function V is the strong solution of the Hamilton-Jacobi inclusion (14.2.11).

In the remaining part of the proof all elements from $\mathcal{H}$ having continuous versions are identified with those versions. The functions h and g are approximated by more general ones.

Step 2. The functions h and g are in $C_b(H)$.
For natural $n > \frac{1}{2}\omega + a$, define

$$\mathcal{G}_n \eta = \int_0^{+\infty} e^{-ns} T_s \eta \, ds, \quad \eta \in \mathcal{H}.$$

The functions

$$h_n = n \mathcal{G}_n h, \quad g_n = n, \quad \mathcal{G}_n g, \quad n > \frac{(\omega + a)}{2},$$

are in $D(\mathcal{L})$ and $h_n \to h$, $g_n \to g$ as $n \to +\infty$ both in $\mathcal{H}$ and uniformly on compact subsets of H. Moreover the corresponding continuous functions (V_n) converge to the value function V uniformly on compact subsets of $[0, +\infty) \times H$. This in turn implies, that V_n, regarded as $\mathcal{H}$-valued functions, converge uniformly on bounded intervals of $\mathbb{R}_+$ to V.

Step 3. The functions h, g satisfy assumption (14.3.1) .

By Step 2 it is enough to show that there exist sequences (h_n), (g_n) of functions from $C_b(H)$ such that $h_n \to h$, $g_n \to g$ as $n \to +\infty$ in $\mathcal{H}$, $h_n \leq g_n$ and the corresponding value functions V_n, $n \in N$, converge to V, as $\mathcal{H}$-valued functions, uniformly on bounded intervals. Moreover V is the value function correponding to the data (h, g).

It turns out that it is enough to choose h_n, g_n, $n \in \mathbb{N}$, identical to h, g, on balls B_n with a fixed center x_0 and radius n and such that on H

$$|h_n| \leq |h|, \ |g_n| \leq |g|, \ n \in \mathbb{N}. \quad \square$$

In some situations the domain $D(\mathcal{L})$ of the generator $\mathcal{L}$ can be completely described and the concept of strong solution of (14.2.10) becomes very explicit. This is the case when for instance X is an Ornstein-Uhlenbeck process satisfying

$$dX = AX dt + B dW, \ X(0) = x,$$

for which the semigroup e^{tA} generated by A has an exponential decay,

$$\|e^{tA}\| \leq M e^{-\gamma t}, \ t \geq 0,$$

for some positive γ and the operators $e^{tA}Q$, $t \geq 0$, are self-adjoint with $Q = BB^*$. Then the Gaussian measure $\mu = N_{Q_\infty}$ with

$$Q_\infty = \int_0^{+\infty} e^{tA} Q e^{tA^*} dt$$

is invariant for the corresponding transition semigroup (R_t), see §10.1. The domain $D(\mathcal{M}_2)$ of its generator $\mathcal{M}_2$ is given by (Theorem 10.2.6)

$$D(\mathcal{M}_2) = W_Q^{2,2}(H, \mu) \cap W_{AQ}^{1,2}(H, \mu)$$

and the operator $\mathcal{M}_2$ is an extension of the differential operator

$$\frac{1}{2} \mathrm{Tr}[Q D^2 \varphi] + \langle x, A^* D\varphi \rangle, \ x \in H.$$

The abstract theorem is, therefore, applicable.

14.4 Excessive measures for Wiener processes

The results obtained in the previous section are applicable if, in particular, the measure μ is invariant for X. For instance, if X is an Ornstein-Uhlenbeck process discussed in Chapter 9. However, invariant measures, even with infinite total mass, do not exist for all Ornstein-Uhlenbeck processes. This is so for the most important case of the Wiener process as we will show now. We have in fact the following result.

Theorem 14.4.1 *If μ is an invariant measure for a nondegenerate Wiener process, on an infinite dimensional Hilbert space H, finite on bounded sets, then $\mu(\Gamma) = 0$ for all $\Gamma \in \mathcal{B}(H)$.*

The following proof, due to S. Kwapien [153], is based on two lemmas.

Lemma 14.4.2 *If μ is an invariant measure for a nondegenerate Wiener process, finite on bounded sets, and f is a bounded Borel function with bounded support then*

$$h(x) = \int_H f(x + y)\mu(dy), \quad x \in H,$$

is harmonic for the heat semigroup.

Proof. Note that

$$
\begin{aligned}
P_t h(x) &= \int_H h(x + z) N_{tQ}(dz) \\
&= \int_H \left[\int_H f(x + z + y)\mu(dy) \right] N_{tQ}(dz) \\
&= \int_H f(x + u)\mu * N_{tQ}(du), \quad x \in H, \; t \geq 0.
\end{aligned}
$$

Since $\mu * N_{tQ} = \mu$, $t \geq 0$, the result follows. $\square$

Lemma 14.4.3 *If h is a nonnegative harmonic function for the heat semigroup then for arbitrary $x, a \in H$,*

$$\frac{1}{2}\left(h(x + Q^{1/2}a) + h(x - Q^{1/2}a) \right) \geq h(x).$$

Proof. By the Cameron-Martin formula we have

$$h(x + Q^{1/2}a) = P_t h(x + Q^{1/2}a) = \int_H h(x + Q^{1/2}a + z) N_{tQ}(dz)$$

$$= \int_H h(x + z) N_{Q^{1/2}a, tQ}(dz)$$

$$= \int_H h(x + z) e^{-\frac{1}{2}|a|^2 + \langle (tQ)^{-1/2}z, (tQ)^{-1/2}Q^{1/2}a\rangle} N_{tQ}(dz)$$

$$= e^{-\frac{|a|^2}{2t}} \int_H h(x + z) e^{\frac{1}{\sqrt{t}}\langle (tQ)^{-1/2}z, a\rangle} N_{tQ}(dz),$$

Consequently

$$\frac{1}{2}\left(h(x+Q^{1/2}a)+h(x-Q^{1/2}a)\right)$$

$$=\frac{1}{2}e^{-\frac{|a|^2}{2t}}\int_H h(x+z)\left[e^{\frac{1}{\sqrt{t}}\langle(tQ)^{-1/2}z,a\rangle}+e^{-\frac{1}{\sqrt{t}}\langle(tQ)^{-1/2}z,a\rangle}N_{tQ}\right](dz).$$

Since, for any $\sigma \in \mathbb{R}$, $\frac{1}{2}(e^\sigma+e^{-\sigma}) \geq 1$ and h is a nonnegative function we have

$$\frac{1}{2}\left(h(x+Q^{1/2}a)+h(x-Q^{1/2}a)\right)$$

$$\geq e^{-\frac{|a|^2}{2t}}\int_H h(x+z)N_{tQ}=e^{-\frac{|a|^2}{2t}}P_t h(x) \geq e^{-\frac{|a|^2}{2t}}h(x).$$

Letting $t \to +\infty$ we obtain the result. □

Proof of Theorem 14.4.1. Assume that $\mu \neq 0$ and that for some $r > 0$, $\mu(B(0,r)) > 0$. Let $f = 1_{B(0,r)}$ and (e_n) be an orthonormal basis of eigenvectors of Q. Define

$$h(x)=\int_H f(x+y)\mu(dy), \ x \in H.$$

Then $h(0)=\mu(B(0,r)) > 0$, and by Lemma 14.4.3

$$\frac{1}{2}\left(h(2re_i)+h(-2re_i)\right) \geq \mu(B(0,r)), \ i \in \mathbb{N}.$$

Consequently, there exists a sequence (ε_i) with $\varepsilon_i=1$ or -1 such that

$$h(2r\varepsilon_i e_i) \geq \mu(B(0,r)).$$

But

$$h(2r\varepsilon_i e_i) = \int_H 1_{B(0,r)}(2r\varepsilon_i e_i + z)\mu(dz)$$

$$= \mu(B(2r\varepsilon_i e_i, r)) \geq \mu(B(0,r)),$$

and the balls $B(2r\varepsilon_i e_i, r)$, $i \in \mathbb{N}$, are disjoint. Therefore

$$\mu(B(0, 3r)) \geq \mu\left(\bigcup_{i=1}^{\infty} B(2r\varepsilon_i e_i, r)\right)$$

$$\geq \sum_{i=1}^{\infty} \mu(B(2r\varepsilon_i e_i, r))$$

$$\geq \sum_{i=1}^{\infty} \mu B(0, r)) = +\infty,$$

a contradiction. $\square$

Thus the $L^2(H, \mu)$ theory cannot be developed if one restricts the considerations to μ being an invariant measure. A plausible possibility would be to choose μ as an ω-excessive measure. Unfortunately μ cannot be chosen to be a Gaussian measure.

Theorem 14.4.4 *Gaussian measures are never ω-excessive for the heat semigroups with $Q \neq 0$.*

Proof. Assume that a measure $N_{m,R}$ is ω-excessive for the heat semigroup

$$P_t\varphi(x) = \int_H \varphi(x + z)N_{tQ}(dz), \ t \geq 0, \ x \in H.$$

Then

$$N_{tQ} * N_{m,R} \leq e^{\omega t} N_{m,R}, \ t \geq 0.$$

Consequently, for any $\lambda \in H$

$$\int_H e^{\langle \lambda, x \rangle} N_{tQ} * N_{m,R}(dx) \leq e^{\omega t} \int_H e^{\langle \lambda, x \rangle} N_{m,R}(dx),$$

and therefore

$$e^{\frac{1}{2}\langle tQ\lambda, \lambda \rangle + \langle \lambda, m \rangle + \frac{1}{2}\langle R\lambda, \lambda \rangle} \leq e^{\omega t} e^{\langle \lambda, m \rangle + \frac{1}{2}\langle R\lambda, \lambda \rangle},$$

or equivalently, for arbitrary $\lambda \in H$, $\frac{1}{2}\langle Q\lambda, \lambda \rangle \leq \omega$, a contradiction if $Q \neq 0$.
$\square$

It is easy to check that the measure $\mu = \int_0^{+\infty} N_{tQ} dt$ is bounded on bounded sets and is excessive for the heat semigroup.

Here is another positive result, see J. Zabczyk [221].

Theorem 14.4.5 *Assume that Q is a diagonal operator on $H = \ell^2$ with positive eigenvalues $(\lambda_n)_{n\in\mathbb{N}}$ such that $\sum_{n=1}^{\infty} \lambda_n < +\infty$. There exists an ω-excessive measure μ for the corresponding heat semigroup of the form $\mu = \mathop{\times}\limits_{n=1}^{\infty} \mu_n$, where μ_n are measures on $\mathbb{R}$, if and only if*

$$\sum_{n=1}^{\infty} \sqrt{\lambda_n} < +\infty. \tag{14.4.1}$$

If (14.4.1) is satisfied one can choose

$$\mu_n(dx) = \frac{1}{2}\lambda_n^{-1/4}e^{\lambda_n^{-1/4}|x|}dx, \quad n \in \mathbb{N}.$$

Part IV

APPENDICES

Appendix A

Interpolation spaces

This appendix is devoted to the basic *interpolation theorem*, which is proved in §A.1. In §A.2 we present a characterization of some interpolation spaces between a Banach space and the domain of a linear operator. We recall that the basic definition of interpolation spaces is given in §2.3.

A.1   The interpolation theorem

The main result of this section is the following.

Theorem A.1.1 *Let X, X_1, Y, Y_1 be Banach spaces such that $Y \subset X$, $Y_1 \subset X_1$ with continuous embeddings. Let moreover T be a linear mapping $T : X \to X_1$, $T : Y \to Y_1$, such that for some $M, N > 0$*

$$\|Tx\|_{X_1} \le M\|x\|_X, \ \|Ty\|_{Y_1} \le N\|y\|_Y.$$

Then T maps $(X, Y)_{\theta, \infty}$ into $(X_1, Y_1)_{\theta, \infty}$, and

$$[Tx]_{(X_1, Y_1)_{\theta, \infty}} \le M^{1-\theta} N^\theta \, [x]_{(X,Y)_{\theta, \infty}}, \quad x \in (X, Y)_{\theta, \infty}.$$

Proof. Let $x \in (X, Y)_{\theta, \infty}$. Then we have

$$[Tx]_{(X_1, Y_1)_{\theta, \infty}} = \sup_{t \in (0,1]} t^{-\theta} \inf\{\|a_1\|_{X_1} + t\|b_1\|_{Y_1} : a_1 + b_1 = Tx\}$$

$$\le \sup_{t \in (0,1]} t^{-\theta} \inf\{\|Ta\|_{X_1} + t\|Tb\|_{Y_1} : a + b = x\}$$

$$\leq \sup_{t\in(0,1]} t^{-\theta}\inf\{M\|a\|_X + tN\|b\|_Y : a+b=x\}$$

$$= M\sup_{t\in(0,1]} t^{-\theta}\inf\left\{\|a\|_X + t\,\frac{N}{M}\|b\|_Y : a+b=x\right\}$$

$$= M\sup_{s\in(0,N/M]}\frac{N^\theta}{M^\theta}s^{-\theta}\inf\{\|a\|_X + s\|b\|_Y : a+b=x\}$$

$$\leq M^{1-\theta}N^\theta[x]_{(X,Y)_{\theta,\infty}}. \quad \square$$

A.2   Interpolation between a Banach space X and the domain of a linear operator in X

We are given a linear closed operator $A : D(A)\subset X \to X$, generator of a strongly continuous semigroup in X, which we denote by e^{tA}. We set

$$\|x\|_X = \|x\|, \quad x \in X, \quad \|x\|_{D(A)} = \|x\| + \|Ax\|, \quad x \in D(A),$$

and

$$(X, D(A))_{\theta,\infty} = D_A(\theta,\infty), \quad \theta \in (0,1). \tag{A.2.1}$$

Proposition A.2.1 *Let $\theta \in (0,1)$. Then we have*

$$D_A(\theta,\infty) = \left\{ x \in X : \sup_{t\in(0,1]} t^{-\theta}\|e^{tA}x - x\| < +\infty \right\}. \tag{A.2.2}$$

Moreover there exists $C_A > 0$ such that for all $x \in D_A(\theta,\infty)$ it holds that

$$\frac{1}{C_A}\|x\|_{D_A(\theta,\infty)} \leq \|x\| + \sup_{t\in(0,1]} t^{-\theta}\|e^{tA}x - x\| \leq C_A\|x\|_{D_A(\theta,\infty)}. \tag{A.2.3}$$

Proof. We set
$$\lambda(x) = \sup_{t\in(0,1]} t^{-\theta}\|e^{tA}x - x\|,$$

and proceed in two steps.

Step 1. If $\lambda(x) < +\infty$ then $x \in D_A(\theta,\infty)$ and there exists $C_1 > 0$ such that $\|x\|_{D_A(\theta,\infty)} \leq C_1\lambda(x)$.

We have in fact
$$\|e^{tA}x - x\| \leq \lambda(x)t^\theta, \quad t \in (0,1).$$

For all $t > 0$ we set

$$a_t = x - \frac{1}{t}\int_0^t e^{sA}x\,ds = \frac{1}{t}\int_0^t (x - e^{sA}x)\,ds,$$

$$b_t = \frac{1}{t}\int_0^t e^{sA}x\,ds.$$

Consequently $b_t \in D(A)$ and $Ab_t = \frac{e^{tA}x-x}{t}$. It follows that

$$\|a_t\| \le \frac{1}{t}\int_0^t s^\theta ds\lambda(x) \le \frac{t^\theta}{\theta+1}\lambda(x),$$

and

$$\|b_t\|_{D(A)} \le \|b_t\| + \|Ab_t\| \le \|x\| + \gamma(x)t^{\theta-1}.$$

So $K(t,x) \le Ct^\theta$ and the conclusion follows.

Step 2. If $x \in D_A(\theta,\infty)$, then $\lambda(x) < +\infty$ and there exists $C_2 > 0$ such that $\lambda(x) \le C_2\|x\|_{D_A(\theta,\infty)}$.

Let $x \in D_A(\theta,\infty)$ and $\varepsilon > 0$. For all $t \in (0,1)$ there exist $a_t \in X$ and $b_t \in D(A)$ such that

$$\|a_t\| + t(\|b_t\| + \|Ab_t\|) \le t^\theta(\|x\|_{D_A(\theta,\infty)} + \varepsilon).$$

For all $\xi > 0, t > 0$ it holds that

$$e^{\xi A}x - x = e^{\xi A}a_t - a_t + e^{\xi A}b_t - b_t = e^{\xi A}a_t - a_t + \int_0^\xi e^{\xi A}Ab_t\,d\xi.$$

It follows that

$$\|e^{\xi A}x - x\| \le \left[2\xi^\theta + \xi t^{\theta-1}\right](\|x\|_{D_A(\theta,\infty)} + \varepsilon),$$

and the conclusion follows since ε is arbitrary. $\square$

Appendix B

Null controllability

B.1   Definition of null controllability

Let H and U be Hilbert spaces. We are given a linear operator $A : D(A) \subset H \to H$, infinitesimal generator of a strongly continuous semigroup (e^{tA}) such that, for some $M > 0$, $\|e^{tA}\| \leq Me^{\omega t}$, $t \geq 0$, and a linear bounded operator $B \in L(U; H)$. We assume for simplicity that the kernel of BB^* is equal to $\{0\}$. We are concerned with the differential equation

$$\begin{cases} y'(t) = Ay(t) + Bu(t), & t \geq 0, \\ y(0) = x \in H, \end{cases} \tag{B.1.1}$$

where $u \in L^2_{\text{loc}}(0, +\infty, U)$.

Problem (B.1.1) has a unique mild solution $y(\cdot) = y(\cdot, x; u)$ given by

$$y(t) = e^{tA}x + L_t u, \quad t \geq 0,$$

where

$$L_t u = \int_0^t e^{(t-s)A} Bu(s) ds.$$

We say that system (B.1.1) is *null controllable in time* $T > 0$, if for all $x \in H$ there exists $u \in L^2(0, T; U)$ such that $y(T, x; u) = 0$. If the system is null controllable in any time, it is called *null controllable*.

Example B.1.1 Assume that $U = H$, and $B = I$. Then system (B.1.1) is null controllable. In fact, given $T > 0, x \in H$, and setting

$$u(t) = -\frac{1}{T} e^{tA}x, \quad t \in [0, T], \tag{B.1.2}$$

we have $y(T, x; u) = 0$.

B.2 Main results

We will need a result on linear operator theory; see e.g. G. Da Prato and J. Zabczyk [101, Appendix A]

Proposition B.2.1 *Let H and Z be Hilbert spaces, $T \in L(U; H)$ and $S \in L(H)$. Then the following statements are equivalent.*

(i) $S(H) \subset T(U)$.

(ii) There exists $C > 0$ such that $|S^*x| \leq C|T^*x|, \quad x \in H$.

Now we can prove the following result.

Theorem B.2.2 *System* (B.1.1) *is null controllable in time $T > 0$, if and only if one of the following statements holds.*

(i) $e^{TA}(H) \subset L_T(L^2(0, T; U))$.

(ii) $e^{TA}(H) \subset Q_T^{1/2}(H)$, *where*

$$Q_T x = \int_0^T e^{sA} BB^* e^{sA^*} x \, ds, \quad x \in H. \tag{B.2.1}$$

Moreover, if system (B.1.1) *is null controllable in any time, we have*

$$L_T(L^2(0, T; U)) = Q_T^{1/2}(H) = \quad constant, \quad T > 0. \tag{B.2.2}$$

Proof. It is obvious that statement (i) is equivalent to null controllability in time T. Now notice that, as easily checked,

$$(L_T^* x)(t) = B^* e^{(T-t)A^*} x.$$

It follows that

$$L_T L_T^* x = \int_0^T e^{sA} BB^* e^{sA^*} x \, ds, \quad x \in H. \tag{B.2.3}$$

Consequently

$$\|L_T^* x\|_{L^2(0,T;U)}^2 = |Q_T^{1/2} x|^2, \tag{B.2.4}$$

from which, in view of Proposition B.2.1,

$$L_T(L^2(0, T; U)) = Q_T^{1/2}(H). \tag{B.2.5}$$

Thus (i) and (ii) are equivalent.

Let us prove the last statement. First notice that, setting

$$S_T u = \int_0^T e^{sA} Bu(s)ds, \quad u \in L^2(0,T;U),$$

we have

$$L_T(L^2(0,T;U)) = S_T(L^2(0,T;U)).$$

This implies

$$L_T(L^2(0,T;U)) \subset L_{T+\varepsilon}(L^2(0,T;U)), \quad \varepsilon > 0.$$

Finally, let us prove the converse inclusion. Let $x \in L_T(L^2(0,T;U)) = S_T(L^2(0,T;U))$, and let $u \in L^2(0,T;U)$ be such that $L_T u = x$. Then we have

$$x = \int_0^{T-\varepsilon} e^{sA} Bu(s)ds + e^{(T-\varepsilon)A} \int_0^{\varepsilon} e^{sA} Bu(s+T-\varepsilon)ds.$$

Since system (B.1.1) is null controllable in any times, there exists $z \in L^2(0,T-\varepsilon;U)$ such that

$$e^{(T-\varepsilon)A} \int_0^{\varepsilon} e^{sA} Bu(s+T-\varepsilon)ds = \int_0^{T-\varepsilon} e^{sA} Bz(s)ds.$$

Thus

$$x = \int_0^{T-\varepsilon} e^{sA} B(u(s)+z(s))ds,$$

and consequently, $x \in L_T(L^2(0,T-\varepsilon;U))$ as required. $\square$

B.3 Minimal energy

We assume here that system (B.1.1) is null controllable. We fix $T > 0$ and $x \in H$, and we consider the affine hyperplane

$$V_x = \{u \in L^2(0,T;U) : L_T u = -e^{TA}x\}.$$

Then we denote by $\hat{u}$ the projection of 0 on V_x; $\hat{u}$ is called the strategy of *minimal energy driving* x *to* 0 and it is the element of $L^2(0,T;U)$ characterized by the following conditions:

$$\begin{cases} \text{(i)} \quad L_T\hat{u} = 0, \\[2ex] \text{(ii)} \quad L_T\eta = 0 \Rightarrow \langle \hat{u}, \eta \rangle_{L^2(0,T;U)} = 0. \end{cases} \tag{B.3.1}$$

Now, we want to find an expression for $\hat{u}$. For this we need some preliminary considerations. First of all we set

$$\Lambda_T = Q_T^{-1/2} e^{TA}, \quad T > 0. \tag{B.3.2}$$

By condition (ii) of Theorem B.2.2 and the closed graph theorem, it follows that Λ_T is bounded for all $T > 0$. Moreover, in view of (B.2.4) and Proposition B.2.1, it follows that $L_T^*(H) = Q_T^{1/2}(H)$, and $|L_T^* Q_T^{-1/2} y| = |y|$, $\forall y \in Q_T^{1/2}(H)$. Since $Q_T^{1/2}(H)$ is dense in H, because Ker $BB^* = \{0\}$, it follows that $L_T^* Q_T^{-1/2}$ is uniquely extendible to a linear bounded operator on H, denoted by $\overline{L_T^* Q_T^{-1/2}}$, such that

$$\|\overline{L_T^* Q_T^{-1/2}}\| = 1. \tag{B.3.3}$$

We can now prove the following result.

Theorem B.3.1 *Assume that system* (B.1.1) *is null controllable in time* $T > 0$, *and let* $x \in H$. *Then the control of minimal energy driving* x *to* 0 *is given by*

$$\hat{u} = -\overline{L_T^* Q_T^{-1/2}} \Lambda_T x. \tag{B.3.4}$$

Moreover, the minimal energy $E(\hat{u}) := |\hat{u}|^2_{L^2(0,T;H)}$ *is given by*

$$E(\hat{u}) = |\Lambda_T x|^2. \tag{B.3.5}$$

Proof. Let $\hat{u}$ be defined by (B.3.4). We are going to check that conditions (i) and (ii) of (B.3.1) are fulfilled. Condition (i) holds since

$$L_T \hat{u} = -L_T \overline{L_T^* Q_T^{-1/2}} \Lambda_T x = -\overline{L_T L_T^* Q_T^{-1/2}} \Lambda_T x = -e^{TA} x.$$

Let us prove (ii). Assume that $L_T \eta = 0$. Then we have

$$\langle \hat{u}, \eta \rangle_{L^2(0,T;H)} = -\left\langle \overline{L_T^* Q_T^{-1/2}} \Lambda_T x, \eta \right\rangle_{L^2(0,T;H)} = \left\langle \Lambda_T x, Q_T^{-1/2} L_T \eta \right\rangle = 0.$$

It remains to prove (B.3.5). Recalling (B.3.3) we have

$$|\hat{u}|^2_{L^2(0,T;H)} = \left| \overline{L_T^* Q_T^{-1/2}} \Lambda_T x \right|^2_{L^2(0,T;H)} \leq |\Lambda_T x|^2.$$

The proof is complete. $\square$

Example B.3.2 We continue here Example B.1.1, assuming $U = H$ and $B = I$. Let u be defined by (B.1.2). Then we have

$$|\Lambda_T x|^2 \leq T^{-2} \int_0^T |e^{tA} x|^2 dt \leq \begin{cases} MT^{-1/2} \text{ if } \omega \leq 0, \\ \\ Me^{\omega T} T^{-1/2} \text{ if } \omega > 0. \end{cases}$$

Example B.3.3 Let $U = H$ and $B = Q^{1/2}$ with Q symmetric and non-negative. Assume that system (B.1.1) is null controllable, and moreover that

$$e^{tA}(H) \subset Q^{1/2}(H), \ t > 0, \tag{B.3.6}$$

and that

$$\int_0^T |Q^{1/2} e^{tA} x|^2 dt < +\infty, \quad t > 0, \ x \in H. \tag{B.3.7}$$

Then the control

$$u(t) = -T^{-1} Q^{-1/2} e^{tA} x, \quad t \geq 0, \tag{B.3.8}$$

drives x to 0 in time T, and so

$$|\Lambda_T x|^2 \leq T^{-2} \int_0^T |Q^{1/2} e^{tA} x|^2 dt. \tag{B.3.9}$$

Assume in addition that the semigroup (e^{tA}) is analytic,

$$\|e^{tA}\| \leq M, \ \|Ae^{tA}\| \leq Nt^{-1}, \ t > 0, \tag{B.3.10}$$

and that there exists $\alpha \in (0, 1/2)$, such that $Q = (-A)^{-\alpha}$. Then by (B.3.9) it follows that

$$|\Lambda_T x| \leq C_\alpha T^{-\alpha-1/2} |x|, \quad x \in H, \tag{B.3.11}$$

for some positive constant C_α.

The following result is a particular case of [196].

Theorem B.3.4 *Assume that system* (B.1.1) *is null controllable; then*

$$\lim_{T \to +\infty} \Lambda_T x = 0, \quad x \in H,$$

if and only if the algebraic Riccati equation

$$PA + A^* P - PBB^* P = 0, \quad P \geq 0, \tag{B.3.12}$$

has a unique solution $P = 0$.

Equation (B.3.12) should be understood in the sense that

$$2\langle PAx, x\rangle - |B^*Px|^2 = 0, \quad x \in D(A).$$

Proof of Theorem B.3.4. For arbitrary operator $R \geq 0$ and $t \geq 0$ denote by $P(t)^R$ the unique solution of the Riccati equation

$$\begin{cases} \frac{d}{dt} P(t) = P(t)A + A^*P(t) - P(t)BB^*P(t), \ t \geq 0, \\ P(0) = R. \end{cases} \tag{B.3.13}$$

Then

$$\langle P(t)^R x, x\rangle = \inf_u \left[\int_0^t |u(s)|^2 ds + \langle Ry^{x,u}(t), y^{x,u}(t)\rangle \right], \tag{B.3.14}$$

where $y(t) = y^{x,u}(t), t \geq 0$, is the solution to the equation

$$\frac{dy}{dt} = Ay + Bu, \quad y(0) = x. \tag{B.3.15}$$

We proceed now in two steps.

Step 1. For arbitrary $t > 0$, $s \geq 0$, $P(s)^{\Lambda_t^*\Lambda_t} = \Lambda_{t+s}^*\Lambda_{t+s}$.

Let $\hat{u}(\sigma)$, $\sigma \in [0, t+s]$, be the control which transfers a given state x to 0 in time $t + s$ with minimal energy. Then

$$\langle \Lambda_{t+s}^*\Lambda_{t+s} x, x\rangle = \int_0^{t+s} |\hat{u}(\sigma)|^2 d\sigma$$

$$= \int_0^s |\hat{u}(\sigma)|^2 d\sigma + \int_s^{t+s} |\hat{u}(\sigma)|^2 d\sigma$$

$$= \int_0^s |\hat{u}(\sigma)|^2 d\sigma + \langle \Lambda_t^*\Lambda_t y^{x,\hat{u}}(s), y^{x,\hat{u}}(s)\rangle.$$

Therefore

$$\langle \Lambda_{t+s}^*\Lambda_{t+s} x, x\rangle \geq \langle P(s)^{\Lambda_t^*\Lambda_t} x, x\rangle.$$

On the other hand for arbitrary control u

$$\int_0^s |u(\sigma)|^2 d\sigma + \langle \Lambda_t^*\Lambda_t y^{x,\hat{u}}(s), y^{x,\hat{u}}(s)\rangle$$

$$= \int_0^s |u(\sigma)|^2 d\sigma + \int_s^{t+s} |\tilde{u}(\eta - s)|^2 d\eta,$$

where $\tilde{u}(\eta)$, $\eta \in [0, t]$, transfers the state $y^{x,u}(s)$ to 0 in time t with minimal energy. Consequently the control

$$\bar{u}(\sigma) = \begin{cases} u(\sigma) \text{ if } \sigma \leq s, \\ u(s) \text{ if } s \leq \sigma \leq s + t, \end{cases}$$

transfers x to 0 in time $t + s$. Therefore

$$\int_0^s |u(\sigma)|^2 d\sigma + \langle \Lambda_t^* \Lambda_t y^{x,u}(s), y^{x,u}(s) \rangle \geq \langle \Lambda_{t+s}^* \Lambda_{t+s} x, x \rangle,$$

and thus

$$\langle \Lambda_{t+s}^* \Lambda_{t+s} x, x \rangle \leq \langle P(s)^{\Lambda_t^* \Lambda_t} x, x \rangle.$$

This completes the proof of Step 1.

The operator valued function $\Lambda_t^* \Lambda_t$, $t > 0$, is decreasing and therefore there exists a symmetric nonnegative limit

$$\lim_{t \to +\infty} \Lambda_t^* \Lambda_t := \overline{P}.$$

Now from (B.3.13), it follows easily that, see e.g. [215],

$$\overline{P}A + A^* \overline{P} - \overline{P}BB^* \overline{P} = 0, \quad \overline{P} \geq 0.$$

Therefore, if $\Lambda_t^* \Lambda_t$ does not tend to 0 as t tends to $+\infty$, the equation (B.3.12) has a nonzero solution.

Assume now that $\Lambda_t^* \Lambda_t \to 0$ as $t \to +\infty$. We need another step.

Step 2. For arbitrary $R \geq 0$ and $t > 0$ we have $P(t)^R \leq \Lambda_t^* \Lambda_t$.

Let in fact $u(\sigma)$, $\sigma \in [0, t]$, be a control which tranfers x to 0 in time t, with minimal energy. Since $y^{x,u}(t) = 0$,

$$\langle \Lambda_t^* \Lambda_t x, x \rangle = \int_0^t |u(\sigma)|^2 d\sigma = \int_0^t |u(\sigma)|^2 d\sigma + \langle y^{x,u}(t), y^{x,u}(t) \rangle \geq \langle P(t)^R x, x \rangle,$$

so the proof of Step 2 is complete.

Finally, if $P \geq 0$ is a solution to (B.3.12), then $P(t) = P(t)^R \leq \Lambda_t^* \Lambda_t$ and so $P = 0$ as required.

The proof of the theorem is complete. $\square$

Assume now that the system (B.3.15) is finite dimensional: dim $H < +\infty$, dim $U < +\infty$.

Theorem B.3.5 *Assume that system* (B.1.1) *is null controllable and finite dimensional. Then* $\lim\limits_{t \to +\infty} \Lambda_t^* \Lambda_t = 0$, *if and only if*

$$\max \{\operatorname{Re} \lambda : \ \lambda \in \sigma(A)\} \leq 0.$$

Proof. Assume that $\max \{\operatorname{Re} \lambda : \ \lambda \in \sigma(A)\} > 0$. Then

$$A = \begin{pmatrix} A_{0,0} & 0 \\ A_{1,0} & A_{1,1} \end{pmatrix}, \quad B = \begin{pmatrix} B_0 \\ B_1 \end{pmatrix},$$

where A is in Jordan form with $A_{0,0}$ corresponding to eigenvalues with strictly positive real parts. The pair $(-A_{0,0}, B_0)$ is controllable and the matrix $-A_{0,0}$ is stable (all eigenvalues have negative real parts). Therefore the equation

$$(-A_{0,0})R_0 + R_0(-A_{0,0})^* + B_0 B_0^* = 0, \quad R_0 \geq 0,$$

has a unique solution $R_0 \geq 0$ which is strictly positive definite. Consequently

$$R_0^{-1}(-A_{0,0}) + (-A_{0,0})^* R_0^{-1} + R_0^{-1} B_0 B_0^* R_0^{-1} = 0.$$

Define

$$P = \begin{pmatrix} R_0^{-1} & 0 \\ 0 & 0 \end{pmatrix}.$$

Then $P \neq 0$ and (B.3.12) holds. By Theorem B.3.4,

$$\lim_{t \to +\infty} \Lambda_t^* \Lambda_t \neq 0$$

Assume now that there exists $P \neq 0$ such that (B.3.12) holds. We can assume that P is diagonal of the form

$$P = \begin{pmatrix} P_0 & 0 \\ 0 & 0 \end{pmatrix},$$

where P_0 has positive elements on the diagonal. Let

$$A - \begin{pmatrix} A_{0,0} & A_{0,1} \\ A_{1,0} & A_{1,1} \end{pmatrix}, \quad B = \begin{pmatrix} B_0 \\ B_1 \end{pmatrix}.$$

Then

$$\begin{pmatrix} P_0 & 0 \\ 0 & 0 \end{pmatrix} \begin{pmatrix} A_{0,0} & A_{0,1} \\ A_{1,0} & A_{1,1} \end{pmatrix} + \begin{pmatrix} A_{0,0}^* & A_{0,1}^* \\ A_{1,0}^* & A_{1,1}^* \end{pmatrix} \begin{pmatrix} P_0 & 0 \\ 0 & 0 \end{pmatrix}$$

$$- \begin{pmatrix} R_0 & 0 \\ 0 & 0 \end{pmatrix} \begin{pmatrix} B_0 \\ B_1 \end{pmatrix} \begin{pmatrix} B_0^* & B_1^* \end{pmatrix} \begin{pmatrix} R_0 & 0 \\ 0 & 0 \end{pmatrix} = 0$$

Therefore

$$\begin{pmatrix} P_0 A_{0,0}, & P_0 A_{0,1} \\ 0 & 0 \end{pmatrix} + \begin{pmatrix} A_{0,0}^* P_0 & 0 \\ A_{0,1}^* P_0 & 0 \end{pmatrix} - \begin{pmatrix} P_0 B_0 B_0^* P_0 & 0 \\ 0 & 0 \end{pmatrix}$$

$$= \begin{pmatrix} 0 & 0 \\ 0 & 0 \end{pmatrix},$$

and equivalently

$$P_0 A_{0,0} + A_{0,0}^* P_0 - P_0 B_0 B_0^* P_0 = 0$$

and

$$P_0 A_{0,1}, \ A_{0,1}^* P_0 = 0.$$

Consequently $A_{0,1} = 0$. Moreover

$$(-A_{0,0}) P_0^{-1} + P_0^{-1} (-A_{0,0})^* + B_0 B_0^* = 0,$$

and the pair $(-A_{0,0}, B_0)$ is controllable. Thus, by a classical result on the Lyapunov equation, see J. Zabczyk [215], the matrix $-A_{0,0}$ is stable and therefore

$$\sup \{ \mathrm{Re}\,\lambda : \lambda \in \sigma(A) \} > 0. \quad \square$$

Appendix C

Semiconcave functions and Hamilton-Jacobi semigroups

C.1 Continuity modulus

We use notation introduced in §2.1 and §2.2. In particular, for any $\varphi \in UC_b(H)$ we define the *uniform continuity modulus* ω_φ of φ by setting

$$\omega_\varphi(t) = \sup\{|\varphi(x) - \varphi(y)| : x, y \in H, \ |x - y| \le t\}, \ t \ge 0. \qquad (C.1.1)$$

Proposition C.1.1 *Let* $\varphi \in UC_b(H)$, *then the following statements hold:*

(i) ω_φ *is not decreasing,*

(ii) ω_φ *is subadditive,*

$$\omega_\varphi(t + s) \le \omega_\varphi(t) + \omega_\varphi(s), \ t, s \ge 0, \qquad (C.1.2)$$

(iii) ω_φ *is continuous in* $[0, +\infty)$.

Proof. Assertion (i) is clear, let us prove (ii). Let $t, s \ge 0$, then for any $\varepsilon > 0$ there exist $x_\varepsilon, y_\varepsilon \in H$ such that

$$|x_\varepsilon - y_\varepsilon| \le t + s, \quad \omega_\varphi(t + s) \le \varepsilon + |\varphi(x_\varepsilon) - \varphi(y_\varepsilon)|.$$

Define $z_\varepsilon = \frac{s}{t+s} x_\varepsilon + \frac{t}{t+s} y_\varepsilon$; then we have

$$x_\varepsilon - z_\varepsilon = \frac{t}{t + s} (x_\varepsilon - y_\varepsilon), \ z_\varepsilon - y_\varepsilon = \frac{s}{t + s} (x_\varepsilon - y_\varepsilon),$$

so that $|x_\varepsilon - z_\varepsilon| \le t$, $|z_\varepsilon - y_\varepsilon| \le s$. Consequently

$$\omega_\varphi(t+s) \le \varepsilon + |\varphi(x_\varepsilon) - \varphi(z_\varepsilon)| + |\varphi(z_\varepsilon) - \varphi(y_\varepsilon)|$$

$$\le \varepsilon + \omega_\varphi(t) + \omega_\varphi(s).$$

The conclusion follows from the arbitrariness of ε.

Finally let us prove (iii). We first show continuity of ω_φ at 0. Since φ is uniformly continuous, for any $\varepsilon > 0$ there exists $t_\varepsilon > 0$ such that

$$|x - y| \le t_\varepsilon \Longrightarrow |\varphi(x) - \varphi(y)| \le \varepsilon.$$

Then $t < t_\varepsilon \Longrightarrow \omega_\varphi(t) \le \varepsilon$, so that ω_φ is continuous at 0. Now let us prove right continuity of ω_φ at $t_0 > 0$. If $h > 0$ we have

$$\omega_\varphi(t_0) \le \omega_\varphi(t_0 + h) \le \omega_\varphi(t_0) + \omega_\varphi(h),$$

so that

$$0 \le \omega_\varphi(t_0 + h) - \omega_\varphi(t_0) \le \omega_\varphi(h),$$

and, by the proved continuity of ω_φ at 0, we have

$$\lim_{h \to 0^+} \omega_\varphi(t_0 + h) = \omega_\varphi(t_0).$$

Left continuity of ω_φ can be proved in a similar way. $\square$

C.2   Semiconcave and semiconvex functions

A function $\varphi : H \to \mathbb{R}$ is said to be *semiconvex* (resp. *semiconcave*), if there exists $K > 0$ such that

$$x \to \varphi(x) + \frac{K}{2}|x|^2 \quad \left(\text{resp. } x \to \varphi(x) - \frac{K}{2}|x|^2 \right),$$

is convex (resp. concave). We will need the following result proved in [155].

Proposition C.2.1 *Assume that $\varphi \in C_b^{0,1}(H)$ is both semiconcave and semiconvex, and let $K > 0$ be such that*

$$x \to \varphi(x) + \frac{K}{2}|x|^2 \text{ is convex and } x \to \varphi(x) - \frac{K}{2}|x|^2 \text{ is concave.}$$

Then $\varphi \in C_b^{1,1}(H)$ and $[\varphi]_{1,1} \le K$.

Proof. We first prove that for any $t \in [0, 1]$ the following estimate holds:

$$|\varphi(tx + (1 - t)y) - t\varphi(x) - (1 - t)\varphi(y)|$$

$$\leq \frac{K}{2}t(1 - t)|x - y|^2, \quad x, y \in H. \qquad (C.2.1)$$

In fact, since $\varphi(x) + \frac{K}{2}|x|^2$ is convex, for any $x, y \in H$ we have

$$\varphi(tx + (1 - t)y) + \frac{K}{2}|tx + (1 - t)y|^2$$

$$\leq t\varphi(x) + t\frac{K}{2}|x|^2 + (1 - t)\varphi(y) + (1 - t)\frac{K}{2}|y|^2,$$

which yields

$$\varphi(tx + (1 - t)y) - t\varphi(x) - (1 - t)\varphi(y) \leq \frac{K}{2}t(1 - t)|x - y|^2.$$

In a similar way, using concavity of $\varphi(x) - \frac{K}{2}|x|^2$, we prove that

$$\varphi(tx + (1 - t)y) - t\varphi(x) - (1 - t)\varphi(y) \geq -\frac{K}{2}t(1 - t)|x - y|^2,$$

and so (C.2.1) is proved.

For any $x, y \in H$, we denote by $\varphi'(x) \cdot y$ the derivative (if it exists) of φ at x in the direction y:

$$\varphi'(x) \cdot y = \frac{d}{d\lambda}\varphi(x + \lambda y)\Big|_{\lambda=0}.$$

Now we proceed in several steps.

Step 1. Existence of the directional derivative.

Let $x, y \in H$ be fixed and set

$$F(\lambda) = \frac{\varphi(x + \lambda y) - \varphi(x)}{\lambda}, \quad \lambda > 0.$$

Then, if $\lambda < \mu$ we have

$$F(\lambda) - F(\mu) = \frac{1}{\lambda\mu}[\mu\varphi(x + \lambda y) - \lambda\varphi(x + \mu y) - (\mu - \lambda)\varphi(x)].$$

Setting $\alpha = \frac{\lambda}{\mu}$, it follows that

$$F(\lambda) - F(\mu) = \frac{1}{\lambda}[\varphi((1-\alpha)x + \alpha(x+\mu y)) - \alpha\varphi(x+\mu y) - (1-\alpha)\varphi(x)].$$

Now, from (C.2.1) with $t = \alpha$, it follows that

$$|F(\lambda) - F(\mu)| \le \frac{K}{2}|\mu - \lambda||y|^2. \qquad (C.2.2)$$

Thus F is uniformly continuous on $(0, +\infty)$ and so there exists the limit $\lim_{\lambda \to 0} F(\lambda) = \varphi'(x) \cdot y$.

Step 2. For any $x \in H$ the directional derivative $\varphi'(x)$ belongs to $L(H)$ and $\|\varphi'(x)\| \le \frac{K}{2} + [\varphi]_1$.

We leave to the reader the simple proof of linearity of $\varphi'(x)$. Moreover, from (C.2.2) with $\mu = 1$ we get

$$|\varphi'(x) \cdot y - \varphi(x+y) + \varphi(x)| \le \frac{K}{2}|y|^2.$$

Now, letting λ tend to 0, we find

$$|\varphi'(x) \cdot y| \le \frac{K}{2}|y|^2 + [\varphi]_1|y|,$$

and the conclusion follows.

Step 3. If $\dim H < +\infty$ then the following statements are equivalent.

(i) It holds that

$$\frac{1}{h^2}|\varphi(x+2hy) - 2\varphi(x+hy) + \varphi(x)| \le K|y|^2, \quad x, y \in H.$$

(ii) It holds that

$$|\varphi'(x) - \varphi'(y)| \le K|x - y|, \quad x, y \in H.$$

If $\varphi \in UC_b^2(H)$ then the equivalence of (i) and (ii) is easy and it is left to the reader. Now the conclusion follows from the fact that $UC_b^2(H)$ is dense in $UC_b^1(H)$.

Step 4. If $\dim H < +\infty$ then the conclusion of the proposition holds.

We have in fact

$$\varphi(x + 2hy) - 2\varphi(x + hy) + \varphi(x)$$

$$= -2\left[\varphi\left(\frac{1}{2}(x + 2hy) + \frac{1}{2}x\right) - \frac{1}{2}\varphi(x + 2hy) - \frac{1}{2}\varphi(x)\right].$$

Thus by (C.2.1) it follows that

$$|\varphi(x + 2hy) - 2\varphi(x + hy) + \varphi(x)| \le Kh^2|y|^2,$$

and Step 4 is proved.

Note that we present a different proof when dim $H = +\infty$, since $UC_b^2(H)$ is not dense in $UC_b^1(H)$, see §2.2.

Step 5. Conclusion when dim $H = +\infty$.

Let x, y, z be fixed. Since the subspace spanned by x, y, z is of dimension less than or equal to 3, by Step 4 it follows that

$$|\varphi'(x) \cdot z - \varphi'(y) \cdot z| \le K|x - y||z|,$$

which yields

$$\|\varphi'(x) - \varphi'(y)\| \le K|x - y|, \quad x, y \in H,$$

by the arbitrariness of z. Since, as is well known, continuity of the directional derivatives implies differentiability of φ, we have

$$\|D\varphi(x) - D\varphi'(y)\| \le K|x - y|, \quad x, y \in H.$$

Therefore $\varphi \in UC_b^{1,1}(H)$ as required. $\square$

C.3 The Hamilton-Jacobi semigroups

In this section we introduce two semigroups of nonlinear operators on $UC_b(H)$ that are the key tool for the definition of inf-sup convolutions. For any $\varphi \in UC_b(H)$ and for any $t > 0$ we set

$$U_t\varphi(x) = \inf_{y \in H}\left\{\varphi(y) + \frac{|x - y|^2}{2t}\right\} = \inf_{y \in H}\left\{\varphi(x - y) + \frac{|y|^2}{2t}\right\} \qquad \text{(C.3.1)}$$

and

$$V_t\varphi(x) = \sup_{y \in H}\left\{\varphi(y) - \frac{|x - y|^2}{2t}\right\} = \sup_{y \in H}\left\{\varphi(x - y) - \frac{|y|^2}{2t}\right\}. \qquad \text{(C.3.2)}$$

We set moreover

$$U_0\varphi = V_0\varphi = \varphi \quad \varphi \in UC_b(H). \qquad \text{(C.3.3)}$$

Remark C.3.1 Setting $u(t, x) = U_t\varphi(x)$ and $v(t, x) = V_t\varphi(x)$, $x \in H$, $t \geq 0$, then u and v are respectively the unique *viscosity solutions* of the Hamilton-Jacobi equations

$$\begin{cases} D_t u(t, x) + \frac{1}{2}|Du(t, x)|^2 = 0, \ x \in H, \ t \geq 0, \\ u(0, x) = \varphi(x), \end{cases}$$

and

$$\begin{cases} D_t v(t, x) - \frac{1}{2}|Dv(t, x)|^2 = 0, \ x \in H, \ t \geq 0 \\ v(0, x) = \varphi(x) \end{cases}$$

These results are due to M. G. Crandall and P. L. Lions [59]. We will not use them in what follows.

We are now going to prove several properties of semigroups U_t, $t \geq 0$, and V_t, $t \geq 0$. First of all we notice that, from the very definition of U_t and V_t, we have

$$U_t\varphi(x) \leq \varphi(x) \leq V_t\varphi(x), \quad t \geq 0, \ x \in H, \quad \varphi \in UC_b(H). \tag{C.3.4}$$

It is also clear that U_t and V_t are order preserving. That is if $\varphi, \psi \in UC_b(H)$ are such that $\varphi(x) \leq \psi(x)$, $x \in H$, then we have

$$U_t\varphi(x) \leq U_t\psi(x), \text{ and } V_t\varphi(x) \leq V_t\psi(x), \quad x \in H.$$

We prove now, following J. M. Lasry and P. L. Lions [155], that U_t and V_t are strongly continuous semigroups of contractions on $UC_b(H)$.

Proposition C.3.2 *Let $U_t, V_t, t \geq 0$, be defined by (C.3.1)-(C.3.3). Then the following statements hold.*

(i) *For any $t > 0$, U_t and V_t map $UC_b(H)$ into itself. Moreover $\omega_{U_t\varphi} \leq \omega_\varphi$, $\omega_{V_t\varphi} \leq \omega_\varphi$, $\varphi \in UC_b(H)$, $t \geq 0$.*

(ii) *We have $U_{t+s} = U_t U_s, V_{t+s} = V_t V_s$, $t, s \geq 0$.*

(iii) *For all $\varphi, \psi \in UC_b(H)$ we have*

$$\|U_t\varphi - U_t\psi\|_0 \leq \|\varphi - \psi\|_0, \quad \|V_t\varphi - V_t\psi\|_0 \leq \|\varphi - \psi\|_0.$$

(iv) *We have*

$$|U_t\varphi(x) - \varphi(x)| \leq \omega_\varphi \left(2\sqrt{t\|\varphi\|_0}\right), \ \varphi \in UC_b(H), \quad x \in H,$$

$$|V_t\varphi(x) - \varphi(x)| \leq \omega_\varphi \left(2\sqrt{t\|\varphi\|_0}\right), \ \varphi \in UC_b(H), \quad x \in H,$$

so that

$$\lim_{t\to 0} U_t\varphi = \varphi, \quad \lim_{t\to 0} V_t\varphi = \varphi, \quad \text{in } UC_b(H). \tag{C.3.5}$$

(v) If $\varphi \in C_b^\alpha(H)$ for $\alpha \in (0,1)$, then there exists $C_\alpha > 0$ such that

$$|U_t\varphi(x) - \varphi(x)| \le C_\alpha \, [\varphi]_\alpha^{\frac{2}{2-\alpha}} \, t^{\frac{\alpha}{2-\alpha}}, \; t > 0, \; x \in H,$$

$$\tag{C.3.6}$$

$$|V_t\varphi(x) - \varphi(x)| \le C_\alpha \, [\varphi]_\alpha^{\frac{2}{2-\alpha}} \, t^{\frac{\alpha}{2-\alpha}}, \; t > 0, \; x \in H.$$

Proof. We will prove all the statements concerning U_t since those concerning V_t can be proved in a similar way. In all the proof φ is a fixed element of $UC_b(H)$.

Let us prove (i). Let $t > 0, x, \overline{x} \in H$, then we have

$$
\begin{aligned}
U_t\varphi(x) - U_t\varphi(\overline{x}) &= \inf_{y\in H}\left\{\varphi(x-y) + \frac{|y|^2}{2t}\right\} + \sup_{z\in H}\left\{-\varphi(\overline{x}-z) - \frac{|z|^2}{2t}\right\} \\
&= \sup_{z\in H}\inf_{y\in H}\left\{\varphi(x-y) - \varphi(\overline{x}-z) + \frac{|y|^2 - |z|^2}{2t}\right\} \\
&\le \sup_{z\in H}\inf_{y\in H}\left\{\omega_\varphi(x - \overline{x} + z - y) + \frac{|y|^2 - |z|^2}{2t}\right\}
\end{aligned}
$$

Since ω_φ is subadditive by Proposition C.1.1 it follows that

$$U_t\varphi(x) - U_t\varphi(\overline{x}) \le \omega_\varphi(x - \overline{x}) + \sup_{z\in H}\inf_{y\in H}\left\{\omega_\varphi(z - y) + \frac{|y|^2 - |z|^2}{2t}\right\}.$$

Setting $y = z$, we see that

$$\sup_{z\in H}\inf_{y\in H}\left\{\omega_\varphi(z - y) + \frac{|y|^2 - |z|^2}{2t}\right\} \le 0,$$

and so $|U_t\varphi(x) - U_t\varphi(\overline{x}| \le \omega_\varphi(x - \overline{x})$, and the conclusion follows.

We now prove the semigroup law (ii). Let $t, s > 0, x \in H$, then we have

$$
\begin{aligned}
U_t U_s\varphi(x) &= \inf_{y\in H}\left\{\inf_{z\in H}\left[\varphi(z) + \frac{|y - z|^2}{2s}\right] + \frac{|x - y|^2}{2t}\right\} \\
&= \inf_{z\in H}\left\{\varphi(z) + \inf_{y\in H}\left[\frac{|y - z|^2}{2s} + \frac{|x - y|^2}{2t}\right]\right\}.
\end{aligned}
$$

Since

$$\min_{y \in H} \left[\frac{|y - z|^2}{2s} + \frac{|x - y|^2}{2t} \right] = \frac{|x - z|^2}{2(t + s)},$$

we have

$$U_t U_s \varphi(x) = \inf_{z \in H} \left\{ \varphi(z) + \frac{|x - z|^2}{2(t + s)} \right\} = U_{t+s} \varphi(x).$$

Let us prove (iii). Let $x \in H, t > 0$, $\varphi, \psi \in UC_b(H)$. For any $\varepsilon > 0$ there exists $y_\varepsilon \in H$ such that

$$U_t \psi(x) + \varepsilon > \psi(y_\varepsilon) + \frac{|x - y_\varepsilon|^2}{2t}.$$

It follows that

$$U_t \varphi(x) - U_t \psi(x) \le \inf_{y \in H} \left\{ \varphi(y) + \frac{|x - y|^2}{2t} \right\} - \psi(y_\varepsilon) - \frac{|x - y_\varepsilon|^2}{2t} + \varepsilon.$$

Setting $y = y_\varepsilon$ we have

$$U_t \varphi(x) - U_t \psi(x) \le \varphi(y_\varepsilon) - \psi(y_\varepsilon) - \varepsilon \le \|\varphi - \psi\|_0 - \varepsilon.$$

The proof follows now from the arbitrariness of ε.

Let us prove (iv). Let $x \in H, t > 0$. For any $\varepsilon > 0$ there exists $y_\varepsilon \in H$ such that

$$U_t \varphi(x) + \varepsilon > \varphi(y_\varepsilon) + \frac{|x - y_\varepsilon|^2}{2t}.$$

Taking into account (C.3.4) we obtain

$$0 \le \varphi(x) - U_t \varphi(x) \le \varphi(x) - \varphi(y_\varepsilon) - \frac{|x - y_\varepsilon|^2}{2t} + \varepsilon. \tag{C.3.7}$$

This implies that $\frac{|x - y_\varepsilon|^2}{2t} \le 2 \|\varphi\|_0 + \varepsilon$, and so $|x - y_\varepsilon| \le \sqrt{4t\|\varphi\|_0 + 2t\varepsilon}$. Consequently, by (C.3.7) it follows that

$$\varphi(x) - U_t \varphi(x) \le \varphi(x) - \varphi(y_\varepsilon) + \varepsilon$$

$$\le \omega_\varphi(|x - y_\varepsilon|) + \varepsilon$$

$$\le \omega_\varphi \left(\sqrt{4t\|\varphi\|_0 + 2t\varepsilon} \right) + \varepsilon.$$

Now the conclusion follows from the arbitrariness of ε.

Let us prove, finally, (v). By (C.3.7) it follows that

$$\frac{|x - y_\varepsilon|^2}{2t} \le \varphi(x) - \varphi(y_\varepsilon) + \varepsilon \le [\varphi]_\alpha |x - y_\varepsilon|^\alpha + \varepsilon,$$

which implies

$$|x - y_\varepsilon|^2 \le 2t[\varphi]_\alpha |x - y_\varepsilon|^\alpha + 2\varepsilon t.$$

By Young's inequality we have for any $C > 0$

$$|x - y_\varepsilon|^2 \le \frac{\alpha}{2} C^{\frac{2}{\alpha}} |x - y_\varepsilon|^2 + \left(1 - \frac{\alpha}{2}\right) \left[\frac{2t[\varphi]_\alpha}{C}\right]^{\frac{2}{2-\alpha}} + 2\varepsilon t.$$

Setting $C = (1/\alpha)^{2/\alpha}$ we find that, for a suitable constant C_α,

$$|x - y_\varepsilon|^2 \le C_\alpha^2 [\varphi_\alpha]^{\frac{2}{2-\alpha}} t^{\frac{2}{2-\alpha}} + 4\varepsilon t.$$

Now by (C.3.7) it follows that

$$\varphi(x) - U_t\varphi(x) \le [\varphi]_\alpha |x - y_\varepsilon|^\alpha + \varepsilon$$

$$\le \left(C_\alpha^2 [\varphi_\alpha]^{\frac{2}{2-\alpha}} t^{\frac{2}{2-\alpha}} + 4\varepsilon t\right)^{\frac{\alpha}{2}} + \varepsilon.$$

The conclusion follows again from the arbitrariness of ε.□

It is well known that the functions $U_t\varphi$ and $V_t\varphi$ are not differentiable even if φ is regular. One can prove, however, following J. M. Lasry and P. L. Lions [155], that they are Lipschitz continuous.

Proposition C.3.3 (i). *For any* $\varphi \in UC_b(H)$, $x, \bar{x} \in H$, *and any* $t > 0$ *we have*

$$\begin{cases} \dfrac{|U_t\varphi(x) - U_t\varphi(\bar{x})|}{|x - \bar{x}|} \le \dfrac{1}{2t}\left[4\sqrt{t}\,\|\varphi\|_0 + |x - \bar{x}|\right], \\[4mm] \dfrac{|V_t\varphi(x) - V_t\varphi(\bar{x})|}{|x - \bar{x}|} \le \dfrac{1}{2t}\left[4\sqrt{t}\,\|\varphi\|_0 + |x - \bar{x}|\right]. \end{cases} \tag{C.3.8}$$

(ii). *For any* $\varphi \in C_b^\alpha(H)$, $\alpha \in (0, 1)$, $x, \bar{x} \in H$, *and any* $t > 0$ *there exists* $C_\alpha > 0$ *such that*

$$\begin{cases} \dfrac{|U_t\varphi(x) - U_t\varphi(\bar{x})|}{|x - \bar{x}|} \le \dfrac{1}{2t}\left[2C_\alpha [\varphi]_\alpha^{\frac{1}{2-\alpha}} t^{\frac{1}{2-\alpha}} + |x - \bar{x}|\right], \\[4mm] \dfrac{|V_t\varphi(x) - V_t\varphi(\bar{x})|}{|x - \bar{x}|} \le \dfrac{1}{2t}\left[2C_\alpha [\varphi]_\alpha^{\frac{1}{2-\alpha}} t^{\frac{1}{2-\alpha}} + |x - \bar{x}|\right]. \end{cases} \tag{C.3.9}$$

Proof. (i) For any $\varepsilon > 0$ there exists $y_\varepsilon \in H$ such that

$$U_t\varphi(\overline{x}) + \varepsilon > \varphi(y_\varepsilon) + \frac{|\overline{x} - y_\varepsilon|^2}{2t}. \qquad (C.3.10)$$

It follows that

$$U_t\varphi(x) - U_t\varphi(\overline{x}) \leq \frac{|x - y_\varepsilon|^2}{2t} - \frac{|\overline{x} - y_\varepsilon|^2}{2t} + \varepsilon$$

$$\leq \frac{1}{2t}\left[|x - \overline{x}|^2 + 2 < x - \overline{x}, \overline{x} - y_\varepsilon > \right] + \varepsilon. \qquad (C.3.11)$$

But from (C.3.10) it follows that $\frac{|\overline{x}-y_\varepsilon|^2}{2t} \leq 2\|\varphi\|_0 + \varepsilon$ so that $|\overline{x} - y_\varepsilon| \leq \sqrt{2t(2\|\varphi\|_0 + \varepsilon)}$. Substituting in (C.3.11), and taking into account the arbitrariness of ε, gives (C.3.8). Let us prove (C.3.9). By (C.3.10) we have $\varphi(\overline{x}) + \varepsilon > \varphi(y_\varepsilon) + \frac{|\overline{x}-y_\varepsilon|^2}{2t}$, which yields $\frac{|\overline{x}-y_\varepsilon|^2}{2t} \leq \varepsilon + [\varphi]_\alpha|\overline{x} - y_\varepsilon|^\alpha$. Arguing as in the proof of Proposition C.3.2(v) we see that there exists $C_\alpha > 0$ such that $|\overline{x}-y_\varepsilon| \leq C_\alpha[\varphi]_{\overline{\alpha}}^{\frac{2}{2-\alpha}} t^{\frac{1}{2-\alpha}} + 4\varepsilon t$. Substituting in (C.3.11), and taking into account the arbitrariness of ε we find (C.3.9). $\square$

Let $t > 0$. By Proposition C.3.3 it follows that for any $\varphi \in UC_b(H)$, $U_t\varphi$ and $V_t\varphi$ are Lipschitz continuous. We prove now that $U_t\varphi$ is semiconcave, and that $V_t\varphi$ is semiconvex.

Proposition C.3.4 *Let $t > 0$ and $\varphi \in UC_b(H)$. Then $x \to U_t\varphi(x) - \frac{|x|^2}{2t}$ is concave, and $x \to V_t\varphi(x) + \frac{|x|^2}{2t}$ is convex.*

Proof. We have

$$U_t\varphi(x) - \frac{|x|^2}{2t} = \inf_{y \in H}\left\{\varphi(y) + \frac{|x - y|^2}{2t} - \frac{|x|^2}{2t}\right\}$$

$$= \inf_{y \in H}\left\{\varphi(y) - \frac{1}{t}\langle 2x - y, y\rangle\right\}.$$

Since the infimum of a set of affine functions is a concave function, the conclusion follows. The statement concerning $V_t\varphi$ is proved similarly. $\square$

We end this section by proving that if φ is semiconcave and $t > 0$, then $V_t\varphi$ is also semiconcave. We first need a lemma, due to J. M. Lasry and P. L. Lions [155].

Lemma C.3.5 *Let $F : H \times H \to \mathbb{R}$, $(x, y) \to F(x, y)$, be concave. Then the function $\psi : H \to \mathbb{R} : \psi(x) = \sup_{y \in H} F(x, y)$ is concave.*

Proof. Let $x_1, x_2 \in H$, and $\varepsilon > 0$. Then there exist $y_{1,\varepsilon}, y_{2,\varepsilon} \in H$ such that

$$\psi(x_1) \le F(x_1, y_{1,\varepsilon}) + \varepsilon, \quad \psi(x_2) \le F(x_2, y_{2,\varepsilon}) + \varepsilon. \tag{C.3.12}$$

Let $\theta \in [0, 1]$. Then we have clearly

$$\psi(\theta x_1 + (1 - \theta)x_2) \ge F(\theta x_1 + (1 - \theta)x_2, \theta y_{1,\varepsilon} + (1 - \theta)y_{2,\varepsilon}).$$

Since F is concave, it follows that

$$\psi(\theta x_1 + (1 - \theta)x_2) \ge \theta F(x_1, y_{1,\varepsilon}) + (1 - \theta)F(x_2, y_{2,\varepsilon}).$$

Finally, from (C.3.12) it follows that

$$\psi(\theta x_1 + (1 - \theta)x_2) \ge \theta\psi(x_1) + (1 - \theta)\psi(x_2) - \varepsilon,$$

which yields the conclusion, due to the arbitrariness of ε. $\square$
     We can now prove the result.

Proposition C.3.6 *Let $\varphi \in UC_b(H)$ and $t > 0$ be such that $x \to \varphi(x) - \frac{|x|^2}{2t}$, is concave. Then for any $t \in (0, 1)$ and any $s \in (0, t)$ the function $x \to V_s\varphi(x) - \frac{|x|^2}{2(t-s)}$ is concave.*

Proof. We have

$$V_s\varphi(x) - \frac{|x|^2}{2(t - s)} = \sup_{y \in H}\{f_1(y) + F_2(x, y)\},$$

where $f_1(y) = \varphi(y) - \frac{|y|^2}{2t}$, and

$$F_2(x, y) = \frac{|y|^2}{2t} - \frac{|x - y|^2}{2s} - \frac{|x|^2}{2(t - s)}.$$

Now f_1 in concave in H by hypothesis and F_2 is concave in $H \times H$ by a direct verification. Then the conclusion follows from Lemma C.3.5. $\square$

Bibliography

[1] N. U. Ahmed, M. Fuhrman and J. Zabczyk, On filtering equations in infinite dimensions, *J. Functional Anal.*, **143**, 180-204, 1997.

[2] A. Albanese and F. Kühnemund, Trotter-Kato approximations theorems for locally equicontinuous semigroups, preprint.

[3] S. Albeverio and R. Høegh-Krohn, Dirichlet forms and diffusion processes on rigged Hilbert spaces, *Z. Wahrsch. verw. Geb.*, **40**, 1-57, 1977.

[4] S. Albeverio and M. Röckner, New developments in the theory and applications of Dirichlet forms in stochastic processes, Physics and Geometry, S. Albeverio et al. (editors), World Scientific, 27-76, 1990.

[5] S. Albeverio and M. Röckner, Stochastic differential equations in infinite dimensions: solutions via Dirichlet forms, *Probab. Th. Relat. Fields*, **89**, 347-86, 1991.

[6] W. Arendt, A. Driouich and O. El-Mennaoui, On the infinite product of C_0 semigroups, *J. Functional Anal.*, **160**, 524-542, 1998.

[7] J. P. Aubin, *Mathematical methods of game and economic theory*, North-Holland, 1979.

[8] D. Bakry, L'hypercontractivité et son utilisation en théorie des semi-groupes, Lectures on Probability Theory, **1581**, Springer-Verlag, 1994.

[9] V. Barbu and G. Da Prato, *Hamilton-Jacobi equations in Hilbert spaces*, Research Notes in Mathematics, Pitman, 1982.

[10] Ph. Bénilan, Une remarque sur la convergence des semigroupes non linéaires, *C. R. Acad. Sci. Paris*, **272**, 1182-1184, 1971.

[11] A. Bensoussan and J. L. Lions, *Applications des inéquations variationnelles en contrôle stochastique*, Dunod, 1978.

[12] Y. M. Berezansky and Y. G. Kondratiev, *Spectral methods in infinite-dimensional analysis*, two volumes, Kluwer, 1995.

[13] P. Besala, On the existence of a fundamental solution for a parabolic equation with unbounded coefficients, *Ann. Polon. Math.*, **29**, 403-409, 1975.

[14] A. Bielecki, Une remarque sur la méthode de Banach-Caccioppoli-Tikhonov dans la théorie des équations différentielles ordinaires, *Bull. Acad. Polon. Sci.*, **4**, 261-264, 1956.

[15] P. Billingsley, *Probability and measure*, Wiley, 1979.

[16] J. M. Bismut, *Large deviations and the Malliavin calculus*, Birkhäuser, 1984.

[17] R. M. Blumenthal and R. K. Getoor, *Markov processes and potential theory*, Academic Press, 1968.

[18] V. Bogachev, *Gaussian measures*, Mathematical Surveys and Monographs, **62**, American Mathematical Society, 1998.

[19] V. Bogachev, G. Da Prato and M. Röckner, Regularity of invariant measures for a class of perturbed Ornstein-Uhlenbeck operators, *Nonlinear Diff. Equations Appl.*, **3**, 261-268, 1996.

[20] V. Bogachev and M. Röckner, Regularity of invariant measures on finite and infinite dimensional spaces and applications, *J. Functional Anal.*, **133**, 168-223, 1995.

[21] V. Bogachev and M. Röckner, Elliptic equations for measures on infinite dimensional spaces and applications, *Probab. Th. Relat. Fields*, **120**, 445-496, 2001.

[22] V. Bogachev, M. Röckner and B. Schmuland, Generalized Mehler semigroups and applications, *Probab. Th. Relat. Fields*, **114**, 193-225, 1996.

[23] V. Bogachev, M. Röckner and T. S. Zhang, Existence and uniqueness of invariant measures: an approach via sectorial forms, *Appl. Math. Optim.*, **41**, 87-109, 2000.

[24] S. Bonaccorsi and G. Guatteri, A variational approach to evolution problems with variable domains, *J. Diff. Equations*, **175**, 51-70, 2001.

[25] H. Brézis, *Operatéurs maximaux monotones*, North-Holland, 1973.

[26] H. Brézis, M. Crandall and A. Pazy, Perturbations of nonlinear maximal monotone sets, *Comm. Pure Appl. Math.*, **23**, 123-144, 1970.

[27] P. Cannarsa and G. Da Prato, On a functional analysis approach to parabolic equations in infinite dimensions, *J. Functional Anal.*, **118**, 22-42, 1990.

[28] P. Cannarsa and G. Da Prato, Second-order Hamilton-Jacobi equations in infinite dimensions, *SIAM J. Control Optim.*, **29**, 474-492, 1991.

[29] P. Cannarsa and G. Da Prato, A semigroup approach to Kolmogoroff equations in Hilbert spaces, *Appl. Math. Letters*, **4**, 49-52, 1991.

[30] P. Cannarsa and G. Da Prato, Direct solution of a second order Hamilton-Jacobi equation in Hilbert spaces, *Stochastic partial differential equations and applications*, G. Da Prato and L. Tubaro (editors), Research Notes in Mathematics, **268**, Pitman, 72-85, 1992.

[31] P. Cannarsa and G. Da Prato, Infinite dimensional elliptic equations with Hölder continuous coefficients, *Advances Diff. Equations*, **1**, 425-452, 1996.

[32] P. Cannarsa and G. Da Prato, Schauder estimates for Kolmogorov equations in Hilbert spaces, *Progress in elliptic and parabolic partial differential equations*, A. Alvino, P. Buonocore, V. Ferone, E. Giarrusso, S. Matarasso, R. Toscano and G. Trombetti (editors), Research Notes in Mathematics, **350**, Pitman, 100-111, 1996.

[33] P. Cannarsa and G. Da Prato, Potential theory in Hilbert spaces, *Proc. Symposia Appl. Math.*, **54**, 27-51, 1998.

[34] P. Cannarsa and V. Vespri, Generation of analytic semigroups by elliptic operators with unbounded coefficients, *SIAM J. Math. Anal.*, **18**, 857-872, 1985.

[35] S. Cerrai, A Hille-Yosida theorem for weakly continuous semigroups, *Semigroup Forum*, **49**, 349-367, 1994.

[36] S. Cerrai, Weakly continuous semigroups in the space of functions with polynomial growth, *Dyn. Syst. Appl.*, **4**, 351-372, 1995.

[37] S. Cerrai, Elliptic and parabolic equations in $\mathbb{R}^n$ with coefficients having polynomial growth, *Comm. Partial Diff. Equations*, **21**, 281-317, 1996.

[38] S. Cerrai, Kolmogorov equations in Hilbert spaces with non smooth coefficients, *Comm. Appl. Anal.*, **2**, 271-297, 1998.

[39] S. Cerrai, Differentiability of Markov semigroups for stochastic reaction-diffusion equations and applications to control, *Stoch. Processes Appl.*, **83**, 15-37, 1999.

[40] S. Cerrai, Ergodicity for stochastic reaction-diffusion systems with polynomial coefficients, *Stoch. Stoch. Reports*, **67**, 17-51, 1999.

[41] S. Cerrai, Smoothing properties of transition semigroups relative to SDEs with values in Banach spaces, *Probab. Th. Relat. Fields*, **113**, 85-114, 1999.

[42] S. Cerrai, Optimal control problems for stochastic reaction-diffusion systems with non Lipschitz coefficients, *SIAM J. Control Optim.*, **39**, 1779-1816, 2001.

[43] S. Cerrai, *Second order PDE's in finite and infinite dimensions. A probabilistic approach*, Lecture Notes in Mathematics, **1762**, Springer-Verlag, 2001.

[44] S. Cerrai, Stationary Hamilton-Jacobi equations in Hilbert spaces and applications to a stochastic optimal control problem, *SIAM J. Control Optim.*, to appear.

[45] S. Cerrai, Classical solutions for Kolmogorov equations in Hilbert spaces, *Proceedings of the Ascona Conference 1999 on Stochastic Analysis, Random Fields and Their Applications*, to appear.

[46] S. Cerrai and F. Gozzi, Strong solutions of Cauchy problems associated to weakly continuous semigroups, *Diff. Integral Equations*, **8**, 465-486, 1994.

[47] A. Chojnowska-Michalik, Transition semigroups for stochastic semilinear equations in Hilbert spaces, *Dissertationes Mathematicae (Rozprawy Mat.)*, **396**, Warsaw, 2001.

[48] A. Chojnowska-Michalik and B. Goldys, Existence, uniqueness and invariant measures for stochastic semilinear equations, *Probab. Th. Relat. Fields*, **102**, 331-356, 1995.

[49] A. Chojnowska-Michalik and B. Goldys, On regularity properties of nonsymmetric Ornstein-Uhlenbeck semigroups in L^p spaces, *Stoch. Stoch. Reports*, **59**, 183-209, 1996.

[50] A. Chojnowska-Michalik and B. Goldys, Nonsymmetric Ornstein-Uhlenbeck semigroup as a second quantized operator, *J. Math. Kyoto Univ.*, **36**, 481-498, 1996.

[51] A. Chojnowska-Michalik and B. Goldys, On Ornstein-Uhlenbeck generators, preprint, School of Mathematics, The University of New South Wales, 1996.

[52] A. Chojnowska-Michalik and B. Goldys, Non symmetric Ornstein-Uhlenbeck generators, *Canadian Mathematical Society Conference Proceedings*, **28**, 99-116, 2000.

[53] A. Chojnowska-Michalik and B. Goldys, Generalized Ornstein-Uhlenbeck semigroups: Littlewood-Paley-Stein inequalities and the P. A. Meyer equivalence of norms, *J. Functional Anal.*, **182**, 243-279, 2001.

[54] A. Chojnowska-Michalik and B. Goldys, *Symmetric Ornstein-Uhlenbeck semigroups and their generators,* preprint.

[55] P. L. Chow, Infinite-dimensional Kolmogorov equations in Gauss-Sobolev spaces, *Stoch. Anal. Appl.*, to appear.

[56] P. L. Chow and J. L. Menaldi, Variational inequalities for the control of stochastic partial differential equations, *Stochastic partial differential equations and applications*, Proceedings, Trento 1988, Lecture Notes in Mathematics **1390**, G. Da Prato and L. Tubaro (editors), Springer-Verlag, 42-52, 1989.

[57] P. L. Chow and J. L. Menaldi, Infinite dimensional Hamilton-Jacobi-Bellman equations in Gauss-Sobolev spaces, *Nonlinear Analysis TMA*, **29**, 415-426, 1997.

[58] M. G. Crandall, H. Ishii and P. L. Lions, User's guide to viscosity solutions of second order partial differential equations, *Bull. Amer. Math. Soc. (N.S.)*, **27**, 1-67, 1992.

[59] M. G. Crandall and P. L. Lions, Hamilton-Jacobi equations in infinite dimensions, Part I: uniqueness of viscosity solutions, *J. Functional Anal.*, **62**, 379-396, 1985.

[60] M. G. Crandall and P. L. Lions, Hamilton-Jacobi equations in infinite dimensions, Part II: existence of viscosity solutions, *J. Functional Anal.*, **68**, 368-405, 1986.

[61] M. G. Crandall and P. L. Lions, Hamilton-Jacobi equations in infinite dimensions, Part IV, Unbounded linear terms, *J. Functional Anal.*, **90**, 237-283, 1990.

[62] Yu. Daleckij, Differential equations with functional derivatives and stochastic equations for generalized random processes, *Dokl. Akad. Nauk SSSR*, **166**, 1035-1038, 1996.

[63] Yu. Daleckij and S. V. Fomin, *Measures and differential equations in infinite-dimensional space*, Kluwer, 1991.

[64] G. Da Prato, *Applications croissantes et équations d'évolution dans les espaces de Banach*, Academic Press, 1976.

[65] G. Da Prato, Some results on Bellman equation in Hilbert spaces, *SIAM J. Control Optim.*, **23**, 61-71, 1985.

[66] G. Da Prato, Some results on Bellman equation in Hilbert spaces and applications to infinite dimensional control problems, Lecture Notes in Control and Information Sciences, **69**, M. Metivier and E. Pardoux (editors), Springer-Verlag, 270-280, 1985.

[67] G. Da Prato, Some results on parabolic evolution equations with infinitely many variables, *J. Diff. Equations*, **68**, 281-297, 1987.

[68] G. Da Prato, Some results on Kolmogoroff equations for infinite dimensional stochastic systems, *Stochastic Differential Systems, Stochastic Control Theory and Applications, IMA*, **10**, W. Fleming and P. L. Lions (editors), Springer-Verlag, 87-97, 1988.

[69] G. Da Prato, Some results on parabolic equations in Hilbert spaces, *Proceedings of the International Conference on Theory and Applications of Differential Equations*, A. Aftabizadeh (editor), Vol. II, Ohio University Press, 170-178, 1989.

[70] G. Da Prato, Smoothing properties of heat semigroup in infinite dimensions, *Semigroups of linear and nonlinear operators and applications*, J. A. Goldstein and A. Goldstein (editors), Kluwer, 129-141, 1993.

[71] G. Da Prato, Parabolic equations in infinitely many variables, *Rend. Circolo Mat. Palermo, Ser. II*, **33**, 25-38, 1993.

[72] G. Da Prato, Null controllability and strong Feller property of Markov transition semigroups, *Nonlinear Analysis TMA*, **25**, 941-949, 1995.

[73] G. Da Prato, Some results on elliptic and parabolic equations in Hilbert spaces, *Rend. Mat. Accad. Lincei*, **9**, 181-199, 1996.

[74] G. Da Prato, Regularity results for Kolmogorov equations on $L^2(H, \mu)$ spaces and applications, *Ukrainian Math. J.*, **49**, 448-457, 1997.

[75] G. Da Prato, Characterization of the domain of an elliptic operator of infinitely many variables in $L^2(\mu)$ spaces, *Rend. Mat. Accad. Lincei*, **8**, 101-105, 1997.

[76] G. Da Prato, *Stochastic evolution equations by semigroups methods*, Centre de Recerca Matematica, Barcelona, **11**, 1998.

[77] G. Da Prato, The Ornstein-Uhlenbeck generator perturbed by the gradient of a potential, *Bollettino UMI*, **8**, 501-519, 1998.

[78] G. Da Prato, Monotone gradient systems in L^2 spaces, *Proceedings of the Ascona Conference 1999 on Stochastic Analysis, Random Fields and Their Applications*, Progress in Probability, Birkhäuser, 73-88, 2000.

[79] G. Da Prato, Elliptic operators with unbounded coefficients: construction of a maximal dissipative extension, *J. Evolution Equations*, **1**, 1-18, 2001.

[80] G. Da Prato, Lipschitz perturbations of Ornstein-Uhlenbeck semigroups, preprint, Scuola Normale Superiore di Pisa, 2001.

[81] G. Da Prato, Some properties of monotone gradient systems, *Dyn. Contin. Discrete Impuls. Syst. Ser. A Math. Anal.*, **8**, n. 3, 401-414, 2001.

[82] G. Da Prato, Bounded perturbations of Ornstein-Uhlenbeck semigroups, *Progress in Nonlinear Differential Equations and Their applications*, **50**, 97-114, Birkhäuser, 2002.

[83] G. Da Prato and A. Debussche, Control of the stochastic Burgers model of turbulence, *SIAM J. Control Optim.*, **37**, 1123-1141, 1999.

[84] G. Da Prato and A. Debussche, Dynamic Programming for the stochastic Navier-Stokes equations, *Math. Modelling Numerical Anal.*, **34**, 459-475, 2000.

[85] G. Da Prato and A. Debussche, Dynamic Programming for the stochastic Burgers equation, *Ann. Mat. Pura Appl.*, **178**, 143-174, 2000.

[86] G. Da Prato and A. Debussche, Maximal dissipativity of the Dirichlet operator corresponding to the Burgers equation, *Canadian Mathematical Society Conference Proceedings*, **28**, 145-170, 2000.

[87] G. Da Prato and A. Debussche, 2*D*-Navier-Stokes equations driven by a space-time white noise, *J. Functional Anal.*, to appear.

[88] G. Da Prato, A. Debussche and B. Goldys, Invariant measures of non symmetric dissipative stochastic systems, *Probab. Th. Relat. Fields*, to appear.

[89] G. Da Prato, D. Elworthy and J. Zabczyk, Strong Feller property for stochastic semilinear equations, *Stoch. Anal. Appl.*, **13**, 35-45, 1995.

[90] G. Da Prato, M. Fuhrman and J. Zabczyk, A note on regularizing properties of Ornstein-Uhlenbeck semigroups in infinite dimensions, *Stochastic partial differential equations and application*, G. Da Prato and L. Tubaro (editors), Dekker, to appear.

[91] G. Da Prato and B. Goldys, On symmetric gaussian diffusions, *Stoch. Anal. Appl.*, **17**, 369-381, 1999.

[92] G. Da Prato, B. Goldys and J. Zabczyk, Ornstein-Uhlenbeck semigroups in open sets of Hilbert spaces, *C. R. Acad. Sci. Paris*, **325**, 433-438, 1997.

[93] G. Da Prato, S. Kwapien and J. Zabczyk, Regularity of solutions of linear stochastic equations in Hilbert spaces, *Stoch.*, **23**, 1-23, 1987.

[94] G. Da Prato and A. Lunardi, On the Ornstein-Uhlenbeck operator in spaces of continuous functions, *J. Functional Anal.*, **131**, 94-114, 1995.

[95] G. Da Prato, P. Malliavin and D. Nualart, Compact families of Wiener functionals, *C. R. Acad. Sci. Paris*, **315**, 1287-1291, 1992.

[96] G. Da Prato and M. Röckner, Singular dissipative stochastic equations in Hilbert spaces, preprint, Scuola Normale Superiore di Pisa, 2001.

[97] G. Da Prato and L. Tubaro, Self-adjointness of some infinite dimensional elliptic operators and application to stochastic quantization, *Probab. Th. Relat. Fields*, **118**, 131-145, 2000.

[98] G. Da Prato and L. Tubaro, On a class of gradient systems with irregular potentials, *Infin. Dimens. Anal. Quantum Probab. Relat. Top.*, **4**, 183-193, 2001.

[99] G. Da Prato and L. Tubaro, Some results about dissipativity of Kolmogorov operators, *Czechoslovak Math. J.*, **51**, 4, 685-699, 2001.

[100] G. Da Prato and J. Zabczyk, Smoothing properties of transition semigroups in Hilbert Spaces, *Stoch. Stoch. Reports*, 35, 63-77, 1991,

[101] G. Da Prato and J. Zabczyk, *Stochastic equations in infinite dimensions*, Cambridge University Press, 1992.

[102] G. Da Prato and J. Zabczyk, *Ergodicity for infinite dimensional systems*, London Mathematical Society Lecture Notes, **229**, Cambridge University Press, 1996.

[103] G. Da Prato and J. Zabczyk, Differentiability of the Feynman-Kac semigroup and a control application, *Rend. Mat. Accad. Lincei*, **8**, 183-188, 1997.

[104] E. B. Davies, *One parameter semigroups*, Academic Press, 1980.

[105] W. Desh and A. Rhandi, On the norm continuity of transition semigroups in Hilbert spaces, *Arch. Math.*, **70**, 52-56, 1998.

[106] J. D. Deuschel and D. Stroock, *Large deviations*, Academic Press, 1984.

[107] N. Dunford and J. T. Schwartz, *Linear operators*, Vol. II, Interscience, 1964.

[108] E. B. Dynkin, *Markov processes*, Vol. I, Springer-Verlag, 1965.

[109] A. Eberle, *Uniqueness and non-uniqueness of singular diffusion operators*, Lecture Notes in Mathematics, **1718**, Springer-Verlag, 1999.

[110] K. D. Elworthy, Stochastic flows on Riemannian manifolds, *Diffusion processes and related problems in analysis*, Vol. II, M. A. Pinsky and V. Wihstutz (editors), 33-72, Birkhäuser, 1992.

[111] K. Engel and R. Nagel, *One-parameter semigroups for linear evolution equations*, with contributions by S. Brendle, M. Campiti, T. Hahn, G. Metafune, G. Nickel, D. Pallara, C. Perazzoli, A. Rhandi, S. Romanelli and R. Schnaubelt, Graduate Texts in Mathematics, **194**, Springer-Verlag, 2000.

[112] S. N. Ethier and T. G. Kurtz, *Markov processes, characterization and convergence*, Wiley, 1986.

[113] F. Flandoli and F. Gozzi, Kolmogorov equation associated to a stochastic Navier-Stokes equation, *J. Functional Anal.*, **160**, 312-336, 1998.

[114] W. H. Fleming and H. M. Soner, *Controlled Markov processes and viscosity solutions*, Springer-Verlag, 1993.

[115] A. Friedman, *Partial differential equations* of parabolic type, Prentice-Hall, 1964.

[116] A. Friedman, *Stochastic differential equations and applications, probability and mathematical statistics*, Academic Press, 1975.

[117] M. Fuhrman, Densities of Gaussian measures and regularity of non-symmetric Ornstein-Uhlenbeck semigroups in Hilbert spaces, preprint, Institute of Mathematics, Polish Academy of Sciences, 1994.

[118] M. Fuhrman, Hypercontractivité des semigroupes de Ornstein-Uhlenbeck non symétriques, *C. R. Acad. Sci. Paris*, **321**, 929-932, 1995.

[119] M. Fuhrman, A note on the nonsymmetric Ornstein-Uhlenbeck process in Hilbert spaces, *Appl. Math. Letters*, **8**, 19-22, 1995.

[120] M. Fuhrman, Analyticity of transition semigroups and closability of bilinear forms in Hilbert spaces, *Studia Math.*, **115**, 53-71, 1995.

[121] M. Fuhrman, Hypercontractivity properties of non symmetric Ornstein-Uhlenbeck semigroups, *Stoch. Anal. Appl.*, **16**, 241-260, 1998.

[122] M. Fuhrman and M. Tessitore, Nonlinear Kolmogorov equations in infinite dimensional spaces: the backward stochastic differential equations approach and applications to optimal control, *Ann. Probab.*, to appear.

[123] M. Fuhrman and M. Tessitore, The Bismut-Elworthy formula for backward SDE's and applications to nonlinear Kolmogorov equations and control in infinite dimensional spaces, Preprint Politecnico di Milano, 2001.

[124] M. Fukushima, *Dirichlet forms and symmetric Markov processes*, North-Holland, 1980.

[125] D. Gątarek and B. Goldys, Existence, uniqueness and ergodicity for stochastic quantization equation, *Studia Math.*, **119**, 179-193, 1996.

[126] I. Gikhman and A. V. Shorokhod, *The theory of stochastic processes*, Vols. I, II and III, Springer-Verlag, 1974, 1975, 1979.

[127] B. Goldys, On analyticity of Ornstein-Uhlenbeck semigroups, *Rend. Mat. Accad. Lincei*, **10**, 131-140, 1999.

[128] B. Goldys and F. Gozzi, Second order parabolic HJ equations in Hilbert spaces and stochastic control: L^2 approach, preprint, Università di Pisa, 2000.

[129] B. Goldys, F. Gozzi and J. van Neerven, Closability of directional gradients, *Potential Anal.*, to appear.

[130] B. Goldys and M. Kocan, Diffusion semigroups in spaces of continuous functions with mixed topology, *J. Diff. Equations*, **173**, 17-39, 2001.

[131] B. Goldys and B. Maslowski, Ergodic control of semilinear stochastic equations and the Hamilton-Jacobi equation, *J. Math. Anal. Appl.*, **234**, 592-631, 1999.

[132] V. Goodman, A divergence theorem for Hilbert spaces, *Trans. Amer. Math. Soc.*, **164**, 411-426, 1972.

[133] F. Gozzi, Regularity of solutions of a second order Hamilton-Jacobi equation and application to a control problem, *Comm. Partial Diff. Equations*, **20**, 775-826, 1995.

[134] F. Gozzi, Global regular solutions of second order Hamilton-Jacobi equations in Hilbert spaces with locally Lipschitz nonlinearities, *J. Math. Anal. Appl.*, **198**, 399-443, 1996.

[135] F. Gozzi and E. Rouy, Regular solutions of second order stationary Hamilton-Jacobi equations, *J. Diff. Equations*, **130**, 201-234, 1996.

[136] F. Gozzi, E. Rouy and A. Swiech, Second order Hamilton-Jacobi equations in Hilbert spaces and stochastic boundary control, *SIAM J. Control Optim.*, **38**, 400-430, 2000.

[137] F. Gozzi and A. Swiech, Hamilton-Jacobi-Bellman equations for the optimal control of the Duncan-Mortensen-Zakai equation, *J. Functional Anal.*, **172**, 466-510, 2000.

[138] L. Gross, Potential theory on Hilbert spaces, *J. Functional Anal.*, **1**, 123-181, 1967.

[139] L. Gross, Logarithmic Sobolev inequalities, *Amer. J. Math.*, **97**, 1061-1083, 1976.

[140] P. Guiotto, Non-differentiability of heat semigroups in infinite dimensional Hilbert space, *Semigroup Forum*, **55**, 232-236, 1997.

[141] P. R. Halmos, *Measure theory*, Van Nostrand, 1961.

[142] K. Itô, On a stochastic integral equation, *Proc. Imp. Acad. Tokyo*, **22**, 32-35, 1946.

[143] S. Itô, Fundamental solutions of parabolic differential equations and boundary value problems, *Japanese J. Math.*, **27**, 5-102, 1957.

[144] C. Knoche and K. Frieler, Solutions of stochastic differential equations in infinite dimensional Hilbert spaces and their dependence on initial data, Diplomarbeit, Fakultät für Mathematik, Universität Bielefeld, 2001.

[145] A. N. Kolmogorov, Über die analytischen Methoden in der Wahrscheinlichkeitsrechnung, *Math. Ann.*, **104**, 415-458, 1931.

[146] N. V. Krylov, *Introduction to the theory of diffusion processes*, *Translations of Mathematical Monographs*, **142**, American Mathematical Society, 1995.

[147] N. V. Krylov, *Lectures on elliptic and parabolic equations in Hölder spaces*, American Mathematical Society, 1996.

[148] N. V. Krylov, On Kolmogorov's equations for finite-dimensional diffusions, Lecture Notes in Mathematics, **1715**, G. Da Prato (editor), 1-64, Springer-Verlag, 1999.

[149] H. H. Kuo, *Gaussian measures in Banach spaces*, Lecture Notes in Mathematics, **463**, Springer-Verlag, 1975.

[150] H. H. Kuo and M. A. Piech, Stochastic integrals and parabolic equations in abstract Wiener space, *Bull. Amer. Math. Soc.*, **79**, 478-482, 1973.

[151] F. Kühnemund, Bi-continuous semigroups on spaces with two topologies: theory and applications, Dissertation der Mathematischen Fakultät der Eberhard-Karls-Universität Tübingen zur Erlangung des Grades eines Doktors der Naturwissenschaften, 2001.

[152] J. Kurtzweil, On approximation in real Banach spaces, *Studia Math.*, **14**, 213-231, 1954.

[153] S. Kwapien, Invariant measures for infinite dimensional Wiener process, manuscript, 2001.

[154] O. A. Ladyzhenskaja, V. A. Solonnikov and N. N. Ural'ceva, *Linear and quasilinear equations of parabolic type*, Translations of Mathematical Monographs, American Mathematical Society, 1968.

[155] J. M. Lasry and P. L. Lions, A remark on regularization in Hilbert spaces, *Israel J. Math.*, **55**, 257-266, 1986.

[156] P. Lévy, *Problèmes concrets d'analyse fonctionnelle*, Gauthier-Villars, 1951.

[157] J. L. Lions and J. Peetre, Sur une classe d'espaces d'interpolation, *Publ. Math. IHES*, **19**, 5-68, 1964.

[158] P. L. Lions, *Generalized solutions of Hamilton-Jacobi equations*, Pitman, 1982.

[159] V. Liskevich and M. Röckner, Strong uniqueness for a class of infinite dimensional Dirichlet operators and application to stochastic quantization, *Ann. Scuola Norm. Sup. Pisa Cl. Sci.*, **27**, 69-91, 1999.

[160] H. Long and I. Simão, Kolmogorov equations in Hilbert spaces with application to essential self-adjointness of symmetric diffusion operators, *Osaka J. Math.*, **37**, 2000.

[161] G. Lumer and R. S. Phillips, Dissipative operators in a Banach space, *Pac. J. Math.*, **11**, 679-698, 1961.

[162] A. Lunardi, *Analytic semigroups and optimal regularity in parabolic problems*, Birkhäuser, 1995.

[163] A. Lunardi, On the Ornstein-Uhlenbeck operator in L^2 spaces with respect to invariant measures, *Trans. Amer. Math. Soc.* **349**, 155-169, 1997.

[164] A. Lunardi, Schauder theorems for linear elliptic and parabolic problems with unbounded coefficients in $\mathbb{R}^n$, *Studia Math.* **128**, 171-198, 1998.

[165] Z. M. Ma and M. Röckner, *Introduction to the theory of (non-symmetric) Dirichlet forms*, Springer-Verlag, 1992.

[166] P. Malliavin, *Stochastic analysis*, Springer-Verlag, 1997.

[167] G. Metafune, L^p spectrum of Ornstein-Uhlenbeck operators, *Ann. Scuola Norm. Sup. Pisa*, **30**, 97-124, 2001.

[168] G. Metafune and G. Pallara, Discreteness of the spectrum for some differential operators with unbounded coefficients in $\mathbb{R}^n$, *Rend. Mat. Accad. Lincei*, **11**, 9-19, 2000.

[169] G. Metafune, G. Pallara and E. Priola, Spectrum of Ornstein-Uhlenbeck operators in L^p spaces, preprint, Università di Torino, 2001.

[170] G. Metafune, G. Pallara and M. Wacker, *Feller semigroups on $\mathbb{R}^n$*, preprint, 2001.

[171] G. Metafune, G. Pallara and M. Wacker, *Compactness properties of Feller semigroups*, preprint, 2001.

[172] G. Metafune, J. Prüss, A. Rhandi and R. Schnaubelt, The domain of the Ornstein-Uhlenbeck operator on an L^p space with invariant measure, *Ann. Scuola Norm. Sup. Pisa*, to appear.

[173] G. Metafune, A. Rhandi and R. Schnaubelt, Spectrum of the infinite dimensional Laplacian, *Arch. Math. (Basel)*, **75**, 280-282, 2000.

[174] J. van Neerven and J. Zabczyk, Norm discontinuity of Ornstein-Uhlenbeck semigroups, *Semigroup Forum*, **3**, 389-403, 1999.

[175] E. Nelson, A quartic interaction in two dimensions, *Mathematical theory of elementary particles*, R. Goodman and I. Segal (editors). MIT Press, 69-73, 1966.

[176] A. S. Nemirovski and S. M. Semenov, The polynomial approximation of functions in Hilbert spaces, *Mat. Sb. (N.S.)*, **92**, 257-281, 1973.

[177] J. Neveu, Sur l'espérance conditionnelle par rapport à un mouvement Brownien, *Ann. Inst. H. Poincaré*, **12**, 105-109, 1976.

[178] D. Nualart and A. S. Ustunel, Une extension du laplacien sur l'espace de Wiener et la formule d'Itô associée, *C. R. Acad. Sci. Paris*, I, **309**, 383-386, 1984.

[179] K.R. Parthasarathy, *Probability measures on metric spaces*, Academic Press, 1967.

[180] A. Pazy, *Semigroups of linear operators and applications to partial differential equations*, Springer-Verlag, 1983.

[181] S. Peszat, On a Sobolev space of functions of infinite numbers of variables, *Bull. Pol. Acad. Sci.*, **40**, 55-60, 1993.

[182] S. Peszat, Large deviation principle for stochastic evolution equations, *Probab. Th. Relat. Fields*, **98**, 113-136, 1994.

[183] S. Peszat, Private communication.

[184] S. Peszat and J. Zabczyk, Strong Feller property and irreducibility for diffusions on Hilbert spaces, *Ann. Probab.*, **23**, 157-172, 1995.

[185] R. R. Phelps, Gaussian null sets and differentiability of Lipschitz map on Banach spaces, *Pac. J. Math.*, **77**, 523-531, 1978.

[186] A. Piech, A fundamental solution of the parabolic equation on Hilbert space, *J. Functional Anal.*, **3**, 85-114, 1969.

[187] A. Pietsch, *Nuclear locally convex spaces*, Springer-Verlag, 1972.

[188] E. Priola, π-semigroups and applications, preprint, Scuola Normale Superiore di Pisa, 1998.

[189] E. Priola, On a class of Markov type semigroups in spaces of uniformly continuous and bounded functions, *Studia Math.*, **136**, 271-295, 1999.

[190] E. Priola, Uniform approximation of uniformly continuous and bounded functions on Banach spaces, *Dyn. Syst. Appl.*, **9**, 181-198, 2000.

[191] E. Priola, On a Dirichlet problem involving an Ornstein-Uhlenbeck operator, preprint Scuola Normale Superiore di Pisa, 2000.

[192] E. Priola, The Cauchy problem for a class of Markov type semigroups, *Comm. Appl. Anal.*, **5**, 49-76, 2001.

[193] E. Priola, Schauder estimates for a homogeneous Dirichlet problem in a half space of a Hilbert space, *Nonlinear Analysis TMA*, **44**, 679-702, 2001.

[194] E. Priola, A counterexample to Schauder estimates for elliptic operators with unbounded coefficients, *Rend. Atti Accad. Lincei*, **12**, 15-25, 2001.

[195] E. Priola, Dirichlet problems in a half space of a Hilbert space, *Infin. Dimens. Anal. Quantum Probab. Relat. Top.*, to appear.

[196] E. Priola and J. Zabczyk, Null controllability with vanishing energy, preprint, Università di Torino, 2001.

[197] E. Priola and L. Zambotti, New optimal regularity results for infinite dimensional elliptic equations, *Bollettino UMI*, **8**, 411-429, 2000.

[198] M. Röckner, L^p-analysis of finite and infinite dimensional diffusions, Lecture Notes in Mathematics, **1715**, G. Da Prato (editor), Springer-Verlag, 65-116, 1999.

[199] I. Simão, Regular fundamental solution for a parabolic equation on an infinite-dimensional space, *Stoch. Anal. Appl.*, **11**, 235-247, 1993.

[200] I. Simão, Regular transition densities for infinite-dimensional diffusions, *Stoch. Anal. Appl.*, **11**, 309-336, 1993.

[201] B. Simon, The $P(\phi)_2$ euclidean (quantum) field theory, Princeton University Press, 1974.

[202] W. Stannat, (Nonsymmetric) Dirichlet operators on L^1: existence, uniqueness and associated Markov processes, *Ann. Scuola Norm. Sup. Pisa*, Ser. IV, **28**, 99-140, 1999.

[203] W. Stannat, The theory of generalized Dirichlet forms and its applications in analysis and stochastics, *Memoirs Amer. Math. Soc.*, **678**, 1999.

[204] L. Stettner and J. Zabczyk, Strong envelopes of stochastic processes and a penalty method, *Stoch.*, **4**, 267-280, 1981.

[205] D. W. Stroock, *Probability theory: an analytic view*, Cambridge University Press, 1993.

[206] A. Swiech, "Unbounded" second order partial differential equations in infinite dimensional Hilbert spaces, *Comm. Partial Diff. Equations*, **19**, 1999-2036, 1994.

[207] A. Talarczyk, Dirichlet problem for parabolic equations on Hilbert spaces, *Studia Math.*, **141**, 109-142, 2000.

[208] G. Tessitore, Infinite horizon, ergodic and periodic control for a stochastic infinite-dimensional affine equation, *J. Math. Syst. Estim. Control*, **8** (electronic), 1998.

[209] G. Tessitore and J. Zabczyk, Comments on transition semigroups and stochastic invariance, preprint, Scuola Normale Superiore di Pisa, 1998.

[210] G. Tessitore and J. Zabczyk, Invariant measures for stochastic heat equations, *Probab. Math. Statist.*, **18**, 271-287, 1998.

[211] G. Tessitore and J. Zabczyk, Trotter's formula for transition semigroups, *Semigroup Forum*, **63**, 114-126, 2001.

[212] H. Triebel, *Interpolation theory, function spaces, differential operators*, North-Holland, 1978.

[213] F. A. Valentine, A Lipschitz condition preserving extension for a vector function, *Amer. J. Math.*, **67**, 83-93, 1945.

[214] K. Yosida, *Functional analysis*, Springer-Verlag, 1965.

[215] J. Zabczyk, Linear stochastic systems in Hilbert spaces: spectral properties and limit behavior, Report **236**, Institute of Mathematics, Polish Academy of Sciences, 1981. Also in Banach Center Publications, **41**, 591-609, 1985.

[216] J. Zabczyk, Symmetric solutions of semilinear stochastic equations, *Stochastic partial differential equations and applications*, Proceedings, Trento 1985, Lecture Notes in Mathematics, **1390**, G. Da Prato and L. Tubaro (editors), Springer-Verlag, 237-256, 1989.

[217] J. Zabczyk, *Mathematical control theory. An introduction*, Birkhäuser, 1992.

[218] J. Zabczyk, Stopping problems on Polish spaces, *Ann. Univ. Marie Curie-Sklodowska*, **51**, 181-199, 1997.

[219] J. Zabczyk, Infinite dimensional diffusions in modelling and analysis, *Jber. Dt. Math.-Verein.*, **101**, 47-59, 1999.

[220] J. Zabczyk, Parabolic equations in Hilbert spaces, *Second order partial differential equations in Hilbert spaces*, Lecture Notes in Mathematics, **1715**, G. Da Prato (editor), Springer-Verlag, 117-213, 1999.

[221] J. Zabczyk, Excessive weights, manuscript, 1999.

[222] J. Zabczyk, Bellman's inclusions and excessive measures, *Probab. Math. Statist.*, **21**, 101-112, 2001.

[223] L. Zambotti, A new approach to existence and uniqueness for martingale problems in infinite dimensions, *Probab. Th. Relat. Fields*, **118**, 147-168, 2000.

[224] L. Zambotti, A reflected stochastic heat equation as symmetric dynamics with respect to 3-d Bessel bridge, *J. Functional Anal.*, **180**, 195-209, 2001.

[225] L. Zambotti, Integration by parts on the δ-d Bessel bridge, $d > 3$, and related SPDEs, *Ann. Probab.*, to appear.

Index

Printed in the United States
By Bookmasters